Ecological Systems Integrity

T0227464

Environmental law and governance are the cornerstones of global efforts to conserve the environment, protect resources and ensure fair and equitable outcomes for all of the planet's inhabitants. This book presents a series of thought-provoking chapters which consider the place of governance and law in the defence against imminent and ongoing threats to ecological, social and cultural integrity.

Written by an international team of both established and early-career scholars from various disciplines and backgrounds, the chapters cover the most pressing and contemporary issues in environmental law and governance. These include access and benefit-sharing; the right to food and water; climate change coping and adaptation; human rights; the rights of indigenous communities; public and environmental health; and many more. The book has a general focus on environmental governance and law in the European Union, and offers points of comparison with Canada and North and South America.

Laura Westra is professor emerita (philosophy), University of Windsor, Canada, a sessional instructor at the Faculty of Law, University of Milano (Bicocca), Italy, and visiting professor at the University of Trento, Italy.

Janice Gray is senior lecturer at UNSW Law, University of New South Wales, Australia.

Vasiliki Karageorgou is assistant professor in European administrative and environmental law at the Panteion University of Social and Political Science in Athens, Greece.

Ecological Systems Integrity

Governance, law and human rights

**Edited by Laura Westra,
Janice Gray and
Vasiliki Karageorgou**

Routledge
Taylor & Francis Group

LONDON AND NEW YORK

from Routledge

First published 2015
by Routledge

2 Park Square, Milton Park, Abingdon, Oxfordshire OX14 4RN
711 Third Avenue, New York, NY 10017

Routledge is an imprint of the Taylor & Francis Group, an informa business

First issued in paperback 2017

© 2015 Laura Westra, Janice Gray and Vasiliki Karageorgou, selection and
editorial material; individual chapters, the contributors

The right of the editors to be identified as the authors of the editorial material,
and of the authors for their individual chapters, has been asserted in accordance
with sections 77 and 78 of the Copyright, Designs and Patents Act 1988.

All rights reserved. No part of this book may be reprinted or reproduced or
utilised in any form or by any electronic, mechanical, or other means, now
known or hereafter invented, including photocopying and recording, or in any
information storage or retrieval system, without permission in writing from the
publishers.

Trademark notice: Product or corporate names may be trademarks or registered
trademarks, and are used only for identification and explanation without intent to
infringe.

British Library Cataloguing-in-Publication Data
A catalogue record for this book is available from the British Library

Library of Congress Cataloging in Publication Data
Ecological systems integrity : governance, law, and human rights / edited by
Laura Westra, Janice Gray, and Vasiliki Karageorgou.
    pages cm
    Includes index.
    1. Environmental law.   2. Biodiversity conservatio—Law and legislation.
    3. Ecology.   4. Convention on Biological Diversity (1992 June 5)
    5. Agreement on Trade-Related Aspects of Intellectual Property Rights
    (1994 April 15)   I. Westra, Laura, editor.   II. Gray, Janice, 1956– editor.
    III. Karageorgou, Vasilike I, editor.
    K3585.E26 2015
    344.04'6—dc23   2014048549

ISBN: 978-1-138-88510-3 (hbk)
ISBN: 978-1-138-57487-8 (pbk)

Typeset in Baskerville
by Keystroke, Station Road, Codsall, Wolverhampton

Contents

Contributors

Aristotelis B. Alexopoulos, PhD
Country of affiliation: Greece
Professor (maritime policy and marine environmental law), Department of
    Shipping, Transport and Logistics, BCA College of Athens
abagr@yahoo.com
aalexopoulos@bca.edu.gr

Susana Borràs, PhD (Law)
Country of affiliation: Spain
Assistant professor in international environmental law, Centre of Environmental
    Law Studies of Tarragona (CEDAT), Rovira i Virgili University
www.cedat.cat
susana.borras@urv.cat

Klaus Bosselmann, PhD
Countries of affiliation: New Zealand, Germany
Professor of law; director, New Zealand Centre for Environmental Law, University
    of Auckland
www.law.auckland.ac.nz/uoa/os-klaus-bosselmann
k.bosslmann@auckland.ac.nz

Tilemachos Bourtzis, BA, MSc, PhD Can
Country of affiliation: Greece
Researcher, EKEPEK – Panteion University of Social and Political Sciences
tbourtzis@gmail.com

Benjamin A. Brown
Country of affiliation: USA
Student, Widener University School of Law, Harrisburg, PA

**Donald A. Brown, Juris Doctor, MA (liberal studies, philosophy
and art)**
Country of affiliation: USA
Scholar in residence and professor, School of Law, Widener University
Part-time professor, Nanjing University of Information Science and Technology

Visiting professor, Nagoya University
Ethicsandclimate.org
Dabrown57@gmail.com

Margherita Brunori
Country of affiliation: Italy
PhD student, Scuola Superiore Sant'Anna – Dirpolis Institute
www.sssup.it

Antonio D'Aloia
Country of affiliation: Italy
Full professor, Constitutional Law, University of Parma
antonio.daloia@unipr.it

Joseph W. Dellapenna, BBA, JD, LLM (International and Comparative Law), LLM (Environmental Law)
Country of affiliation: USA
Professor of law, Villanova University School of Law
dellapen@law.villanova.edu

Rose A. Dyson, EdD
Country of affiliation: Canada
Media education consultant, President, Canadians Concerned about Violence in Entertainment
www.C-CAVE.com
rdyson@oise.utoronto.ca
rose.dyson@alumni.utoronto.ca

Anastasia Fotinakopoulou
Country of affiliation: Greece
PhD candidate, researcher, European Centre of Environmental Research and Training, Panteion University of Social and Political Sciences
natfotin@yahoo.com

Constanze Katharina Frank-Oster
Country of affiliation: Germany
UN
Constanze.frank@gmx.de

Geoffrey Garver, BS (Chem Eng), JD, LLM
Countries of affiliation: Canada, United States
PhD candidate (geography), sessional instructor, McGill University
gginmont@videotron.ca

Janice Gray
Country of affiliation: Australia
Senior lecturer, Faculty of Law, UNSW Australia
Editor-in-Chief, *Australasian Journal of Natural Resources Law and Policy*
j.gray@unsw.edu.au

Donato Gualtieri
Country of affiliation: Italy
PhD candidate in international environmental law, Department of Law, University
    of Pavia
donato.gualtieri01@ateneopv.it

Kathryn Gwiazdon, JD
Countries of affiliation: USA, Japan
Consultant, international environmental law and ethics
kakintzele@gmail.com

Sandra Jen
Country of affiliation: Belgium
Consultant, environmental policy and law
Post-graduate diploma, Environmental Law, University of Strasbourg (F)
Master degrees, Institute of Political Sciences, and International and European
    Law, University of Toulouse (F)
sjen@sjenconsult.org

Vasiliki (Vicky) Karageorgou, PhD (Law)
Country of affiliation: Greece
Assistant professor, Panteion University of Social and Political Sciences
IUNC Environmental Law Commission, European Environmental Law Forum
vkaragiorgou@yahoo.gr

Yuliya Lyamzina, PhD, MBA
Country of affiliation: Czech Republic
Johannes Gutenberg-Universitaet Mainz
lyamzina@yahoo.com

Kathleen Mahoney, QC, FRSC, LLM, JD
Country of affiliation: Canada
Professor, University of Calgary
kmahoney@ucalgary.ca

Massimiliano Montini, LLM, LLB
Country of affiliation: Italy
Associate professor, and director of Environmental Legal Team, University of
    Siena
Co-director, R4S Regulation for Sustainability Research Group
massimiliano.montini@gmail.com
www.elt.unisi.it, www.r4s.unisi.it

Antoni Pigrau, PhD (Law)
Country of affiliation: Spain
Professor, public international law, University Rovira i Virgili
antoni.pigrau@urv.cat
www.cedat.cat

Gerasimos Rodotheatos, BA, MA, PhD Can
Country of affiliation: Greece
Researcher, EKEPEK – Panteion University of Social and Political Sciences
yrodo@panteion.gr

Colin L. Soskolne, PhD (Epidemiology)
Countries of affiliation: Canada and Australia
Professor emeritus, University of Alberta, Edmonton, Canada
Adjunct professor, Faculty of Health, University of Canberra
www.colinsoskolne.com
colin.soskolne@ualberta.ca

Grigoris Tsaltas, PhD
Country of affiliation: Greece
Professor (international law), Department of International, European and Area
    Studies, Panteion University of Social and Political Sciences
gtsaltas@panteion.gr

Peter Venton, BA (Econ), MA (Econ)
Country of affiliation: Canada
President, JPV Associates
Former senior economist, Ministry of Finance, Government of Ontario
Peter.venton@bell.net

Laura Westra, PhD, PhD (Law)
Countries of affiliation: Canada, Italy
Professor emerita (philosophy), sessional instructor, Faculty of Law, University of
    Windsor
www.ecointegrity.net
www.globalecointegrity.net

Preface

The annual meeting of the Global Ecological Integrity Group met for the 22nd time on the Island of Rhodes in Greece, from 21 to 26 June 2014. This time the group included a larger number of European scholars than usual, probably attracted by the incredible historical location, and the sunshine and blue skies that blessed our meeting throughout.

This increased presence introduced, if not a new note, at least a renewed emphasis to our discussion (see Chapter 1). It also served to initiate, or perhaps renew, a somewhat novel aspect in our collective work: the increasing need for system integrity to develop global forms of governance, beyond the present systems.

In other words, beyond extending and enlarging our ongoing critique of failed organizations and institutions, from the standpoint of public health, environmental legal regimes, from the effectiveness of democratic institutions to the attacks on legal protests, our emphasis is now primarily on the 'global' aspects of the Global Ecological Integrity Group's work. This emphasis entails the need for new forms of governance to implement that move.

This volume represents only the beginning, as we look forward to continue our conversation next year, once again in Italy, at the University of Parma.

Finally, for this past meeting, thanks are due to many who made this fruitful meeting possible, starting with Vicky Karageorgou, our local organizer, who arranged both our seaside hotel and our outstanding location in the old town, as well as an unforgettable buffet reception in the old town, in a fantastic restaurant.

I owe both thanks and an apology to my two co-editors, Vicky Karageorgou and Janice Gray, who had to carry a heavier editing burden than they should have, because of unexpected pressures of teaching on my part, which rendered me less efficient than I should have been. Many thanks are due also to Luc Quenneville (University of Windsor) who, once again, managed to get us organized and finished on time.

Part I

Biological integrity, ecology and the law

1 Environmental and ecological concerns in Europe and North America contrasted

Laura Westra

Introduction

This volume combines the work of authors from both Europe and North America. Aside from those addressing global issues in international law, there is a decided difference between the approach taken by those considering international or European laws from a European perspective, and those writing from a North American perspective: in short, the Europeans appear far more ready to trust the effectiveness of the environmental law regimes they discuss than their North American counterparts.

This difference applies not only to ecological/biological issues, but extends to the very basis of legal governance and human rights, as there is only one clear Canadian case where the author demonstrates that the law eventually prevailed in defence of human rights against the state itself (see Chapter 15 by Kathleen Mahoney). But that was a *sui generis* case, that is, one which involved Canadian First Nations, who supposedly enjoy a special relationship with the Federal Government of Canada, given the fiduciary duty of the latter regarding the former.

At any rate this is by no means a new development, as many legal scholars working and publishing in the United States, Canada and New Zealand appear primarily concerned with changing the approach to governance as a whole, rather than focusing on this or that specific legal regime. I include myself in this group, as the whole impetus behind the Global Ecological Integrity Project, from the start, was more than an effort to achieve a specific change in environmental law, perhaps (Westra 1994; Westra 1998), or even to achieve the elimination of environmental racism and the general disregard of human rights (Westra and Lawson 2001). Rather, it was and has been an ongoing effort to effect a radical revision of current forms of governance, with their flaws and *lacunae*, viewed from the perspective of a principled defence of human rights, starting from the biological integrity of all organisms. That principle is, in turn, grounded on the ecological integrity of the habitat required by each natural organism.

Thus, broadly speaking, it would seem that, outside of Europe, the emphasis (at least in our group's experience for the last 22 years) has been one of demanding a radical reconstruction of systems of governance founded on the radical destruction of the ongoing systems, which give primacy to trade, capital and unsustainable growth instead of ecologically sound reasoning.

In contrast, European scholars approach the overwhelming problems of the present systems of governance from at least a moderately optimistic position. They seem to consider each legal regime as a 'work in progress', or at least as the basis for present and future emendations, intended to render each instrument closer to a more desirable position, closer to one respectful of human environmental rights.

Thus their efforts towards *lex ferenda* start most decidedly from the present *lex lata*, which they appear to view as worthy of considered re-elaboration, not only as practically necessary. Clearly, even those who move directly to a discussion of international law must still use it as a basis for the future vision they propose, but there seems to be far less interest in doing what Bill Rees famously termed 'rearranging the deck chairs on the Titanic' at one of our meetings some years back.

I believe that this contrast reflects more than different temperaments, perhaps even different age groups, present in the generalizations I am suggesting. It seems that there is a basic difference in the way governance actually functions in North America, in contrast with the way it functions in Europe (perhaps including a country such as New Zealand): the basic difference is the amount and the extent of corporate power present in the respective legal systems in these continents.

That power is, undoubtedly, present everywhere today due to globalization. But there is a substantive difference in the ways the intrusive corporate power is felt, as it operates in North America or in Europe.

From *Santa Clara* to the Monsanto Immunity Act: the ascent of corporate power

> [T]he rise of the giant corporations during the past century has been the principal influence in the creation of a second – the secret – constitution. Corporations, at least those of giant size, are private governments and should be recognized as such.
>
> (Miller 1987: 242)

Although Miller's argument pertains exclusively to the US Constitution, the fact that the 'giants' to which he refers are indeed the supranational powers that intrude in global governance, entails that his position supports the argument of this work, even beyond its application to US governance.

It seemed obvious that the court in *Santa Clara* did not want to discuss the personhood question, and were quite satisfied taking it for granted as a starting point, without any philosophical or political argument in its support. Some of the commentators suggested personal motives, while others assumed simply carelessness, even negligence.

In contrast, Miller proposes that in fact, the *Santa Clara* court knew precisely what it was doing, although it could not predict all the ramifications and consequences of its decision over time. Miller says:

> *Santa Clara*, accordingly, is best seen as one of a clutch of decisions in which the Supreme Court was a willing ally of the property-owning class in the United States – those that Alexander Hamilton called 'the rich and well-born' in the Constitution Convention of 1787.
>
> (Miller 1987: 243)

The attack on the Fourteenth Amendment is considered to be especially heinous, as it transformed a landmark document designed to protect freed slaves, into one for the protection of those that needed it the least, as 'the state action doctrine added to corporate privilege, rendering the firms immune from constitutional norms' (*ibid.*).

Nor do we need to wait for the racially motivated repressions in South Carolina or Florida (Bullard 1994), or the repression of Indigenous revolts against mining and extracting industries today (Westra 2007). August 2012 saw the violent battle of South African policemen against the miners at America mine, who protested for non-payment of wages and poor working conditions, resulting in four murdered individuals.

But the use of federal troops to repress protesting workers, in the US, dates back to 1895[1] (Miller 1987: 243), when a strike by Pullman workers in 1894 'stopped trains and the mail', killing 'some of the mob': a result that was seen as acceptable as a solution to the problem (*ibid.*). Not only does the working class, one that includes African Americans, suffer from the constitutional dualism Miller decries: every US citizen is affected by it.

The US is 'governed by a type of democratic elitism' (*ibid.*: 246) but the secret institutions of the corporate-directed second constitution 'have marked antidemocratic authoritarian tendencies' (*ibid.*). Miller does not discuss the actual detailed results of this duality, but he does state that the separation between the economy and the polity is only functional, as they are complementary instead. Hence a truthful understanding of the situation would be to term it the 'corporate state'.

This intimate relation entails that the most important aspects of democratic citizenship are eliminated. Mark Kesselman notes that 'The hidden face of power is exercised not so much by the suppression of specific issues from the political agenda as by the exclusion of the most fundamental matters of public concern from the political sphere' (Kesselman 1982: 571).

The corporate control here acknowledged could not persist without an additional control, one we have neglected in this work given its wide reach and the public awareness of its results: the corporate control of the mass media. That is the final form of supranational control: neither the control of the law, of governance, nor of the markets exerts such widespread power. It is the ultimate social control of people's wants and preferences, so that unhealthy and even injurious choices become the 'real' preferences of US citizens to begin with but of people everywhere eventually, as the 'better way' is spread, from harmful McDonald's hamburgers and Coca-Cola, to mind-warping computer games and other such 'necessities', promoted to the public from childhood on.

Thus what sets North America apart is the presence of multiple aspects of legal corporate dominance, from the effortless acquisition of personhood in the *Santa Clara* decision in 1886 to the 2012 Monsanto Immunity act, signed into law by President Barack Obama; which, although expired in 2013, remains as a precedent and as an ominous indication of the incestuous relationship existing between the US government and Big Business (Westra 2014). The main advantage of being a legal person is to avoid responsibility for one's actions (Neoclous 2003: 147–65), as the ability to avoid the legal consequences that a natural person would endure derive from the corporate powers acquired over time, both legally and illegally.

The corporations, associations and institutions that are part of globalization are not open to any consideration of morality or human rights, beyond their own 'rights' and interests. Invoking the criminal law is therefore not only difficult, but often futile: 'the corporation, as a disembodied jural entity . . . has no physical body capable of incarceration, or corporal or capital punishment' (Grear 2010: 91).

The same thought is expressed even more forcefully by an anonymous protester from the Occupy movement in New York, as he waved a placard saying, 'I'll believe corporations are people when Texas executes one' (Collins 2013). Grear acknowledges that 'the new trade-related market friendly paradigm calls upon the state . . . to save capital by pursuing "de-regulation", de-nationalization and disinvestment' (Grear 2010: 14). But the borderless flow of capital (*ibid.*: 13) no longer needs the state, which has become, as we have seen, nothing but the agent, the tool of corporate power, for the most part.

Multiple cases simply repeat and reinforce our own case for overwhelming, multifaceted corporate power, no matter how acquired and exercised. But that power demonstrates the underlying criminality, the illegality pursued with impunity, both for acquiring it and for exercising it. In sum, corporate power, as it now exists, fails to reflect law or justice in several, connected ways. This lack includes the following aspects:

1   persistent lobbying of governments and corresponding intrusion in the drafting of treaties and regulations in international law;
2   intrusion in the political process through large donations to parties and individual candidates, especially in the US, home of the most powerful multinational corporations (MNCs);
3   ongoing control of domestic legal regimes to ensure impunity for activities that affect basic human rights to life, health and normal development;
4   arrogating traditional state powers for the protection of citizens, by the imposition of supranational constitutions, through the secret decisions of economic tribunals, such as those of NAFTA or the WTO, as those supersede the decisions arising from state constitutions;

5   the continued production of the chemical and other industrial products that affect human rights as in (3), with full knowledge of their composition and eventual effects, with the impunity guaranteed by both lobbying (1) and their intrusion in the political process (2); and

6   collusion with states' illegal aggressions, oppressions and attacks, and by the ongoing manufacture of weapons both legal and illegal.

Some of these activities would be viewed simply as criminal were they perpetrated by natural persons. Others, perhaps, are borderline between illegality and immoral legal behaviour. The intrusion in the political, democratic process is particularly worrisome despite the US court decisions that have legitimized that behaviour. The results of that intrusion ensure that elections now range quite far from the ideal of 'one man one vote'.

When these corporate activities are viewed together, it reinforces the impression of a total lack of concern for the public interest, the public order, and even the legitimacy of domestic governments, and the soundness of international law.

Beyond the state/corporate power grid: present and ongoing issues in Europe

> The most daunting challenge of adapting to the realities of the Anthropocene era is achieving a soft transition from state-centric world order, to a geocentric reconfiguring of the political community to enable the emergence of effective and humane global governance.
>
> (Falk 2014)

The corporate control discussed in the previous section entails that the expected connection between democracy and state governance has been broken, as especially North American states' main alliance is with corporate power not the aspirations, let alone the needs of their citizens. The political community is now formed primarily of transnational citizens (Sassen 2006), whose alliance is to global communities, 'geo-centric' movements and organizations, by-passing the local/national allegiance, now no longer in sync with their expectations.

Yet interest in international law and global compacts persists especially in the European Union, where the excesses of corporate power are somewhat less obvious, perhaps more subdued, at least in the amount of power they can display. In 2004 Jeremy Rifkin explained that the 'American Dream' of old was tied to an old system that no longer functions (Rifkin 2004: 196). Europe's 'dream' is not tied to American particularism, with its increasingly unacceptable exceptionalism, but the EU constitution is a document that resonates strongly with the citizens of many countries:

> If we were to sum up the gist of the document [the EU Constitution], it would be a commitment to respect human diversity, promote inclusiveness, champion human rights and the rights of nature, foster quality of life, free

the human spirit for deep play, build a perpetual peace, and nurture a global consciousness.

(Rifkin 2004: 213)

These principles indeed resonate with the global community, as do several of the European practices that render some of these principles and aspirations operational. Some of these practices reflect a level of actual concern and respect for the human right to health and for the environment, that are not present in North America. Evidence of this approach can be found especially in two areas where Europe's defence of the related human rights is clearly in evidence, in contrast with the negligence and disinterest present in North America. The first is the area of agriculture, which combines pesticide exposure and the promotion of GMOs; the second is the disregard for issues connected to climate change. These issues will be discussed briefly in sequence.

Food safety and corporate power in Europe

However in 1999 a group of EU Member States decided to systematically block the EU's authorization of any new GM products (both import and cultivation), until a legislative framework would be in place that would guarantee consumer choice (labeling and traceability), food and environmental safety (novel food and feed regulation, deliberate release directive), and liability in case of damage to the environment because of GMOs.

(Corporate Europe Observatory 2004)

A wide variety of questions were raised in Brussels regarding GM products primarily because after the 'first generation of GM producers', there was no clear 'consumer benefit' and a series of 'food scandals' primarily regarding the emergence of BSE and other diseases of domestic animals showed clearly the downside of GMOs (*ibid.*). In addition, the tired arguments from biotech corporate propaganda, about the 'fact' that 'GM is good for the environment, is safer because more scientific, is beneficial for hungry people' (*ibid.*) were not acceptable to the European people. Perhaps only the argument about possible employment was still considered with some favour, and may still continue to gain ground. But many of the proposed laws are being designed and implemented, if slowly.

For instance, the 'Deliberate Release Directive and Novel Food and Feed' is intended to assess the risk possibility each new product may pose. But the European Food Safety Authority (EFSA), which should decide on human and animal health and safety, still continues to rely on 'data provided by the company' (*ibid.*). A new Directive (EC MEMO 04/16) has introduced the labelling of animal feed, but animal products fed on GM food do not have to be labelled, although 'The rules on traceability provide means to trace products containing or produced from GMOs, through the food production and distribution chain' (Corporate Europe Observatory 2004: 2).

A new 'Environmental Liability Directive' was designed and implemented to cover 'damage to biodiversity, even limited to protected species and habitats' (*ibid.*). But damage to health or to organic products is not covered by the directive, defined by environmentalists as an 'empty shell' (*ibid.*: 3). More hopeful was the Cartagena Protocol on Biosafety, including a Biosafety Regulation purporting to control 'the export of GM products from EU to other parts of the world' (*ibid.*). That document introduces the precautionary principle (it is noteworthy that the US refused to sign the document over their objections to that point) but it also had a clause added stating that the provisions of the Protocol would not affect such international agreements as the WTO (*ibid.*).

However, many specific EU countries worked on establishing so-called 'co-existence': the Italian Biotechnology Association (Assobiotec) reported to EuoBio four cases of regional co-existence legislation (in Tuscany, Basilicata, Apulia and the Marche) that basically meant a ban on the use of GM in those regions' (*ibid.*: 8).

As well, at least three Austrian regions (Upper Austria, Salzburg and Carinthia) have effected a ban on GM cultivation (*ibid.*: 8), whereas The Netherlands, the UK and Belgium are pro-GM instead. Leaving aside the specifics of various EU regulatory bodies and preferences, some welcome developments appear:

> In any case, some GM companies, like Bayer CropScience, need to start making some money soon – with GM. Otherwise investors might demand to move out of it. Syngenta has announced to withdraw from all GM research activities from Europe following similar decisions earlier by Monsanto, Dupont and others. Reasons named were public resistance, high authorization hurdles and the lack of market opportunities. Ultimate proofs that overall the biotech lobby has not been a great success. At least the biotech climate in Europe is not changing.
>
> (Corporate Europe Observatory 2004: 9)

At any rate, this brief examination of the practical aspects of GM promotion and lobbying, somewhat dated though it is, cements further our conclusions regarding:

• corporate power and its intrusion into public life and policy decisions, despite the generally better environmental regulatory regimes present in the EU (see previous section); and
• the total lack of concern on the part of corporate giants, even those affiliated with medicine, regarding human rights to health and a safe environment, contrasted with their total and exclusive focus on their economic advancement.

Hence the question posed earlier – that is, how can interventions such as those by Greenpeace and Friends of the Earth (Europe), as well as by local groups such as IFOAM (organic farming) and Eurocommerce (retail industry), be permitted to have equal footing (to say the least) with MNCs like Monsanto and Bayer, who, by their words and actions, clearly indicate their total disregard for human rights, ecological concerns, let alone the presence of democratic institutions?

After the Treaty of Amsterdam entered into force on 1 May 1999, the 'democratic deficit' in the EU was increasingly considered to be acute, as the liberalization of trade reached new heights, and as multinational accords (MAIs) were secretly negotiated under the directives of intergenerational industry with the main purpose of eliminating any measures intended for the protection of local economies, or for the rights of workers and of the environment itself (Balanya *et al.* 2005: 199).

The march away from democracy and human rights was taking place while, according to the World Bank, rich countries' average income was US$26,500 a year, five times higher than the global average of US$5,120 and 62 times higher than the income of the poorest countries (*ibid.*: 203). But the economic crisis did not inspire MNCs to provide measures capable of redressing the gross imbalances and desperate world poverty. In contrast, the situation whetted their appetite to capture the 'emergent markets' in the East and in Latin America.

The EU moratorium on GMOs persisted:

> On September 29, 2006, a World Trade Organization (WTO) dispute settlement panel issued its final decision (the EC-Biotech decision) in the complaint brought by the United States, Canada and Argentina against the European Communities (EC), over EC's alleged moratorium on the approval and marketing of agricultural and food products containing genetically modified organisms (GMOs).
>
> (Gonzales 2007: 583)

The WTO panel concluded that the moratorium on biotech products (1999–2003), which 'resulted in undue delay' in the 'premarket approval procedures in violation of the WTO Agreement on the Application of Sanitary and Phytosanitary Measures' (*ibid.*: 584). The conflict was based on the approach of the US, which could be defined as 'product oriented', as it assumed that to transfer genes from one species to another was no more dangerous or novel than the traditional hybridization technologies. Thus there was no need for any stricter oversight than in the past.

In contrast, the 'EC has adopted a process oriented approach, which assumes that genetically altered products may pose novel or unique human health and environmental risks as a consequence of genetic modification' (*ibid.*). Hence the EU expects risk evaluations and 'public input' before issuing any approval of GMOs, and it insists on an extensive tracing system. It is worthy of note that the panel involved in the decision did not even attempt to consider the safety of the GMOs or the rights of countries to follow the democratic decisions of their citizens regarding their position on GMOs, nor did the panel discuss the precautionary principle and its possible impact on the controversy.

The conflict's effects involved problems beyond Europe and North America, as several countries in Africa were offered 'aid' in the form of GM seeds, which were refused. The EU accused the US of exploiting a humanitarian crisis in order to expand the market for GMOs (World Trade Organization 2006: 4; Nuffield Council on Bioethics 2004). One of the reasons for Africa's refusal of the 'aid' proposed by the Bush administration was precisely the difficulty of continuing these countries' lucrative trade with Europe, due to the difficulties deliberately posed by the EU on the sale of GMOs.

Capitalism and climate change in North America and Europe

> Finding new ways to privatize the commons and profit from disaster is what our current system is built to do; left to its own devices, it is capable of nothing else.
>
> (Klein 2014: 9)

The current system Naomi Klein refers to is the forward march of American capitalism and the globalized primacy of trade it fosters. As it seeks 'free trade deals', it continues to ensure that neither health nor environment are given serious consideration. In contrast, MNCs are increasingly free 'to produce their goods as cheaply as possible and sell them with as few regulations as possible – while paying as little in taxes as possible' (*ibid.*: 21).

This approach ensures that regulatory regimes intended to take over the control of the commons, with no concern for the ongoing looming disasters created by climate change, remain committed to 'market fundamentalism' (*ibid.*). And disasters include more than tsunamis, recurring floods and other extreme events. They include slowly but inexorably melting ice in the Arctic and other famous world glaciers; they also include the terrible human rights abuses that accompany the mining and extractive operations that support the fuel overconsumption which, in turn, exacerbates climate change.

Nor are the corporate giants that have fostered and supported climate change moving to correct their mistakes. On the contrary, they are actively engaged in what Michael Klare terms 'the race for what's left':

> Most of the world's conventional, easy-to-extract reserves of oil and natural gas have now been exhausted. The major energy companies are doing everything they can to prolong the life of existing fields and to develop new deposits in previously inaccessible areas, such as the Arctic and the deep waters of the gulf of Mexico; increasingly, however, they will be forced to rely on unconventional hydro carbon sources to sustain the flow of gas and oil.
>
> (Klare 2014: 127)

Hence, rather than acknowledging the damage they have wrought through their promotion of overconsumption of a non-renewable resource, the corporations continue to pursue the relentless quest for profit, using even riskier tactics and procedures, and extending their activities to the most vulnerable areas in the globe:

> With oil companies bidding on more and more exploration blocks and sending additional drillships to the area, Greenland is coming to encapsulate many features of what is being called the Arctic oil rush . . . exploiting the Arctic will not be easy, extremely low temperatures, frequent storms, ubiquitous sea-ice and growing concern over endangered species and a fragile environment all pose significant obstacles.
>
> (Klare 2014: 74)

Other options for continuing exploitation include the Athabasca tar sands in the boreal forest of northeastern Alberta. However, given the climate, the bitumen, which is like a thick syrup in warmer climates, solidifies in the cold, and must be either mined or 'heated underground in order to make it fluid' (*ibid.*: 100). The latter option involves multiple environmental risks, and it is so difficult that, for the most part now, the many oil producers dig it up from open-pit mining, like coal or iron, which entails cutting down large areas of boreal forest and removing the topsoil, leaving behind hazardous pools of tailings. As well, the '[c]hemical laced water from the ponds, is leaking into the Athabasca river and endangering the health of downstream communities' (*ibid.*: 102; see also Kinzig 2009: 37–59).

In addition to this method of extraction, another in use is 'steam-assisted gravity drainage' (SAGD), a practice which is highly energy intensive (using natural gas), but also uses large quantities of water, eventually left in tailing ponds, laced with multiple chemicals (Klare 2014: 103). This practice destroys the health of the land and the survival of the animals required by the Indigenous Peoples living a traditional lifestyle, such as the First Nations of Canada.

Added to the growing list of environmental disasters that characterize the seemingly unstoppable march of corporations 'racing' for whatever is left, with complete disregard for the effects their previous activities have caused, is the determination to extract gas from shale formations below the surface of the earth:

> In a typical application energy firms use conventional drills to dig a vertical well down to the level of the shale, usually a mile or so underground . . . once numerous channels have been drilled and encased in concrete, small explosive charges are set off to puncture the concrete and shatter the rock. Finally millions of gallons of water laced with sand, lubricants, and other chemicals are forced into these fissures . . . allowing gas to escape toward the central well and then be collected on the surface.
>
> (Klare 2014: 117)

But hydrofracking is not a benign operation: each well uses as much as five million gallons of water, which is mixed with toxic chemicals and the remaining water, 'the flowback' (*ibid.*), is stored in caverns to be eventually treated. However, both unlined storing ponds and the unsupervised 'treatment' itself pose grave danger to public health and the environment (*ibid.*: 119; Gray 2014: 129–47). Social movements and protests are everywhere in Europe against the spread of this particular corporate activity, and they extend to Australia and New Zealand. In Canada an ongoing case involves Jessica Ernst, whose water comes from a private well presently contaminated by the fracking activities of Encana near Rosebud, Alberta. Kathleen Mahoney (author of Chapter 15 in this volume) is active in the case, as she reports that the judge did not accept the harms inflicted on Ms Ernst, but protected instead the Energy Resources Conservation Board's immunity in its decisions regarding allowing Encana to continue fracking despite the severe contamination of her well water and other harms (Kathleen Mahoney, personal communication).

Concluding thoughts

It is noteworthy that the ongoing, unrepentant attacks on the environment and human and animal life are the result of the quest for profit by MNCs based in North America. This brief additional analysis of corporate power in Canada and the US indicates there is a disproportion between the unrestrained corporate power in the North American continent, and the somewhat muted effects in Europe that often run aground because of the very visible social resistance. Citizens march against fracking, against exploration and drilling in the Arctic, and they are clearly reported in the news media, often succeeding in achieving some changes in public policy or at least delaying the harmful practices and obtaining impact assessment and research before projects are finally accepted.

The extent of those protests would be unthinkable, not only in the US where such activities are viewed as unpatriotic and even forms of ecoterrorism, but even in Canada, where a largely unresponsive government has no apparent interest in listening to the beliefs and wishes of that country's citizens.

The dissonance between the lack of commitment to democracy and the rule of law evidenced by most North American academics should be compared to the interest of many European scholars to work within the system, to improve current regimes, evident in the chapters of Part I of this book.

This is not a hard and fast separation, to be sure; Chapters 6, 7 and 15 (by D'Aloia, Dellapenna and Mahoney, respectively) all discuss the possibilities existent in international law, while Montini (Chapter 9), from Italy, in contrast, addresses an issue that, while desirable, could not really be addressed in the Canadian context, for the reasons mentioned (see also Chapter 17 on Canadian democracy, by Peter Venton). Far more typical of the strong critique of present government's practices in North America are the chapters by Soskolne (Chapter 16), Dyson (Chapter 18), Gwiazdon (Chapter 20), and Borràs and Pigrau (Chapter 21), all of which indicate the grave difficulties faced by those who attempt to protect public health and the environment, in a general climate that is decidedly unfriendly to the support of human rights and the environment and to restrain corporate profit seeking. Despite the fact that Borràs and Pigrau's chapter pertains to a European country, it is the only one that does indicate that public protests do not always prevail even in Europe.

From a global perspective – across Europe and North America – Klaus Bosselmann (Chapter 22) reflects on the book's general theme: systems integrity, and what it may mean for ecology, governance and law. His concluding chapter shows the gap between the status quo of state-centred governance and the prospects of achieving ecological integrity. The gap may be widening, but the reverse may also be true: the normative qualities of ecological integrity and their potential for institutionalizing trusteeship are persuasive. Earth governance, as Bosselmann defines it, could provide a unifying framework across all nations and cultures.

Note

1 In re Debs, 158 US 564 (1895).

References

Balanya, Belen, Ann Doherty, Olivier Hoedeman, Adam Maamet and Erik Wesselius (2005) *Europe, Inc.: Comment les Multinationales Consruirent l'Europe et l'Economie Mondiale*. Marseille: Éditions Agone.

Bullard, Robert (1994) *Dumping in Dixie*. Westview Press, Boulder, CO.

Collins, Sheila (2013) Can Occupy Wall Street Make the New World Possible? In L. Westra, P. Taylor and A. Michelot (eds), *Confronting Ecological and Economic Collapse: Ecological Integrity in Law, Policy and Human Rights*. Abingdon: Routledge.

Corporate Europe Observatory (2004) Power Struggle over Biotech in Brussels. December. Available at http://archive.corporateeurope.org/biotechbrussels.html.

Falk, Richard (2014) Changing the Political Climate: A Transitional Imperative. Available at www.greattransition.org/publication/changing-the-political-climate-a-transitional-imperative.

Gonzales, Carmen G. (2007) Genetically Modified Organisms and Justice: The International Environmental Justice Implications of Biotechnology. *Georgetown International Environmental Law Review* 19: 583–636.

Gray, Janice (2014) Frack off! Law, Policy, Social Resistance, Coal Seam Mining and the Earth Charter. In L. Westra and M. Vilela (eds), *The Earth Charter, Ecological Integrity and Social Movements*, 129–47. Abingdon: Routledge.

Grear, Anna (2010) *Redirecting Human Rights: Facing the Challenge of Corporate Legal Humanity*. Basingstoke: Palgrave Macmillan.

Kesselman, Mark (1982) The Conflictual Evolution of American Political Science: From Apologetic Pluralism to Trilateralism and Marxism. In J. Greenstone (ed.), *Public Values and Private Power in American Politics*, 571. Chicago, IL: University of Chicago Press.

Kinzig, Robert (2009) The Canadian Oil Boom: Scraping Bottom. *National Geographic* (March): 37–59.

Klare, Michael T. (2014) *The Race for What's Left*. New York: Henry Holt & Company.

Klein, Naomi (2014) *This Changes Everything: Capitalism versus the Climate*. New York: Allen Lane.

Miller, Arthur S. (1987) Corporations and Our Two Constitutions. In W. J. Samuels and A. S. Miller (eds), *Corporations and Society*, 241–62. New York: Greenwood Press.

Neoclous, M. (2003) Staging Power: Marx, Hobbes, and the Personification of Capital. *Law and Critique* 1(14): 147–65.

Nuffield Council on Bioethics (2004) The Use of Genetically Modified Crops in Developing Countries. Available at http://nuffieldbioethics.org/project/gm-crops-developing-countries.

Rifkin, J. (2004) *The European Dream*. New York: Arches/Penguin.

Sassen, Saskia (2006) *Territory Authority Rights: From Medieval to Global Assemblages*. Princeton, NJ: Princeton University Press.

Westra, Laura (1994) *The Principle of Integrity*. Lanham, MD: Rowman & Littlefield.

Westra, Laura (1998) *Living in Integrity*. Lanham, MD: Rowman & Littlefield.

Westra, Laura (2007) *Environmental Justice and the Rights of Indigenous Peoples*. London: Earthscan.

Westra, Laura (2014) *Revolt against Authority*. Leiden: Brill.

Westra, Laura, and Bill Lawson (2001) *Faces of Environmental Racism*, 2nd edn. Lanham, MD: Rowman & Littlefield.

World Trade Organization (2006) *European Communities – Measures Affecting the Approval and Marketing of Biotech Products*. Panel Report, WT/DS/291/R, WT/DS/292/R, WT/DS/293/R, September 29. Available at www.wto.org/english/tratop_e/dispu_e/cases_e/ds291_e.htm.

2 Access to genetic resources and traditional knowledge and the fair sharing of benefits

The way forward in the European Union

Sandra Jen

On 16 April 2014, the European Union (EU) adopted Regulation No 511/2014 on Compliance Measures for Users from the Nagoya Protocol on Access to Genetic Resources and the Fair and Equitable Sharing of Benefits Arising from their Utilization in the Union (hereafter the EU ABS Regulation).[1] Its aim is to fulfil the European Union international obligations as a contracting party to the Nagoya Protocol. The final text adopted is the result of two years of debates and negotiations between the European Parliament, EU member states and the European Commission in the framework of the EU co-decision procedure. Following a public consultation between 24 October 2011 and 30 December 2011[2] and a stakeholder workshop, the European Commission published on 4 October 2012[3] a proposal for a Regulation on Access to Genetic Resources and the Fair and Equitable Sharing of Benefits Arising from their Utilization in the Union.

The Nagoya Protocol on Access to Genetic Resources and the Fair and Equitable Sharing of Benefits Arising from their Utilization to the Convention on Biological Diversity (hereafter the Protocol or NP) was adopted at the very end of the 10th Conference of the Parties to the Convention on Biological Diversity (CBD) on 29 October 2010, after long and intense negotiations (Wallbott *et al.* 2014: 36). The negotiations at CBD COP 10 were the culmination of a ten-year process to set up a legally binding framework to better implement Article 15 and Article 8j of the CBD. The adoption of the Bonn Guidelines[4] at CBD COP 6 in 2002 marked one step in this process. However, as the guidelines were a non-binding instrument and focused on access to biodiversity genetic resources rather than on measures obliging users, developing and provider countries kept pressing to have a legally binding protocol (Wallbott *et al.* 2014: 36). Following a recommendation at the World Summit on Sustainable Development in 2002[5] and UN General Assembly resolution 57/260, negotiations were relaunched for an 'international ABS regime'. They resulted in the adoption of the Nagoya Protocol.

In the very last steps of the negotiations, some important elements for the effectiveness of the regime were still being debated and remained as open issues for discussion and agreement between the contracting parties. This includes procedures and mechanisms to promote compliance and address cases of non-compliance (Article 30 NP), the Access and Benefit-sharing Clearing-House (Article 14 NP), and the global multilateral benefit-sharing mechanism (Article 10 NP). In July 2014, the Nagoya Protocol reached 50 ratifications, the number required for its subsequent entry into force 90 days later, on 12 October 2014.

Through the participation of both the Members States and the European Commission, the EU has been a key player in the negotiations of the Protocol, working to reach an agreement but also to promote user-country measures and international access standards. With the US not a party to the CBD, the EU is the largest user of biodiversity genetic resources under the Protocol (Oberthür and Rabitz 2014: 79). Similarly, the EU is the second-largest player in global biotechnology behind the US.

The EU ABS Regulation No 511/2014 is only one part of the implementation system for the Nagoya Protocol in the 28 EU member states. It sets up a number of obligations and procedures for users to comply with the Nagoya Protocol (i.e. for companies from different sectors using genetic resources from biodiversity from the rest of the world for research, development and commercialization, for suppliers and for academic, university and non-commercial researchers). This regulation will still have to be completed by implementing acts from the European Commission, in particular to define concepts such as 'stage of final development of a product', the procedure for due diligence declaration, or the recognition of best practices.

As regards access to biodiversity genetic resources on the territory of EU member states, this remains the competence of national governments: if one of the EU member states intends to regulate access to biodiversity genetic resources from their territory and organize the fair and equitable sharing of benefits arising from the biodiversity genetic resources and associated traditional knowledge originating from its territory, specific legislation has to be adopted at national level to that end. Some EU countries had already done so in application of Article 15 of the CBD.[6] This is the case with Bulgaria's Biological Diversity Act (Article 66) of 2002,[7] Croatia's Nature Protection Act of 2003 (Articles 89–91), and Malta's Flora, Fauna and Natural Habitats Protection Regulations No. 311 of 2006 (Articles 36 and 37). Following the adoption of the Nagoya Protocol and related EU Regulation, a number of countries like Germany and the Netherlands have chosen to pursue their policy of not regulating access and benefit sharing. A number of others at the time of writing were still considering the need to adopt access legislation and the choice between a system where free access would be the principle and regulation for specific areas would be the exception or the opposite. France with its overseas departments and territories is one of the few EU member states, if not the only one, to set up a comprehensive system of controlled access and benefit sharing.[8]

This chapter considers selected provisions from the EU ABS Regulation as key elements of the EU system set up for users to comply with the Nagoya Protocol in the EU. It does not seek to address issues concerning the temporal and material scope of the Protocol and the Regulation, nor links with other international agreements addressing ABS, in particular the International Treaty on Plant Genetic Resources for Food and Agriculture. It compares and analyses the initial text proposed by the European Commission, the final text adopted in 2014 and corresponding provisions in the Nagoya Protocol in order to try understand the approach chosen in the final text.

Due diligence as core principle for EU users to comply with ABS regulatory systems

Rights and obligations of Nagoya Protocol Parties regulating access to genetic resources and associated traditional knowledge are stipulated in Articles 6 and 7 of the Nagoya Protocol. Whenever a Party chooses to regulate access in application of the Protocol, prior informed consent (PIC) is the key procedure for operators to comply with. For cases when indigenous and local communities have to give their consent for access to genetic resources and/or associated traditional knowledge the protocol refers to 'PIC or approval and involvement'.

The EU system set up to comply with NP provisions on user compliance relies on due diligence to be exercised by users to ascertain that the material they use, as well as associated traditional knowledge, has been accessed in conformity with the legislation and regulations of the provider country. This includes ensuring that benefits are fairly and equitably shared according to mutually agreed terms (MAT) and in accordance with any applicable legislative or regulatory requirements (Article 4.1 EU ABS Regulation). The requirement for users to exercise due diligence to ensure that the resources they use is of good legal status can be compared to the provisions of the EU Illegal Timber Regulation.[9] Article 6 of this regulation provides a detailed description of the 'due diligence system' to be applied by operators placing timber or timber products on the EU market. The 'due diligence system' is described as a framework of procedures and measures to address three elements: access to specific information; risk assessment procedures; and risk mitigation procedures. Article 5 of the same regulation provides a clear obligation for traders to ensure traceability throughout the supply chain. Neither due diligence measures nor best practices to be recognized by the European Commission are defined with the same level of precision in the EU ABS regulation. The provisions for traceability are also weak compared to those of Article 5 of the EU Timber regulation.

Article 4.2 of the EU ABS Regulation reinstates the obligation to respect MAT for the transfer and utilization of genetic resources and associated traditional knowledge in so far as the legislation or regulatory system of the providing country may require it. The wording of this article raises questions regarding the motivation for emphasizing the distinction between Parties to the Nagoya Protocol with a regulatory system requiring the establishment of MAT and those which do not. This text is the result of the co-decision procedure between Council and European Parliament and was not in the Commission initial proposal (see Document COM (2012) 576). The intention may have been to clarify that the

EU is not imposing a general obligation for EU users to obtain PIC and MAT but only to comply with the legislation and regulatory requirements of the provider country. This provision should reassure concerned operators that when they operate in a Party to the Protocol which does not regulate access to genetic resources or does not require the establishment of MAT as part of their basic ABS system, the transfer and utilization of genetic resources and associated traditional knowledge can be done without MAT. The emphasis on the possibility or opportunity to access biodiversity genetic resources and associated traditional knowledge without MAT raises ethical concerns in relation to the objective of 'fair and equitable sharing of benefits arising from the utilization of genetic resources' (Article 1 NP), and even more so when associated traditional knowledge is at stake. These considerations if not properly addressed in the EU ABS Regulation should be considered carefully by companies as part of their corporate responsibility and business ethics policy.

Availability of internationally recognized certificate of compliance

Article 4.3 of the EU ABS Regulation gives further clarifications on the obligations on users to undertake due diligence, making a distinction between situations where an internationally recognized certificate of compliance is available and where it is not. The European Commission initial proposal for Article 4 made no distinction between such situations and listed five types of information to be sought, kept and transferred (Article 4.2a), an obligation to obtain additional information or evidence where uncertainties about the legality of access and use persist (Article 4.2b) and an obligation to obtain a proper access permit, establish MAT, or discontinue the use where it appears that access was not in accordance with applicable access and benefit-sharing legislation or regulatory requirements (Article 4.2c). A system with three stages of due diligence measures was thus proposed but making no consideration for the specific legal value of the internationally recognized certificate of compliance. In the final text adopted the provisions of Article 4.2b and 4.2c have been merged in Article 4.5 with alteration of wording (see comments below). The combination of these provisions in a new article can lead to questions as to whether its scope intends to also cover situations where users have an internationally recognized certificate of compliance or only for situations where such a certificate is not available (Article 4.3b).

Reference to the internationally recognized certificate of compliance in the initial Commission proposal was made only in Article 9.5 on 'Checks on user compliance' stating that these certificates should be accepted by the competent authorities as evidence of compliance with the ABS legislation or regulatory requirements of the Party to the Nagoya Protocol from where it had been accessed. Likewise in the Nagoya Protocol, the term 'internationally recognized certificate of compliance' is introduced for the first time in Article 17.1a(iii),[10] entitled 'Monitoring The Utilization of Genetic Resource'. The certificates which are to become the core elements of the ABS international system of compliance and monitoring are not defined in Article 2 of the Protocol concerning 'Use of term' but in Article 17.2, which states that 'a permit or its equivalent issued in accordance with Article 6, paragraph 3e and made available to the Access and Benefit-sharing Clearing House, shall constitute an internationally recognized certificate of compliance'. These certificates are to be granted in application of Article 6.3e of the Nagoya Protocol. Article 17.3 of the Protocol describes the legal status of these certificates stating that 'an internationally recognized certificate of compliance shall serve as evidence that the genetic resources which it covers have been accessed in accordance with prior informed consent and that mutually agreed terms have been established, as required by the domestic access and benefit sharing legislation or regulatory requirements of the Party providing prior informed consent'. Finally, Article 17.4 lists the minimum information that the certificate of compliance shall contain, with the precision at the end: 'when it is not confidential'.

The requirement in Article 4.3a of EU ABS Regulation that the transfer of the information on MAT can be limited to the content 'relevant for the subsequent users' may lead to situations where information may be reduced through the supply chain since the precedent users may consider that only part of the information it possesses on MAT is relevant for the subsequent user.

Article 4.3b of the EU ABS Regulation provides for situations 'where no internationally-recognized certificate of compliance is available'. In light of Article 2.4 of the EU ABS Regulation, these provisions clearly do not intend to address situations where resources have their origin in a country not party to the Nagoya Protocol or situations where the resources have been accessed in countries not regulating access to biodiversity genetic resources. Article 6.3 of the Protocol specifies obligations of Parties regulating access and requiring Prior Informed Consent. These include in particular 'issuance at the time of access of a permit or its equivalent as evidence of the decision to grant prior informed consent and of the establishment of mutually agreed terms, and notify the Access and Benefit-sharing Clearing House accordingly' (Article 6.3e NP). Combined application of Articles 6.3 and 17 of the Protocol, the establishment of the ABS Clearing-House[11] and support for capacity building should result over time in all Parties regulating access to be able to provide users with the internationally recognized certificate of compliance. However, in the meantime, the list of information and documents required in application of Article 4.3b of the EU Regulation provide users with elements to ascertain legal certainty as regards their obligation of due diligence. The list under Article 4.3b includes 'the presence or absence of rights and obligations relating to access and benefit sharing' 'access permits, where applicable' and 'mutually agreed terms, including benefit-sharing arrangements, where applicable'. Article 4.3 should also be read in conjunction with Article 4.5, which states that 'when the information in their possession is insufficient or uncertainties about the legality of access and utilization persist, users shall obtain an access permit or its equivalent and establish mutually agreed terms, or discontinue utilization'. This article raises questions about the situation envisaged and seems to allow for a large part of discretion concerning the basic legal checks and controls that should be done before starting to utilize any resources. It appears to allow for situations where resources can start being used without proper checks and assurance on the requirements for PIC and MAT from the country of origin and provide for measures to comply a posteriori with the access requirements of the country of origin. While obtaining information on ABS regime from different countries may have been challenging in the past, the adoption and entry into force of the Nagoya Protocol and the establishment of the ABS Clearing House should result in a situation where there are no excuses for not being aware of the legal conditions attached to biodiversity genetic resources from a specific country. In addition in line with Article 15(5) of the CBD Article 6 of the Nagoya Protocol states clearly that 'access . . . shall be subject to prior informed consent . . . unless otherwise determined by that Party'. It should therefore always be assumed that PIC is required unless a country clearly declares that it is not (Greiber *et al.* 2012: 96). Some difficult situations may remain for a while after the entry into force of the Protocol when States have not yet adopted the requisite legislation or regulatory requirements to comply with Article 6 NP. Nevertheless, basic measures of due diligence in that case should be to consult the State authorities and avoid using resources from that country until clarification on the legal status is obtained. Provisions for risk assessment procedures inspired from the EU Timber regulation (Article 5.1b) could have been beneficial to bring in more objective elements to address situations of uncertainty.

Registered collections and best practices to support due diligence

The Regulation provides for a EU-wide system of registered collections of biodiversity genetic resources (Article 5) and recognition of best practices (Article 8) developed by associations of users or other interested parties to comply with users' obligations and monitoring requirements. Users choosing to obtain a genetic resource from a collection included in the register of collections will benefit from a presumption of due diligence with regards to the information required in application of Article 4.3 EU ABS Regulation. The registered collections have to comply with criteria and procedures to ascertain the good legal status of their resources. Control is under the responsibility of member states. In practice these collections should include, for instance, those of networks of botanical gardens such as the International Plant Exchange Network (IPEN) or collections of micro-organisms such as the Belgian Co-ordinated Collections of Micro-Organisms (BCCM).

Implementation of recognized best practices is to be considered by national authorities as a factor to reduce risks of non-compliance (Article 9 EU ABS Regulation), thus reinforcing as well the declarations of due diligence to be done by user at the stage of final development of a product and potentially leading to lighter control procedures on user implementing them.

The combination of best practices and registered collections are important elements to support users in their due diligence procedures to ensure compliance and reduce risks of exposure to criticism of biopiracy.

Declaration of due diligence at the stage of final development of a product

The stage at which users should make their declaration of due diligence has been extremely debated in the adoption process. While the Commission text proposed this declaration should be made when requesting market approval or at the time of commercialization when no market approval is required, amendments were proposed in the European Parliament to have a declaration of compliance also made when applying for patents. This would have significantly strengthened the direct implications of the compliance procedures but, like during the negotiations of the Protocol, this proposal was strongly opposed. In application of Article 7 of the EU ABS Regulation, declaration of due diligence must be made by users at the stage of the final development of a product and by all recipients of research funding whether public or private when research involves the utilization of genetic resources and associated traditional knowledge. The Commission is charged with the task of determining the stage of final development of a product in different sectors. If this sectoral approach could provide the flexibility requested by economic operators, it could make scrutiny and monitoring more difficult for other stakeholders such as civil society, NGOs or representatives of indigenous people and local communities.

Penalties

Finally penalties for infringement of the regulation and their application shall be decided by member states. The French draft legislation establishes a significant deterrent, with penalties of €150,000 and one year imprisonment for utilizing biodiversity genetic resources and/or associated traditional knowledge without document required in application of Article 4.3 of the EU Regulation or for failing to seek, keep and transfer to subsequent users information relevant for the subsequent users in application of Article 4 of the Regulation. The penalty is brought up to €1 million when the 'utilization of resources or traditional knowledge has led to commercial use'.[12]

Reporting and review clause

Member states have to report on the application of the Regulation every five years starting from 11 June 2017. In 2018, based on the member states' reports, the Commission will draw a synthesis report on the EU-wide application of the regulation and a first assessment of its effectiveness. Article 16.3 of the EU ABS Regulation provides for a review every ten years of the functioning and effectiveness of the Regulation in achieving the objectives of the Nagoya Protocol. It should be noted that the reporting timescales envisaged are long. In comparison, the EU Timber Regulation requires member states to report every second year and the Commission to undertake a review every six years and the Regulation on illegal, unreported and unregulated fishing[13] requires member states to report every two years, with the Commission undertaking a report every three years.

Conclusion

The EU ABS Regulation sets the fundamental framework for the implementation of the Protocol in the EU and its effective consideration by the relevant economic operators in Europe. Its application, with a strong emphasis on due diligence measures that users have to take and demonstrate, is likely to imply significant adaptations and changes in the procedures followed so far for bioprospecting, traceability, accountability, transparency, and research and development. The recognized best practices and register of collections will need to be closely scrutinized in order to assess the reliability of the system.

A complete overview of the EU system to comply with the Nagoya Protocol will be possible once all EU Members States have adopted their own implementation legislation concerning access to genetic resources over which they have sovereignty rights and when procedures for implementation of the EU ABS Regulation have been set up. These include national procedures for declaration of due diligence, designation of national competent authorities, checks on user compliance, rules for penalties but also the European Commission system for recognition of best practices, the register of collections within the EU and the definition of 'final development of a product' in different sectors.

Above all, however, addressing biopiracy still depends on the legislative and regulatory regime of the provider countries – in particular, their capacity to put in place and implement robust, clear and efficient procedures to regulate access, apply PIC procedures, regulate and negotiate MAT, build up capacities of authorities, and indigenous and local communities, as well as raise awareness of the new principles and procedures.

Finally, it will be important to see what consideration and participatory procedures will be put in place for indigenous peoples and local communities in national regulatory systems to be established in relevant EU member states (e.g. France for its overseas territories, Denmark for Greenland and the Faroe Islands).

Notes

1 Regulation (EU) No. 511/2014 of the European Parliament and of the Council of 16 April 2014 on compliance measures for users from the Nagoya Protocol on Access to Genetic Resources and the Fair and Equitable Sharing of Benefits Arising from their Utilization, available at http://eur-lex.europa.eu/legal-content/EN/TXT/PDF/?uri=CELEX:32014R0511&from=EN.
2 See http://ec.europa.eu/environment/consultations/abs_en.htm and http://ec.europa.eu/environment/consultations/abs_results_en.htm.
3 COM 2012/0576 Final 4 October 2012 Proposal for a Regulation of the European Parliament and of the Council on Access to Genetic Resources and the Fair and Equitable Sharing of Benefits Arising from Their Utilization in the Union, available at www.ipex.eu/IPEXL-WEB/dossier/document/COM20120576.do.
4 Bonn Guidelines on access to genetic resources and the fair and equitable sharing of benefits arising from their utilization, available at www.cbd.int/doc/publications/cbd-bonn-gdls-en.pdf.
5 Para. 44o, Johannesburg Plan of Implementation.
6 See IEEP, Ecologic and GHK (2012) and Cabrera Medaglia *et al.* (2012).

7 Biological Diversity Act Promulgated in State Gazette No. 77/9.08.200, last amended in State Gazette No. 19/8.03.2011.
8 See Projet de Loi relative à la biodiversité, Title IV, Access to genetic resources and fair and equitable sharing of benefits Articles 18 to 26, 26 March 2014, available at www.assemblee-nationale.fr/14/projets/pl1847.asp.
9 Regulation (EU) No 995/2010 of 20 October 2010 laying down the obligations of operators who place timber products on the market. OJ L 295/23, available at http://ec.europa.eu/environment/forests/timber_regulation.htm.
10 An Explanatory Guide to the Nagoya Protocol on Access and Benefit Sharing, IUCN Environmental Policy and Law Paper No 83, 2012, available at https://cmsdata.iucn.org/downloads/an_explanatory_guide_to_the_nagoya_protocol.pdf.
11 See Article 14, Nagoya Protocol.
12 Projet de Loi relative à la Biodiversité, Article 20, available at www.assemblee-nationale.fr/14/projets/pl1847.asp.
13 Council Regulation (EC) No 1005/2008 of 29 September 2008 establishing a Community system to prevent, deter and eliminate illegal, unreported and unregulated fishing (IUU Regulation), available at http://eur-lex.europa.eu/legal-content/EN/TXT/PDF/?uri=CELEX:02008R1005-20110309&from=EN.

Legislation

COM 2012/0576 Final 4 October 2012 Proposal for a Regulation of the European Parliament and of the Council on Access to Genetic Resources and the Fair and Equitable Sharing of Benefits Arising from Their Utilization in the Union.
Council Regulation (EC) No 1005/2008 of 29 September 2008 establishing a community system to prevent, deter and eliminate illegal, unreported and unregulated fishing (IUU Regulation), OJ L 286, 29 October 2008.
Nagoya Protocol on Genetic Resources and the Fair and Equitable Sharing of Benefits Arising from the Utilization to the Convention on Biological Diversity.
Regulation (EU) No 995/2010 of 20 October 2010 laying down the obligations of operators who place timber products on the market. OJ L 295/31, 12 November 2010.
Regulation (EU) No. 511/2014 of the European Parliament and of the Council of 16 April 2014 on compliance measures for users from the Nagoya Protocol on Access to Genetic Resources and the Fair and Equitable Sharing of Benefits Arising from their Utilization OJ L 150/67, 20 May 2014.
République Française, Projet de Loi relatif à la Biodiversité, 26 March 2014.

References

Cabrera Medaglia, Jorge, Frederic Perron-Welch and Olivier Rukundo (2012) *Overview of National and Regional Measures on Access to Genetic Resources and Benefit Sharing, Challenges and Opportunities in Implementing the Nagoya Protocol*. July. Montreal: Center for International Sustainable Development Law. Available at http://cisdl.org/biodiversity-biosafety/public/CISDL_Overview_of_ABS_Measures_2nd_Ed.pdf.
Greiber, Thomas, Sonia Peña Moreno, Mattias Åhrén, Jimena Nieto Carrasco, Evanson Chege Kamau, Jorge Cabrera Medaglia and Maria Julia Oliva and Frederic Perron-Welch in cooperation with Natasha Ali and China Williams (2012) *An Explanatory Guide to the Nagoya Protocol on Access and Benefit-Sharing*. Gland: IUCN.

IEEP, Ecologic and GHK (2012) *Study to Analyse Legal and Economic Aspects of Implementing the Nagoya Protocol on ABS in the European Union*. Final report for the European Commission, DG Environment. April. Brussels: Institute for European Environmental Policy.

Oberthür, Sebastian, and Florian Rabitz (2014) "The Role of the European Union in the Nagoya Protocol Negotiations: Self-Interested Bridge Building." In Sebastian Oberthür and G. Kristin Rosendal (eds), *Global Governance of Genetic Resources*, 79–95. Abingdon: Routledge.

Wallbott, Linda, Franziska Wolff and Justyna Pozarowska (2014) "The Negotiations of the Nagoya Protocol." In Sebastian Oberthür and G. Kristin Rosendal (eds), *Global Governance of Genetic Resources*, 33–59. Abingdon: Routledge.

3 A regional alternative to the ineffective global response to biological invasions

The case of the European Union

Donato Gualtieri

Introduction

Palm trees constitute a dominant landscape element in various touristic locations on the Italian Adriatic coast, ideally representing a *trait d'union* linking the sandy beaches of Jesolo (Veneto) to the rocky shores of Vasto (Abruzzo), passing through the notorious *Riviera delle Palme* (Marche).[1] Though not part of the indigenous flora,[2] these palms rapidly adapted to the temperate climate of the region and thrived in harmony with native ecosystems.

In recent years, however, palm trees have been threatened by another 'host' species. Since the beginning of the new millennium, in fact, thousands of palms have died following the invasion of an exotic species originating from Asia: the red palm weevil (*Rynchophorus ferrugineus*), part of the family of *Curculionidae*. While adult insects create limited damages, the burrowing of the larvae into the heart of the tree creates the greatest mortality, as they feed on the soft fibres and terminal buds excavating tunnels up to 1 metre through the internal tissue, thus leaving the outside apparently healthy until it is too late to intervene with a conservative action.

Similarly to palms, these insects did not cross the Mediterranean *propria sponte*: paradoxically, the red palm weevils arrived in Italy from the Middle East or northern Africa – where they had already spread in the 1990s – inside cargo ships transporting date palms (*Phoenix dactylifera*). Once established in the coastal areas of Italy, the *Rynchophorus ferrugineus* started its slow but determinate expansion, favoured by the proximity of palms to beaches and urban contexts and by the consequent unfeasibility of chemical control, as insecticides would have posed a risk to human health. The only way to block the invasion has been to cut down, chop into pieces and burn the palms.

This concrete case dramatically emphasizes the role of human agency in the dispersion of exotic species beyond their natural habitats: this phenomenon, known as 'biologic invasion', is a danger for global biodiversity, whose preservation and conservation are in turn considered to be primary objectives for the contemporary international legal order.[3]

This chapter will illustrate the global regulatory framework set up in the last decades at the universal level in order to give a response to the ecological, economical and social challenges posed by biological invasions. The existence of gaps, inconsistencies and the ultimate legal weakness of such a framework will stress the need for an alternative model characterized by the reduction of the scale of intervention to the regional level, a reconfiguration of the role of the state and the presence of a system of judicial implementation and control. In this sense, the European Union represents at the moment a *unicum* at the international stage, and seems to be the most serious candidate to realize this paradigmatic shift. To verify this assumption, the chapter will be concluded by the analysis of a proposed regulation on biological invasions that the European Commission has submitted to the European Parliament and to the Council.

Biological invasions as a part of international environmental law

For the greatest part of the twentieth century, legal responses to the spread of invasive alien species beyond their natural ranges were an exception – in a panorama of general inaction – at the national level.[4] Even more, the issue was hardly detectable at the supranational stage. The substantial economic benefits deriving from the introduction of non-native species for agricultural or recreational purposes, in fact, worked as a powerful driver of procrastination (Shine *et al.* 2005).

Because of this silence, only recently have biological invasions been included in the international regulatory agenda, as a result of the influence of two preconditions and a triggering event. In fact, if the 'migration' of ecological concerns from the field of social activism to the regulatory sphere allowed for an ecologically-backed approach in which the preservation of nature would become an autonomous goal, the progressive specialization of the international legal order widened the horizons of multilateral regulation to a vast array of legal tools and institutional frameworks, each one of them characterized by a sectoral approach and by a definite set of rules and practices (Fischer-Lescano and Teubner 2004). Those factors paved the way for the internationalization of a problem which had started to be increasingly felt at the national level since the second half of the 1980s, when the negotiation for the adoption of a multilateral legal instrument aimed at conserving the existent biological diversity began.

This legal and historical reconstruction finds parallelism if looking at the factual evolution in the international regulatory system: if the International Plant Protection Convention (1951) was still focused on the safeguard of those plants having an immediate economic value – relegating conservation to the role of side effect – the 'international environmental decade' opened by the Convention on Wetlands of International Importance (Ramsar Treaty, 1971),[5] continued with the signing of the Convention on International Trade in Endangered Species of Wild Fauna and Flora (CITES, 1973),[6] and closed by the Convention on the Conservation of Migratory Species of Wild Animals (CMS, 1979)[7] was marked by an impressing move towards autonomization of ecological concerns.

Yet not one of those conventions addressed the problem of biological invasions (if not the CMS, and in a rather narrow and sectoral way): it was only in 1992, through the adoption of the Convention on Biological Diversity (CBD), that legally binding norms were adopted to oblige Parties to prevent the introduction of alien species capable of threatening ecosystems, habitats or species and, in case, to control and/or eradicate those who succeed in establishing in a host environment.[8] This provision can be read in conjunction with others regarding national and integrated strategic and cross-sectoral planning,[9] regulation or management of potentially damaging processes and categories of activities,[10] involvement of local populations and the private sector,[11] incentives and environmental impact assessment,[12] transboundary notification and emergency planning.[13]

All the same, neither the qualitative evolution nor the quantitative proliferation of legal instruments at the universal level has been accompanied by a real improvement in terms of effectiveness, as demonstrated by the failure in achieving the 2010 Biodiversity Target set by the CBD,[14] and as confirmed by recent studies showing how the extinction risk of birds, mammals and amphibians is still increasing over time because of the impact of invasive alien species on native ecosystems.[15] The reasons for such a regulatory fiasco rest upon a plurality of levels, veiled by an allure of confusion in which the most immediate – and perhaps more evident – causes reveal all their superficiality at a more accurate analysis aimed at finding the *causa prima*[16] of the overall failure of the universal legal architecture set up to respond to biological invasions.

The causes of the regulatory fiasco and the need for a paradigmatic shift

It is beyond doubt that biological invasions, as a phenomenon overreaching national boundaries, need transnational regulation. For this reason, it is absolutely necessary to avoid any isolationist tendency aimed at finding solutions within national borders. All the same, the impact of invasive alien species has a significant impact on a large spectrum of different sectors, thus escaping traditional categorization into specific legal ideal-types. Those remarks certainly open the way to a complex analysis of the causes of the assessed ineffectiveness of the international legal framework.

At a more superficial level, the impossibility to regulate biological invasions through an *ad hoc* legal instrument or a dedicated institutional mechanism inevitably implies the acceptance of a vast array of gaps and inconsistencies at the most diverse levels. A non-exhaustive list of the consequences of such a '*gruyere* effect' include:

- the lack of a common glossary for relevant scientific concepts;[17]
- the almost exclusive attention paid to higher *taxa* of alien animals and plants by universal agreements, while coverage below the species level is geographically or functionally limited (Secretariat of the Convention on Biological Diversity 2001);[18]
- the inexistence of binding international standards in relation to the control of pathways and vectors of introduction of invasive alien species;[19] and

- the lack of coordination in remedial measures; the complete absence of operational coherency and of concrete indicators and criteria for practical implementation.

All the same, a more insightful reflection on the theoretical foundations of international law reveals that gaps and inconsistencies are intermediate causes of ineffectiveness and, in turn, effects of a more rooted dysfunction.

As an eminently transterritorial phenomenon overhauling state borders and capacities, in fact, biological invasions stand in a rather dialectic relationship with the state-centric system developed since the times of the Peace of Westphalia (1648), so that the response to the transboundary impacts of invasive alien species can be accommodated within contemporary international law only at the price of a certain degree of ineffectiveness (Biermann and Dingwerth 2004).[20] In fact, the final aim of reducing biodiversity loss introduces an element of discontinuity with respect to precedent legal and intellectual traditions which have placed men – as individuals or collective aggregations – and human needs at the centre of any supranational regulatory effort, while switching the focus to the strictly ecological safeguard. The attention paid to the needs of nature thus proposes an 'ethical cleansing' aimed at dehumanizing the legal framework in the name of an external, superior good – the conservation of the environment – which exceeds human life and wellbeing for its intrinsic value, existing independently of its importance to humans and sociopolitical aggregations (Soulè 1985).

However, this process of transformation which is now *in potentia* at the international universal level needs a further step to be translated *in acto*. In particular, to effectively operate in a human and state-centric scenario such as the international legal framework, a paradigmatic shift is needed.

With a view to the superior good of environmental conservation, in fact, a conceptual reconstruction capable of decoupling sovereignty and territory should give to the territorial element of sovereignty a less 'cartographic' and more ecological meaning (Sassen 2007). Finally, detaching its legal value of possession 'integrally related to the state' from its actual, environmental significance of a reality 'not inherently tied to the state' (Biggs 1999), it would become possible to embrace a different idea of territory: from basis for the exercise of state jurisdiction to 'space for ecological governability'. Under this new paradigm, states would cease to be the only actors, opening new possibilities at different scales of governance. If the role of sub-national levels falls behind the scope of this chapter, the next paragraph will define the ideal scale of environmental governance in the field of biological invasions.

The choice of the regional scale: European Union as a role model

In order to avoid that the choice of the regional level as the optimal scale of environmental governance with regard to biological invasions would become a mere translation of ineffectiveness to a more geographically restricted level, some caveat must be introduced in order to find an organization endowed with peculiar features useful in overcoming the immediate and rooted causes of ineffectiveness at the global level: a significant delegation of sovereignty from national governments in the environmental sector; the legal authority to adopt and enforce regulations binding upon member states; the possibility to refer any lack of compliance to a judicial instance. The combination of those three elements helps in identifying in the European Union a *unicum* in the contemporary international community.

At first, the European Union explicitly included environmental protection as an objective of the integration process and has thus realized a gradual transfer of sovereignty from member states to supranational institutions. In the 1970s, the completion of the common market and the removal of the most direct restrictions on trade brought attention to the need to harmonize national regulatory standards related to ecological and human health protection, as those norms often worked as barriers to transnational exchange (Fligstein and Stone Sweet 2002).[21] In this way, a chain of events started to take place:

- the issue of environmental programmes of action in the 1970s;
- the 'communitarization' of ecological concerns through the clarification of the objectives, principles and evaluation criteria of a common environmental policy set out in the Single European Act in 1987;[22]
- the inclusion of sustainable development as an overall objective of the Community and of the precautionary principle as a headlight of European action in the Treaty of Maastricht (1992);[23] and
- the introduction of the integration principle – according to which 'environmental protection requirements must be integrated into the definition and implementation of the Community policies and activities with the Treaty of Maastricht (1997).[24]

Finally, with the Treaty on the Functioning of the European Union (TFEU, 2007), environment was designed as a matter of shared competence between the states and the Union,[25] so that – in light of the principle of conferral – states are allowed to exercise their competences 'to the extent that the Union has not exercised its competence'[26] or in case the Union should decide to 'cease exercising its competence'.[27]

If the attribution of competence to the Union on environmental matters satisfies the precondition of 'switching sovereignty' and gives to the European institution a legitimate law-making power, the binding value of such norms upon states remains to be demonstrated. At first, Article 258 of TFEU established an infringement procedure, on the basis of which the Court of Justice of the European Union may determine whether a member state has fulfilled the legal obligations deriving from an act of primary or secondary legislation and, in case of breach of European Union law, it can impose on the infringing state to terminate the breach without delay. Furthermore, on the plan of principles, the Court of Justice developed two theories: on the one hand, the theory of the direct effect – illustrated in the *Van Gend and Loos* case (1963) in relation to treaty provisions, and in the *Marshall* case (1986) with regard to secondary legislation (i.e. directives and regulations) – states that any legal act issued by the European institutions has direct effect before the national courts of the member states, thus creating an alternative enforcement pathway; on the other, the principle of supremacy of European law – elaborated in the *Costa v. Enel* case (1964) – sets the primacy of community law over national legislation.

Finally, with particular regard to the legal approach to biological invasions in the European Union, the subsidiarity principle as defined in Article 5 of TFEU seems to be tailored for the individuation of the regional scale as the optimal level of intervention. According to this provision, in fact, the Union has the right to act in place of the state if the objective of a legal act can – 'by reason of the scale or effects of the proposed action' – be better achieved 'at Union level'.[28]

The EU regulation on invasive species: a proposal for paradigmatic shift?

On 9 September 2013, the European Commission issued a proposal for a Regulation on the prevention and management of the introduction and spread of invasive alien species.

The decision to intervene at the Union level came after five years of continuous efforts, on the base of the consideration of the transboundary nature of biological invasions:[29] even if some member states had already dealt with the problem at national level, the lack of action by neighbouring countries – combined with the inexistence of an adequate coordination in controlling the external borders of the Union – could undermine the effectiveness of such provisions and have distortive effects on the internal market and on the free circulation of goods. After consultations with relevant stakeholders and European citizens,[30] which revealed the insufficiency of a maximization of the existing legislation and policies in the field,[31] the Commission set up a comprehensive legal framework at the Union-level. Unconventionally, European institutions decided to tackle environmental risk through a regulation and not through a directive: this choice would have left, in fact, a narrower margin for measures of adaptation to member states, while imposing a more immediate obligation upon national authorities at the central and local level.

Considering that over 12,000 alien species are already present in the EU – out of which 10 to 15 per cent are invasive[32] – the proposed regulation provides for a prioritized and proportionate approach – through an initial capping of the number of priority species to the top 3 per cent of some 1,500 invasive alien species in Europe – and for a shift towards prevention, early warning and rapid response, while management is deemed to be an *extrema* ratio.[33]

Based on Article 192, paragraph 1 of TFEU, the proposal upholds the three-tier approach (prevention, early warning and rapid response, and management) contained in the Guiding Principles on Invasive Species adopted by the Conference of Parties to the CBD in 2002,[34] aiming at establishing a framework for action to prevent, minimize and mitigate the adverse impacts of invasive alien species on biodiversity and ecosystem services, while at the same time limiting their economic and social adverse effects.[35] As far as prevention is concerned, Article 7 introduces a ban on the voluntary or involuntary introduction of invasive alien species of Union concern, even if Article 8 introduces some exceptions for research and *ex situ* conservation on the basis of national permits. Article 9 also obliges member states to take emergency actions in case of evidence concerning the actual or imminent presence into their territories of an invasive alien species, while Article 11 provides for a binding duty to carry out a comprehensive analysis on the pathways of introduction. With regard to early detection and rapid eradication, Chapter III of the proposal binds member states to set up a rigid border control and a dynamic surveillance system in order to detect the appearance in the environment of the member states of any invasive alien species, and provides for a period of three months from the early detection to put in place a rapid eradication.[36] Finally, Chapter IV introduces uniform norms for management and measures aimed at minimizing the impact of established invasive alien species on biodiversity, ecosystem services, human health and the economy through physical, chemical or biological control, and at restoring damaged ecosystems at the highest possible level.

Finally, in relation to enforcement measures, member states will lay down rules on administrative sanctions applicable on any natural or legal person for breaches of the regulation, beside taking all necessary steps to ensure that those measures are enforced in an effective, proportionate and dissuasive way.[37]

At present, the proposal has been adopted in first lecture by the European Parliament in April 2014, while the Council expressed its consent on 29 September.

Pending the entry into force of the regulation in January 2015, some final remarks can be made. In particular, after the agreement between the Parliament and the Council abolishing the initially proposed cap of the list of species of Union concern to 50 species, one main aspect separates the proposed regulation from giving substance to the theorized paradigmatic shift. In particular, the perplexity regards the possibility for member states to issue permits to carry out certain commercial activities involving invasive alien species. Once again, the difference between success and failure will be decided by economic aspects. In this sense, a strict application of the 'polluter pays principle' for the recovery of restorations costs for the damages brought by those species would be a good way to 'fight fire with fire'.

Notes

1 For a more detailed reconstruction of the geography of palm trees in Italy see, *inter alia*, Di Domenico (2013).
2 Different species of palms have been introduced for ornamental purposes between the eighteenth and nineteenth centuries, the most recurring undoubtedly being the Canary Island date palm (*Phoenix canariensis*).
3 For a good overview on biological invasions see, *inter alia*, Simberloff (2013).

4 An exception is represented by a group of States – such as New Zealand, Australia, United States, Canada – whose regulatory frameworks have systematically taken biological invasions into account.

5 The Convention on Wetlands of International Importance, especially as Waterfowl Habitat was signed in Ramsar (Iran) by 18 countries in 1971: not enshrined in the United Nations system, this intergovernmental treaty has as its main goal the 'wise' use of all of the wetlands in members' territories.

6 The Convention on International Trade in Endangered Species of Wild Fauna and Flora, adopted in Washington in 1973, offers a multi-layered protection to more than 35,000 specimens of animals and plants, targeting the degree of protection on the actual risk of extinction.

7 The Convention on the Conservation of Migratory Species of Wild Animals, signed in Bonn in 1979, aims at conserving terrestrial, marine and avian migratory species within their natural range. The CMS operates as an 'umbrella convention', and both binding and non-binding agreements can be signed under its framework in order to give more targeted protection to individual species.

8 Article 8(h), CBD

9 Article 6, CBD.

10 Article 8(l), CBD

11 Article 10, CBD.

12 Article 11, CBD.

13 Article 14, CBD.

14 The 2010 Biodiversity Target was adopted by COP Decision VI/26 and was aimed at reducing the rate of loss of the components of biodiversity. This strategy identified biological invasions as a major threat and prescribed in Goal 6 to adopt stricter measures aimed at preventing the introduction and managing the eradication or control of alien species threatening ecosystems, habitats or species.

15 See, *inter alia*, McGeoch *et al.* (2009).

16 In a famous passage of his *Summa Theologica*, Thomas Aquinas uses five arguments to explain the existence and nature of God. The Second Way, from the nature of efficient cause, admits by plain reason that an infinite regression of causes is not possible when talking about the origin of all that exists in the sensible world.

17 As an example, the Agenda 21 variously refers to invasive alien species as to 'exotic', 'pests', 'foreign'. All those terms, in turn, have a range of intrinsic nuances of significance according to the particular sector in which they are used (i.e. farming or conservation).

18 See *supra*, note 23. A notable example comes from the international sanitary and phytosanitary treaty-system, which potentially covers all taxonomic groups and lower taxonomic categories, but only to the extent that these are injurious to animal or plant health.

19 A fitting example comes from maritime transport, where the 2004 Water Ballast Management Convention has been adopted but never entered into force. Aviation has even more limited and sectorial measures, while land transport and in-land waterways are not formally regulated to minimize transfer risks.

20 Biermann and Dingwerth (2004) propose environmental law as a paradigmatic example of the changing nature of sovereignty.

21 The first interventions of the European Economic Community in the sector – based on Article 100 of the Treaty of Rome – were related to the harmonization of national safety and protection standards in the fields of water quality (i.e. Council Directive 76/464/EEC on pollution caused by discharges of certain dangerous substances), air pollution (i.e. Council Directive 80/779/EEC, setting limit values and guide values for sulphur dioxide and suspended particulates), and the disposal of waste oils (i.e. Council Directive 75/439/EEC).

22   The cornerstone of the new European environmental policy was Article 130R, par. 1 and 2: 'Action by the Community relating to the environment shall have the following objectives: to preserve, protect and improve the quality of the environment; to contribute towards protecting human health; to ensure a prudent and rational utilization of natural resources. Action by the Community relating to the environment shall be based on the principles that preventive action should be taken, that environmental damage should as a priority be rectified at source, and that the polluter should pay. Environmental protection requirements shall be a component of the Community's other policies'.

23   Articles 2 and 174, Treaty establishing the European Community as amended in Maastricht (1992).

24   Article 6, Treaty establishing the European Community as amended in Amsterdam (1997)

25   Article 4, TFEU.

26   Article 2, TFEU.

27   *Ibid.*

28   Article 5, TFEU.

29   See, *inter alia*, the Communication by the European Commission, Towards a Strategy on Invasive Alien Species, 3 December 2008; the Communication by the European Commission, Our Life Insurance, Our Natural Capital: An EU Biodiversity Strategy to 2020, 3 May 2011; and the Commission funded report Assessment to Support Continued Development of the EU Strategy to Combat Invasive Alien Species, issued by the IEEP (Institute for European Environmental Policy) in November 2011.

30   See Response Charts for Invasive Alien Species – An European Concern, 29 May 2008, and Online Public Consultation on Invasive Alien Species, 27 January–12 April 2012.

31   *Inter alia*: the Wildlife Trade Regulation (338/97) restricts imports of endangered species, including imports of seven invasive species; the Regulation on the use of alien and locally absent species in aquaculture (708/2007) addresses the release of alien species for aquaculture purposes; the regulations on plant protection products (1107/2009) and on biocides (528/2012) address the intentional release of micro-organisms as plant protection products or biocide respectively. Finally, the Birds Directive (2009/147/EC) and the Habitats Directive (92/43/EEC), the Water Framework Directive (2000/60/EC) and the Marine Strategy Framework Directive (2008/56/EC) require the restoration of ecological conditions and refer to the need to take invasive species into consideration.

32   Article 3 of the proposed regulation differentiates between alien species (i.e. 'any live specimens of species, subspecies or lower taxon of anima, plants fungi or micro-organisms introduced outside its natural past or present distribution') and invasive alien species (i.e. 'an alien species whose introduction or spread has been found, through risk assessment, to threaten biodiversity').

33   Article 4 of the proposed regulation proposes the adoption by the Commission of a list of invasive alien species of Union concern, and fixes a cap of fifty species.

34   Though non-binding, the Guiding Principles adopted through Decision VI/23 aim to provide all governments and organizations with guidance, clear direction and common goals for developing effective strategies to minimize the spread and impact of invasive alien species through the provision of a three-tier hierarchical approach based on prevention, early detection and management.

35   European Commission, Proposal for a Regulation of the European Parliament and of the Council on the Prevention and Management of the Introduction and Spread of Invasive Alien Species, 9 September 2013.

36   Article 15 of the proposed regulation on invasive alien species.

37   Article 24, proposed regulation on invasive alien species.

References

Biermann, F., and K. Dingwerth (2004) Global Environmental Change and the Nation State. *Global Environmental Politics* 4: 1. Available at www.glogov.org/images/doc/ Biermann_Dingwerth(2004).pdf.

Biggs, M. (1999) Putting the State on the Map: Cartography, Territory, and European State Formation. *Comparative Studies in Society and History* 41: 374–99.

Di Domenico, M. (2013) Invasioni biologiche. Il caso drammatico delle palme e di due specie di insetti, il Punteruolo rosso delle palme (*Rynchophorus ferrugineus*, Coleotteri) e il castnide delle palme (*Paysandisia archon*, Lepidotteri) in Italia. *Altre Modernità* 10: 164–76.

Fischer-Lescano, A., and G. Teubner (2004) Regime Collisions: The Vain Search for Legal Unity in the Fragmentation of Global Law. *Michigan Journal of International Law* 25: 999–1046.

Fligstein, N., and A. Stone Sweet (2002) Constructing Markets and Polities: An Institutionalist Account of European Integration. *American Journal of Sociology* 107: 1206–43.

McGeoch, M. A., D. Spear and E. Marais (2009) *Status of Alien Species Invasion and Trends in Invasive Species Policy*. Summary Report for the Global Invasive Species Programme (GISP). Centre for Invasion Biology. Available at www.bipindicators.net/LinkClick. aspx?fileticket=qe6SbidMiOk=&tabid=80&mid=693

Sassen, S. (2007) *A Sociology of Globalization*. New York: W. W. Norton & Company.

Secretariat of the Convention on Biological Diversity (2001) *Review of the Efficiency and Efficacy of Existing Legal Instruments Applicable to Invasive Alien Species*. CBD Technical Series no. 2. Montreal: Secretariat of the Convention on Biological Diversity.

Shine, C., N. Williams and F. Burhenne-Guilmin (2005) Legal and Institutional Frameworks for Invasive Alien Species. In H. A. Mooney, R. N. Mack, J. A. McNeely, L. E. Neville, P. J. Schei and J. K. Waage (eds), *Invasive Alien Species: a New Synthesis*. Washington, DC: Island Press.

Simberloff, D. (2013) *Invasive Species: What Everyone Needs to Know*. New York: Oxford University Press.

Soulè, M. (1985) What is Conservation Biology? *BioScience* 35(11): 727–34.

4 Redefining the relationship between the Convention on Biological Diversity and the TRIPS Agreement

The first step towards confronting biopiracy?

Anastasia Fotinakopoulou

Introduction

Who owns life? There is a great debate nowadays concerning the status of the legal holders of natural resources. Is it the state where they are situated? Or are they considered an indisputable part of the Common Heritage of Mankind?[1] Who has the jurisdiction to exploit these resources? Is a balance between global economic development and national sovereignty possible? These are some of the main questions that one should take into account in order to address a global challenge such as biopiracy. Nowadays, a common debate about a possible resolution in regard to the issue of biopiracy is centred towards the examination of the relationship between two international agreements: the Convention on Biological Diversity (hereafter CBD) and the Agreement on Trade-Related Aspects of Intellectual Property Rights (hereafter TRIPS Agreement).

The main goal of this chapter is to examine whether or not there is conflict in the mutually supportive application of these two agreements, and how they are interlinked with the issue of biopiracy. The second section will cover some introductory remarks about the context of biopiracy in general, the problem of defining it as well as the main implications caused by this issue in relation to the pillars of sustainable development. The third section will be divided into two parts, with the first covering the CBD, focusing mainly on its third objective, which refers to 'the fair and equitable benefit-sharing arising out of the utilization of genetic resources', as well as its effort to regulate for the first time the issue of biotechnology, mainly through Article 16, and the second focusing on the TRIPS Agreement, and its main contradictions in comparison to the content of CBD. The main question in this part is whether or not the TRIPS Agreement legalizes biopiracy and benefits only the developed countries. The final section will discuss the possible course of action that needs to be taken so as to reach a balance between these agreements as a part of a permanent solution towards a global threat such as biopiracy. It will also establish whether or not they provide the appropriate solid basis needed for the adoption of a new regime that could tackle this emerging new challenge.

Defining biopiracy and its implications

The term 'biopiracy' was originally used by a Canadian NGO, Rural Advancement Foundation International (RAFI), and was attributed to its director, activist Pat Mooney. According to RAFI, biopiracy refers to 'the appropriation of the knowledge and genetic resources of farming and indigenous communities by individuals or institutions who seek exclusive monopoly control (patents or intellectual property) over these resources and knowledge'.[2] In 2004, Dutfield gave a more specific and wider scope to the definition of biopiracy, writing that 'biopiracy normally refers either to the unauthorized extraction of biological resources and/or associated traditional knowledge from developing countries, or to the patenting of spurious "inventions" based on such knowledge or resources without compensation'.[3]

Both definitions became a source of ambiguity as they made reference to another complex and unresolved issue: the subject of appropriately defining what traditional knowledge entailed. There is no single universally accepted definition of 'traditional knowledge' as it is a term that has taken on various meanings through the years.[4] Since 2000, WIPO has been trying to resolve this matter through the establishment of the Intergovernmental Committee on Intellectual Property and Genetic Resources, Traditional Knowledge and Folklore (IGC), as well as the development of a potential international legal instrument that would ensure the effective protection of traditional knowledge, genetic resources and traditional cultural expressions,[5] and would effectively cover the possible deficiencies of other relative frameworks.

The main problem concerning biopiracy lies in the absence of a specific legal context that would provide the solid basis for a possible solution. Developing countries claim that their resources are taken illegally by corporations without acknowledging their sovereign rights and without reference to their origin or the relevant traditional knowledge. Developed countries are the sole beneficiaries of the patenting system and the evolving developments of biotechnology and science in general. There are a lot of important implications should the issue of biopiracy remain unresolved.[6]

Economic implications

Biopiracy has a negative influence on the economic growth of a state's national wealth. The economic implications refer to the empowerment of neocolonialist practices by increasing the dependence on the companies owning the resources, which in most cases, especially those of monopoly, have been acquired illegally. Moreover, there have been certain examples of individuals or groups that are excluded from exporting their products as a result of a patent fortification or are obliged to pay a higher market price for imported patented products originating from the country's own resources. There is also the matter of prohibition of using a resource for an extended period of time or a limited access to the resource. Prime examples of the connection between biopiracy and the way it affects national wealth are the Neem[7] and the Hoodia[8] cases.

Social implications

The social implications of biopiracy concern mainly the violation of rights of indigenous communities as well as the potential loss of their cultural identity and value. Indigenous people have neither access to legal information that would provide them with the necessary background to support their legal rights, nor the financial capability to obtain them. Even though both CBD and the Nagoya Protocol have the potential for holding countries accountable for their Indigenous Peoples rights,[9] the traditional communities have little opportunity to participate in the formation of a national or international framework regarding both the access to genetic resources and intellectual property rights. Moreover, according to the US, traditional knowledge does not incorporate the element of 'newness',[10] a necessary element for a reward to be granted. The most prominent example of the social implications of biopiracy is the basmati case.[11]

Environmental (and ethical) implications

Biopiracy has major ecological impacts as well as potentially negative effects on human health, mainly through the promotion of the biotechnology industry, 'especially via the large scale introduction of genetically engineered organisms in the field of agriculture', as noted by V. Shiva.[12] This results in the loss of biodiversity, a direct consequence arising from the increased application of monoculture, and poses the risk of a potential 'biological pollution', susceptibility to disease, the prominence and dominance of one species in an ecosystem and possible DNA alterations through gene transfer.[13] These actions lead to the direct manipulation of life itself raising a lot of biosafety issues. In most cases the cost-benefit approach of biotechnology is debatable and the potential risks great.

The regulation of access to genetic resources and intellectual property rights: CBD versus TRIPS

Shortly after the adoption of both CBD and the TRIPS Agreement, the issues of whether or not there is conflict[14] in their mutual application and if the TRIPS Agreement should be amended so that it won't legalize biopiracy were widely discussed. The proposed amendment is centred towards the 'determination of the origin of a resource as well as the traditional knowledge used in an invention claimed in a patent application'. This discussion has widened the rift between the North and the South; however, a balance must be found in order to reach a permanent solution towards a global threat such as biopiracy. In the end, the main question that should be addressed is whether the environment precedes trade or the opposite, and what the boundaries that divide are.

CBD and the access and benefit sharing system

The Convention of Biological Diversity was signed in 1992 as a result of the Earth Summit and entered into force in 1993. Its main goal is to regulate biodiversity through the promotion of three pillars: the conservation of biological diversity, the sustainable use of its components and the fair and equitable sharing of the benefits arising out of the utilization of genetic resources.[15] For the first time there is a legal recognition concerning the sovereign rights of the states over their own resources[16] as well as an attempt to broach and regulate the challenging issue of biotechnology.[17] CBD is based on a government-to-government approach[18] through a fair exchange of benefits consolidated by the legal foundation of the access and benefit sharing (ABS) system.[19] The provider states have the authority to develop national laws and policies in order to determine the conditions of access to their resources, while the user states should develop a mechanism to share the benefits arising from the utilization of such resources with the country of origin, mainly through the transfer of derivative technologies.

For a transaction to take place, the user state is obliged to ensure prior informed consent from the provider. Then, in a private law contractual agreement, the provider and the user define the exact (mutually agreed) terms and conditions governing the utilization of said resources as well as the fair and equitable sharing of the benefits. Once the utilization has produced benefits, users share them with the provider country in accordance with the previously agreed terms, supporting the principles of equity, sustainability and biodiversity conservation. This procedural course was considered a substantial first step towards tackling biopiracy, along with the special reference to the role of the indigenous people, who are the key holders of traditional knowledge.[20]

CBD brings forth a contradiction as it is the first legal text trying to bridge the wide gap between the privatization of genetic resources and the globalization of intellectual property rights. This notion results from the wording of Article 16, concerning the role of intellectual property rights in relation to the implementation of the Convention. Although the Article recognizes the importance of IPRs, it is later claimed that these rights 'should be supportive and not run counter to the Convention's objectives.' Through this provision[21] developing countries are given access to patentable biotechnology, which incorporates the use of their resources. This possibility provoked negative reactions from the North, especially from the US, one of the capital users of genetic resources, and was the central factor in their decision to abstain from signing the final text of CBD at first.[22] The US considered the Convention to be a potential threat to the competitiveness of biotechnology corporations as well as a crucial legal obstacle to maximizing access and rights to important resources and, in order to create an alternative course of protection as a 'safety net' against CBD, supported the adoption of the TRIPS Agreement during the Uruguay Round that culminated in the creation of the WTO, in 1994.[23]

Despite the fact that CBD was a crucial step towards bringing important and controversial issues into the spotlight, it became the source of many contrasting views due to the vague and contradictory language used there, as well as the creation of legal uncertainty. That is why in 2010, after a lengthy negotiation process, the Nagoya Protocol was adopted, creating a more specific framework concerning the implementation of the ABS system. The Protocol is based on the general provisions of CBD, including some new mechanisms that could contribute to tackling biopiracy, such as the creation of a Multilateral Benefit-Sharing Mechanism,[24] that would enhance transboundary cooperation. The Nagoya Protocol has not yet entered into force,[25] but there is hope for its potential implementation until the new COP, which is going to take place in October.

The TRIPS Agreement and its contradictions

While CBD promotes collective rights, the TRIPS Agreement was the means to protect private rights. The TRIPS Agreement seeks to impose and universalize the levels and forms of intellectual property protection existing in the North. Article 27 of TRIPs, the most disputed provision in respect to the TRIPS–CBD conflict, provides a broad scope for protection, allowing for the patenting of 'any inventions, whether products or processes, in all fields of technology, provided that they are new, involve an inventive step and are capable of industrial application.' In addition, Article 27.3 (b)[26] relates more to the issue of biotechnology as it clarifies that members are not to exclude from patentability non-biological and microbiological processes. The concrete content of these terms is not addressed properly by the TRIPS Agreement, leaving ground for national interpretation,[27] due to the sensitivity of the subject and the controversy that has already been created. This Article has become the source of many disagreements as the developing countries claim that an amendment is of high importance.[28]

The relationship between CBD and TRIPs is defined as another major part of the North–South conflict, pitting 'the interests of the wealthier developed countries of the North versus those of the resource providing developing countries of the South'.[29] Developing nations claim that the TRIPS Convention benefits mainly industrialized developed countries by granting them, via the patenting system, monopolies of resources gained illegally in most cases and without involving the provision of an inventive step, as most of them are part of a long traditional knowledge practice that either isn't addressed or recognized by the TRIPS Agreement, asserting at the same time that a compliance with the Agreement causes heavy economic burdens to them.[30] No mechanisms are established that grant traditional communities control over their knowledge and innovations, or ensure they reap a fair share of the benefits, as the assignment of IPRs goes to corporations or individuals. Moreover, while CBD requires the granting of prior informed consent of the provider states or the local communities who are identified as key holders of the biodiversity for any use of genetic materials, the TRIPS Agreement does not require the patent holder to disclose the source of genetic material on which a patent may have been granted. It cannot be clearly sustained if a concrete legal rivalry between these two agreements is valid. There is no obvious conflict of interest for a member state to fulfil its obligations under both legal frameworks separately. The real problem is created in relation to biopiracy, especially concerning the status quo of ownership of the accessed genetic resources. It is possible for a multinational company to gain control over a resource and benefit from its use without an obvious legal violation of international obligations. This is especially true in cases where a national legal framework is non-existent.

As Basil Matthew stresses:

> The TRIPS Agreement facilitates the misappropriation of ownership or rights over living organisms, knowledge and processes on the use of biodiversity takes place. The sovereignty of developing countries over their resources, and over their right to exploit or use their resources, as well as to determine access and benefit sharing arrangements, is compromised.[31]

Does that constitute an act of biopiracy?

Is the review of the relationship between CBD and the TRIPS Agreement a first step against biopiracy?

It is widely claimed, and has been almost from the first moment that those two texts entered into force, that they should be implemented in a mutually supportive way in order to avoid possible conflicts and create the possibility of dealing successfully with emerging global challenges such as biopiracy. Ethiopia and India were the first countries to address the need for a review concerning the consistency between the two texts. In 1996, India proposed a possible solution directly to the WTO through a genetic resource disclosure requirement in patent applications, which could be attained through a TRIPS amendment.[32]

This discussion culminated during the next decade along with the negotiations about the new Protocol of CBD concerning the specification of the ABS System. Mainly through the production of soft law texts such as the Doha Declaration (2001)[33] and the Cusco Declaration (2002),[34] the need for an immediate TRIPS review in accordance with the provisions of CBD became obvious. But why is an amendment to the TRIPS Agreement necessary, and what should be contained in it, in order to support successfully the content of CBD? As the main custodians of the Earth's natural wealth, developing countries have made several suggestions. In a paper submitted in 2005 to the WTO, they proposed three types of disclosure requirements:[35]

1   Disclosure of origin of the biological resources and associated traditional knowledge used in developing the invention claimed in the patent application.
2   Disclosure of PIC from the provider country's appropriate authority.
3   Disclosure of a benefit-sharing Agreement (MAT Contract).

Moreover, they have suggested specific amendments in relation to the conflicting TRIPS Articles. The first proposed amendment is an added exception principle to patentability (Article 27) concerning 'products or processes which directly or indirectly include genetic resources or traditional knowledge obtained in the absence of compliance with national and international legislation, including failure to obtain PIC as well as failure to reach a benefit sharing agreement'.

Moreover, in accordance with CBD, this provision will support the government-to-government approach by not preventing the Members from adopting a *sui generis* system in their domestic legislation in accordance with the principles and obligations contained in the CBD. An amendment to Article 29[36] has also been suggested so as to include a disclosure of the country of origin as well as any relevant traditional knowledge used in the invention. Developing countries strongly believe that should these proposals be successfully implemented, there will be more legal certainty and transparency, alleviating the risks of increasing biopiracy.

On the other hand, the US strongly opposes any possible amendments and argues that there is no obvious conflict between the CBD and the TRIPS Agreement. Moreover, it is also argued that most times there is great difficulty in determining the exact source of a genetic material, especially in cases of transit for commercial purposes, and that the determination of the origin is irrelevant with the patenting system.[37] They further claim that the matter of misappropriation of genetic resources is irrelevant with the scope of the TRIPS Agreement and constitutes part of a separate regulatory system. The US proposes a system of contractual agreements that will be more effective and tangible between the parties of the provider and the user, referring to the Merck and InBio case,[38] the first successful example of a bioprospecting agreement. The US has not ratified CBD and doesn't seem interested in doing so in the future. On the other hand, the EU believes that the entry into force of the Nagoya Protocol will have a positive influence as it will give another perspective to future discussions concerning the issue of biopiracy.[39] It is obvious that a TRIPS Amendment will be a first positive step in tackling the challenge of biopiracy, but is it the only one?

A possible TRIPS Amendment is a necessary step towards appropriate protection of countries' sovereign rights over their biological resources, but it is certainly not the only one, as will be explained below.

There is a great need for revised approaches not only at an international, but especially at a regional and national level. The states should develop an appropriate domestic legal framework, by enacting a number of measures concerning the access to their resources. India, Costa Rica and Peru are considered three successful examples on that account. India has played a crucial role in the discussions concerning the review of the relationship between the CBD and the TRIPS Agreement, as it is the holder of approximately 8 per cent of the total global biodiversity and 49,000 species of plants. India has been a victim of biopiracy cases many times. It has also participated in one of the most characteristic patent conflicts that have taken place: The turmeric case.[40] India's steps towards a legislative reform have included the implementation of the Biological Diversity Act 2002 (Biodiversity Act)[41] and the Patents (Amendment) Act 2005 (Patents Act).[42] The Biodiversity Act is based on the pillars established by CBD, including the establishment of a National Biodiversity Authority, responsible for granting PIC before any transaction. Biodiversity Act is said to exceed the expectations set by CBD. While the Biodiversity Act provides for the sustainable use and equitable sharing of biological resources as well as the prevention of possible biopiracy attempts, the Patents Act provides for a broader application of patents law in India.

It logically follows that the Acts must operate together for India's legislative IP regime to be effective. However, neither Act makes reference to the other in resolving potential inconsistencies that may arise between them. And that is a matter that should be seriously addressed in the future.

Another similar example is that of a regional cooperation group, the Andean Community.[43] In 1996, the Commission of the Andean Community introduced Decision 391[44] on a Common Regime on Access to Genetic Resources, which serves to regulate benefit-sharing and access to genetic resources and biologically derived materials within the region. Decision 391 also aims to 'promote conservation of biological diversity and the sustainable use of the biological resources that contain genetic resources,' and to 'promote the consolidation and development of scientific, technological and technical capacities at the local, national and sub-regional levels'. This initiative was a predecessor to the new mechanism suggested later by the Nagoya Protocol about transboundary cases of genetic resources. Although it was considered a progressive move, it was accused of overriding the TRIPS Agreement in many cases.[45] Despite the positive steps taken in establishing an effective legislative scheme dealing with access and benefit-sharing with the aim to meet their CBD obligations, most countries have introduced legislation that appears to be inconsistent with the broad level of patentability provided for under TRIPs. As a result, a broader spectrum of harmonization should be taken into account. Beside the TRIPS Agreement there are other international frameworks dealing with IPRs, such as the Patent Cooperation Treaty (WIPO),[46] which developing countries claim should be amended with the same requirement. Maybe a holistic approach will bring more satisfactory results in tackling biopiracy.

The role of the US, where the most cases of biopiracy stem from, remains a big controversy. Beside their unwillingness to commit to the ratification of CBD, there is a provision in Article 102 of US Patent Law that defines 'prior art'.[47] For the US Patent Law Use in a foreign country does not constitute prior art, which in a way disregards any existence of traditional knowledge. In order to stop biopiracy, it is necessary for the US to revise their patent law or add a relevant rule to the TRIPS Agreement.

Developing countries should also try to explore a combination of other possible options along with the development of their national legislation:[48] Documentation of traditional knowledge and practices, a system proposed by India, creation of a registration system of genetic resources, or developing a regional alliance system consisting of countries with common interests and problems. Unfortunately in many cases, such as the Least Developed Countries, there are other more important development problems, such as poverty, lack of the appropriate institutional structures and zero economic ability to successfully implement both CBD and the TRIPS Agreement.

Concluding remarks

Although there are many disagreements as to whether there is indeed an existing conflict between CBD and the TRIPS Agreement, no one can deny that those texts tend to interrelate. TRIPS and CBD must be mutually supportive and not undermine each other's objectives. Attempting to resolve the TRIPs–CBD conflict requires consideration of a number of issues, all of which are influenced by the broader underlying context of the North–South conflict. The exact content of biopiracy has remained inconclusive long enough, and it does not exist due to the conflict between these two texts. An obvious solution would be the development of a new regime that would regulate biopiracy as a whole. Unfortunately, for this route to be taken, a balance must be found between the already existing frameworks as well as the promotion of effective cooperation between all the appropriate stakeholders based on equitable terms and benefits.

Another significant aspect of this matter concerns the interaction between these two instruments. Most responses have been handled through adaptation mechanisms at a domestic level. Despite that fact that the synergy between these two instruments should be applied in a broader spectrum, it should be taken into account that from one hand there is a Multilateral Environmental Agreement (CBD) without the necessary backup from a World Environmental Organization that could create the appropriate leverage against the promotion of interests through the WTO. Maybe the continued refusal of TRIPS to grant the CBD secretariat observer status at the WTO[49] should be reconsidered in light of emerging challenges that should be dealt with collectively. The matter of legitimacy and legal certainty is a crucial first step towards tackling biopiracy as well as enhancing international equity.

Notes

1   The notion of including the genetic resources in the Common Heritage of Mankind began in 1983, through the adoption of the Resolution 8/83 in FAO, according to which plant genetic resources should be considered a common heritage of mankind and be available without restrictions for plant breeding, scientific and development purposes to all countries and institutions concerned. See International Undertaking on Plant Genetic Resources (Resolution 8/83), available at www.fao.org/docrep/x5563E/X5563e0a.htm#e.%20plant%20genetic%20resources%20(follow%20up%20of%20conference%20resolution%2068. Later, due to objections from developing countries during the negotiations for CBD, biological diversity was affirmed as a common concern of Mankind; see CBD preamble, available at www.cbd.int/convention/articles/default.shtml?a=cbd-00.

2   See D. Robinson, *Confronting Biopiracy: Challenges, Cases and International Debates*, Earthscan, London, 2010, pp. 18–20.

3   G. Dutfield, *Intellectual Property, Biogenetic Resources and Traditional Knowledge*, Earthscan, London, 2004, p. 52.

4   According to Robinson, there are some general criteria concerning the exact content of traditional knowledge: (1) knowledge developed over time; (2) transmitted from generation to generation; (3) typically, transmitted orally; (4) typically, collectively held and owned; (5) typically of a practical nature and relating to natural resources (as has been defined in intellectual property circles, otherwise it falls under folklore or cultural expression); (6) dynamic and evolving with environmental and external influences; (7) often involving elements of innovation and experimentation; (8) typically embedded in specific environmental settings; (9) embedded in customs, language, local practices and cultural heritage; (10) still existing in memory when it is removed from its local,

cultural or environmental setting, but it becomes less 'traditional'; (11) often linked to, or taking the form of, stories, songs, folklore, proverbs, cultural values, beliefs, rituals, community laws, local language and agricultural practices; and (12) having a holistic aspect. See Robinson, *Confronting Biopiracy*, p. 19.

5   The Negotiations started in 2009 and have not been concluded yet, mainly due to disagreements between the North and the South.

6   See V. Shiva, *Biopiracy: The Plunder of Nature and Knowledge*, South End Press, Boston, MA, 1997, pp. 32–37; Robinson, *Confronting Biopiracy*, pp. 102–16.

7   See C. Hamilton, 'Biodiversity, Biopiracy and Benefits: What Allegations of Biopiracy tell us about Intellectual Property', *Developing World Bioethics*, vol. 6, no. 3, 2006, pp. 164–69.

8   See R. Wynberg, 'Hot air over Hoodia', 13 October 2010, available at www.grain.org/article/entries/4047-hot-air-over-hoodia.

9   See M. DeGeer, 'Biopiracy: The Appropriation of Indigenous Peoples' Cultural Knowledge', *New England Journal of International and Comparative Law*, vol. 9, no. 1, 2002, pp. 203–4.

10 Patent law requires that the invention meet three criteria before a patent protection can be applied, namely that the object be (1) new, (2) non-obvious and (3) useful (*ibid.*, p. 183).

11 See J. Ragnar, *Biopiracy, the CBD and TRIPS: The Prevention of Biopiracy*, master's thesis, Faculty of Law, University of Lund, 2004, pp. 13–14, available at http://lup.lub.lu.se/luur/download?func=downloadFile&recordOId=1561387&fileOId=1565619.

12 See Shiva, *Biopiracy*, pp. 33–4.

13 *Ibid.*, p. 39.

14 See J. Carr, 'Agreements that Divide: TRIPS vs CBD and Proposals for Mandatory Disclosure of Source and Origin of Genetic Resources in Patent Applications', *Transnational Law and Policy*, vol. 18, no. 1, pp. 130–54.

15 See CBD, Article 1.

16 Legal Recognition of Principle 21 of the Stockholm Declaration on Human Environment, adopted in 1972. See www.unep.org/Documents.Multilingual/Default.asp?documentid=97&articleid=1503.

17 Although the issue of biotechnology constitutes an important issue of regulation in the Convention, it is not mentioned either in the preamble or in the objectives in order to avoid a possible negative charge of the text due to its conflicting aspects. See P. Grigoriou, G. Samiotis and G. Tsaltas, *The UN Conference on the Environment and Sustainable Development (Rio de Janeiro)*, Papazisis Publications, Athens, 1993, pp. 145–6.

18 'It attempts to reconcile Northern control of biotechnology with Southern control over biodiversity, by creating a framework under which each could benefit from the other's endowment.'; B. Matthew, 'Conflicts and Divergent Perspectives to Protect Traditional Knowledge and Indigenous People', *International Research Journal of Social Sciences*, vol. 2, no. 11, pp. 33–36 (here p. 33).CBD follows a state central approach because it is mainly based on the development of a national framework for establishing an effective system of access and sharing the benefits of biodiversity.

19 Through Article 15 of CBD, 'Access to Genetic Resources'.

20 See Robinson, *Confronting Biopiracy*, p. 28.

21 Article 16 of CBD, 'Access to and Transfer of Technology' par. 3.

22 Although the US was a leader in the negotiations as well as the drafting of the CBD text, they signed CBD in 1993 but never ratified it. Currently, the US has only observer status in the COP. See W. Snape, 'Joining the Convention on Biological Diversity. A Legal and Scientific Overview of Why the United States Must Wake Up', *SDLP*, Spring 2010, pp. 15–16.

23 See K. Rosendal, 'Interacting International Institutions: The Convention on Biological Diversity and TRIPs – Regulating Access to Genetic Resources', paper presented

as part of 'Interaction between International Institutions: Synergies and Conflicts' (Panel FC22), Friday 28 February 2003, 44th Annual Convention of the International Studies Association, Portland, Oregon, available at www.ecologic.eu/download/projekte/850-899/890/isa/isa_convention_on_biodiversity.pdf.

24 Article 10 of the Nagoya Protocol, 'Global Multilateral Benefit-sharing Mechanism'. The establishment of this mechanism was a proposal of the African Group. According to the African Group, 'there is a moral obligation to share benefits arising from continuing uses of material accessed before the protocol's entry into force, and the protocol should "encourage" such benefit-sharing; and there is a legal obligation to share benefits arising from new uses of such material, possibly through a multilateral mechanism.' The function of such a mechanism has been a source of many interpretations as it has raised a lot of conflicting issues, especially regarding its compliance with the Convention's purposes. One of the main problems is what happens in cases where one or more states refuse to grant PIC. See M. Walløe Tvedt, *A Report from the First Reflection Meeting on the Global Multilateral Benefit-Sharing Mechanism*, FNI Report 10/2011, Fridtjof Nansen Institute, Lysaker, pp. 11–12.

25 The Nagoya Protocol on Access to Genetic Resources and the Fair and Equitable Sharing of Benefits Arising from their Utilization will enter into force on 12 October 2014 following its ratification by 51 Parties to the Convention on Biological Diversity (CBD). Ratification of the Nagoya Protocol by 51 Parties to the CBD represents a major step towards achieving Aichi Biodiversity Target 16. See www.africasciencenews.org/en/index.php?option=com_content&view=article&id=1278%3Aaccess-and-benefit-sharing-treaty-receives-required-number-of-ratifications-to-enter-into-force-&catid=52%3Aenvironment&Itemid=115.

26 'Members may also exclude from patentability: (b) plants and animals other than micro-organisms, and essentially biological processes for the production of plants or animals other than non-biological and microbiological processes. However, Members shall provide for the protection of plant varieties either by patents or by an effective sui generis system or by any combination thereof. The provisions of this subparagraph shall be reviewed four years after the date of entry into force of the WTO Agreement'. See www.wto.org/english/docs_e/legal_e/27-trips_04c_e.htm.

27 The common ground between CBD and the TRIPS relates only to the consideration of 'welfare enhancement', as it is claimed by Bowman, as well as the fact that both texts rely both on the national legal and administrative mechanisms of Member States. See M. Smith, 'The Relationship between TRIPS and the CBD: A Way Forward?', *SSRN*, 11 May 2009, available at http://papers.ssrn.com/sol3/papers.cfm?abstract_id=1403000, p. 4.

28 The specific proposals for amendment to the TRIPS Agreement have come from the African Group, the Andean Community, Bolivia, Brazil, China, Columbia, Cuba, Ecuador, India, Indonesia, Kenya, Pakistan, Peru, Thailand, Venezuela and Zimbabwe. See Carr, 'Agreements that Divide', p. 140.

29 See Smith, 'The Relationship between TRIPS and the CBD', p. 5.

30 The establishment of an administration compliant with the TRIPS provisions along with the associated expenses is estimated at $1 million. See Carr, 'Agreements that Divide', p. 136.

31 Matthew, 'Conflicts and Divergent Perspectives to Protect Traditional Knowledge and Indigenous People', p. 34.

32 Carr, 'Agreements that Divide', p. 137.

33 'We instruct the Council for TRIPS, in pursuing its work programme including under the review of Article 27.3(b), the review of the implementation of the TRIPS Agreement under Article 71.1 and the work foreseen pursuant to paragraph 12 of this Declaration, to examine, inter alia, the relationship between the TRIPS Agreement and the Convention on Biological Diversity, the protection of traditional knowledge and folklore, and other relevant new developments raised by Members pursuant to

Article 71.1.' See Doha Declarations, WTO, available at www.wto.org/english/res_e/booksp_e/ddec_e.pdf.

34 Cusco Declaration echoed the importance of paragraph 19 of the 4th WTO Doha Ministerial Declaration of December 20th, 2001 which instructs the TRIPs Council to examine the relationship between the TRIPs Agreement and the CBD and the protection of traditional knowledge and folklore. For the full text see https://www.environment.gov.za/sites/default/files/docs/cuscodeclaration_traditionalknowledge_intellectualpropertyrights.pdf.

35 For the detailed proposals of both developed and developing countries, see www.wto.org/english/tratop_e/trips_e/art27_3b_e.htm.

36 Article 29, Conditions on Patent Applications. See www.wto.org/english/tratop_e/trips_e/t_agm3c_e.htm.

37 See Carr, 'Agreements that Divide', pp. 145–6.

38 According to the terms of a September 1991 agreement, INBio collects and processes plant, insect, and soil samples in Costa Rica and gives them to Merck for evaluation as prospective medicines. The contract provides Costa Ricans with an economically beneficial alternative to deforestation and concurrently advances the research efforts of Merck. See E. Blum, 'Making Biodiversity Conservation Profitable, A Case Study of the Merck/INBio Agreement', *Environment*, vol. 5, no. 4, available at http://qed.econ.queensu.ca/pub/faculty/garvie/eer/blum.pdf.

39 That notion was illustrated through EC's first draft Regulation for the implementation of the Nagoya Protocol on Access to Genetic Resources and the Fair and Equitable Sharing of Benefits Arising from their Utilization (ABS). Its intent was to protect the holders of genetic resources and traditional knowledge from the loss of benefits associated with these resources, a concern known as 'biopiracy', while encouraging research and development. Eventually, the word biopiracy was excluded from the final text of the Regulation, and was replaced by a more generic approach, in Article 9: 'It is also essential to prevent the utilisation in the Union of genetic resources or traditional knowledge associated with genetic resources, which were not accessed in accordance with the national access and benefit-sharing legislation or regulatory requirements of a Party to the Nagoya Protocol'. See EU Regulation No 511/2014 and http://sullivanlaw.net/eu-proposed-access-benefit-sharing-regulation-would-implement-nagoya-protocol.

40 The US Patent Office to revoke the turmeric patent on the basis of a challenge filed by the New Delhi-based Council for Agriculture Research (CSIR). The patent had been granted in March 1995 to two non-resident Indians associated with the University of Mississippi Medical Centre, Jackson, USA. As turmeric has been used for thousands of years for healing wounds and rashes, CSIR challenged the patent on the ground that it lacked novelty. The US Patent Office upheld the objection and withdrew the patent. See V. Shiva, 'The Turmeric Patent is Just the First Step in Stopping Biopiracy', available at www.twnside.org.sg/title/tur-cn.htm.

41 For the full text of India's Biodiversity Act, see http://nbaindia.org/content/25/19/1/act.html.

42 For the full text of India's Patent Act, see www.wipo.int/wipolex/en/details.jsp?id=2407.

43 The Andean Community was established in 1969. It consists of four full members (Bolivia, Colombia, Ecuador, Peru), five associates (Argentina, Brazil, Paraguay, Uruguay, Chile) and one observer country (Spain). See www.comunidadandina.org/ingles/who.htm.

44 For the full text of Decision 391, see www.comunidadandina.org/ingles/normativa/d391e.htm.

45 Mainly on the subject of sovereignty over derivatives, see Smith, 'The Relationship between TRIPS and the CBD: A Way Forward?', pp. 13–17.

46 The Patent Cooperation Treaty was adopted in 1970. See www.wipo.int/pct/en/texts/articles/atoc.htm.

47   'Section 102. Conditions for patentability; novelty and loss of right to patent: A person shall be entitled to a patent unless – (a) the invention was known or used by others in this country, or patented or described in a printed publication in this or a foreign country, before the invention thereof by the applicant for patent.' See www.law.cornell.edu/uscode/text/35/102.
48   See Ragnar, *Biopiracy*, p. 133.
49   See http://biodiversity-l.iisd.org/news/disclosure-requirement-gains-support-at-trips-council-meeting-consultations-to-begin-on-cbd-observer-status.

5 The emerging right to land in new soft law instruments

Margherita Brunori

Introduction

There are two reasons why it is worthwhile to study the new soft law instruments concerning the governance of land and other natural resources for agricultural purposes. The first reason pertains to the contents, the second to the nature of the documents. Land governance deserves attention, since the overlapping of interests of diverse nature and scopes is generating a growing pressure on natural resources as well as complex conflicts between competing subjects, likely to become more and more compelling in the coming decades. Soft law instruments, as currently used, can be portrayed as the expression of a new international way of regulating transnational matters which, despite its unconventionality, requires a careful analysis of its elaboration process and contents, in order to shed light on the evolution of the international legal system. This chapter is a first attempt to outline the emerging legal framework of the topic, offering a look at the possible perspectives that current debates are beginning to shape.

Definition of *land* in international law

In international treaty law there are several references to *land*, the definition or meaning of which changes substantially according to the scope of the convention in question. In this section, I will go through international environmental treaty law and human rights law in order to reconstruct a definition of *land*.

Environmental law

In international environmental law, land is first mentioned by the 1972 United Nations Declaration on Human Environment. Principle 2 of the Declaration says:

> The natural resources of the earth, including the air, water, land, flora and fauna and especially representative samples of natural ecosystems, must be safeguarded for the benefit of present and future generations through careful planning or management, as appropriate.[1]

According to the Declaration, therefore, *land* is a natural resource that must be safeguarded for the benefit of present and future generations through careful planning or management. In addition, in the Preamble of the 1989 Basel Convention on the Control of Transboundary Movements of Hazardous Wastes and its Disposal it is written that:

> 'Area under the national jurisdiction of a State' means any land, marine area or airspace within which a State exercises administrative and regulatory responsibility in accordance with international law in regard to the protection of human health or the environment.[2]

The Basel Convention includes *land* in the area under the national jurisdiction of a State within which it exercises administrative and regulatory responsibility in accordance with international law in regard to the protection of human health and environment. Those definitions mention *land* exclusively for the purpose of defining the category of objects to which the provisions enshrined in the convention are applicable. On the contrary, for the purposes of article 2, para. 2(e) of the United Nation Convention to Combat Desertification (UNCCD), entered into force in 1996, 'Land means the terrestrial bio-productive system that comprises soil, vegetation, other biota, and the ecological and hydrological processes that operate within the system.'[3] In contrast with the previous ones, the UNCCD provides a definition describing the resource land in itself.

Human rights law

In international human rights law, the term *land* does not find any explicit definition and it is cited only in relation to people's rights, or individuals' rights. On the one hand, for what concerns the *people's* relationship to land, we can identify the right of people's sovereignty over their natural resources. The right is affirmed in the first articles of both International Covenant of Political and Civil Rights, and in the International Covenant of Economic, Cultural and Social Rights. People's right of permanent sovereignty over natural resources finds a development in the 1962 United Nations General Assembly Resolution on the Permanent Sovereignty over Natural Resources, where it is declared that peoples and nations have the right to permanent sovereignty over natural resources (i.e. the right to exploit, develop and dispose of natural resources in the interest of their national development and the wellbeing of the people of the state concerned).[4]

On the other hand, the ILO Convention no. 169 affirms the Indigenous and Tribal Peoples' rights over their lands. In the Convention, *land* has a broader meaning and is associated with the concept of *territory*. Article 13, para. 2 says that 'the use of the term lands . . . shall include the concept of territories, which covers the total environment of the areas which the peoples concerned occupy or otherwise use'.[5]

The ontological connection with the concept of territory stresses on its spiritual and cultural function for peoples living on it. Indigenous people's lands have a special spiritual and cultural value and are essential for their livelihood, therefore States have the obligation to protect their property, possession, access or use of their traditionally occupied land, and indigenous peoples have the right to participate in the use, management and conservation of natural resources of their ancestral lands.

The relationship between land and individuals is expressed in human rights law through the possibility of individuals to enjoy property rights or other forms of tenure rights. In an interpretative lecture of human rights law, the right of individuals to own land is found in Article 17 of the 1948 Universal Declaration of Human Rights, in the International Convention on the Elimination of All Racial Discrimination[6] and in the 1979 Convention on Elimination of all forms of Discrimination Against Women.[7] In the latter, there is explicit reference to agricultural land, where Article 14 (g) says that rural women have the right 'to have access to agricultural credit and loans, marketing facilities, appropriate technology and equal treatment in land and agrarian reform as well as in land resettlement schemes'.

Land has an essential connection with the right to housing and the right to adequate food. With reference to the latter, Article 11, para. 2 of the International Covenant on Economic, Social and Cultural Rights (ICESCR) states that:

> 2. The States Parties to the present Covenant, recognizing the fundamental right of everyone to be free from hunger, shall take, individually and through international co-operation, the measures, including specific programmes, which are needed:
>
> (a) To improve methods of production, conservation and distribution of food by making full use of technical and scientific knowledge, by disseminating knowledge of the principles of nutrition and by *developing or reforming agrarian systems in such a way as to achieve the most efficient development and utilization of natural resources*;
>
> (b) Taking into account the problems of both food-importing and food-exporting countries, to ensure an equitable distribution of world food supplies in relation to need.[8]

Article 11 of the Covenant on Economic, Social and Cultural Rights affirms the opportunity to develop and reform agrarian systems in order to achieve the most efficient development and utilization of natural resources and enhance the right to food. Nevertheless, it doesn't make an explicit reference to the concrete way to achieve the 'most efficient' utilization of land and other natural resources, leaving to national governments the power to choose the best solution according to the specific context. The only sure interpretation that can be drawn about the expression 'most efficient development and utilization of natural resources' is that States shall draft agrarian systems that at least do not harm other human rights.

The right to food and the agrarian question

During the period in which the International Covenants were drafted, the agrarian question was attracting the attention of states in many regions. During the 1950s, several countries developed agrarian reforms based on the redistribution of land. The success of the first few (Taiwan, Korea, Japan) encouraged other states to adopt similar land policies, but didn't obtain the expected outcomes, so progressively the strategy of redistribution of land was abandoned in favour of less expensive solutions. Studies of the World Bank also reaffirmed the theory according to which marked-based agrarian reforms were more apt in allocating the resources to whomever would have made the most efficient use of it (de Soto 2000). The diffusion of those studies, together with the new techniques of agricultural production on which the Green Revolution was based, led the issue of land redistribution to be dismissed in favour of attention on investing in techniques to increase productivity (Mazoyer and Roudart 2006).

A revival of the debate on the need for a fairer allocation of the land and the need for protection from the distortion of the market came about in the 1970s, when rising oil prices, a variety of market shocks on both the supply and demand sides, and longer-term pressures on international food commodity markets generated a widespread food crisis in 1972–74. In 1974, the World Food Conference adopted the Universal Declaration on the Eradication of Hunger and Malnutrition.[9] The Declaration does not explicitly talk about land access, but in paragraph it 4 states that 'effective measures of socio-economic transformation by agrarian, tax, credit and investment policy reform and the reorganization of rural structures, such as the reform of the conditions of ownership' are of prime importance for the attainment of the removal of food production obstacles for the eradication of hunger.

Further claims for a new debate on rural development and agrarian reform continued in following years and were brought to the 1979 World Conference of Agrarian Reform and Rural Development (WCARRD), whose final Declaration, called the Peasants' Declaration, strongly recalled the need for equal access to natural resources.[10] However, the ambitious objectives of the Peasants' Declaration did not match with the following decade's policies. In the 1980s, World Bank's loans promoted liberal policies in several domains previously controlled by States, among them the agricultural sector, by pushing for the privatization of state corporations and enhancing free market systems. In general, investments in agriculture decreased in all less-developed countries (Brunori 2013).

The new millennium witnessed a renewed attention to the right to food and to the agricultural sector. In 1999, the General Comment no. 12 of the Committee of Economic, Social and Cultural Rights provided a descriptive interpretation of the right to adequate food, and the year after, the Special Rapporteur on the progressive realization of the right to adequate food was instituted by the Human Right Council. In 2004, the Voluntary Guidelines on the Right to Food were adopted by the FAO council.

Investments in agriculture, in parallel, started to grow after a long period of under-investment. Between 1998 and 2004, official development assistance oscillated around US$3.5–4.5 billion per year, but the renewed focus on agricultural development has since led to increased support to the sector: investments jumped to about US$8 billion in 2008, and even further in 2009, with World Bank lending alone increasing from US$1.5 billion in 2008 to US$3.8 billion in 2009 (World Bank Group 2011).[11] The consequent massive increase of investments into agriculture realized without adequate social, environmental and economic assessment caused severe harm to the livelihood and food security of thousands of indigenous and rural communities. The model promoted by private and public investments privileged individual formalized property rights and large-scale production, in contexts in which small-scale farming based on customary collective rights was the norm.

Forced eviction with low or no compensation led to the loss of land access to thousands families, which had tremendous consequences for their food security and livelihood. Loss of land damaged women particularly, firstly because their land rights enjoy less recognition compared to men, secondly because the possibility to have a direct source of food production represents a fundamental guarantee of minimum subsistence for themselves and their families, and thirdly because in many regions, access for women to salaried work or social security is constrained. Pastoralists, nomadic peoples and tribal peoples suffer equally for the lessening of open access lands. Not least, investments often promote intensive mono-crop production in most fertile lands, with a high need for external inputs and water, therefore generating loss of biodiversity, water and soil consumption and pollution.

The improved sensitivity to the right-based approach to development and in particular to the right to food, together with the need to react to the compelling question of the so-called land grabbing, has pushed international community to agree upon basic shared principles to regulate the matter. A study conducted by the Committee of Food Security pointed out that nowadays, thirty-one transnational documents exist regulating investments in agriculture and land tenure (CFS Open Ended Working Group 2013).

The multifunctionality of land

As the previous paragraphs suggest, land is a crucial element for multiple reasons. Conservation and safeguarding of the quality of land, its components and related natural resources are part of environmental integrity protection. Among other aims, a healthy environment is a precondition for human wellbeing of present and future generations.

Access to (fertile) land and related natural resources is equally instrumental for the fulfilment of basic human rights for the majority of the world's population. Secure land access is an essential component for the realization of the right to adequate food: there are an estimated 500 million smallholder farms in the developing world, supporting almost 2 billion people who depend on them for their livelihood, and small farms produce about 80 per cent of the food consumed in Asia and sub-Saharan Africa. Smallholder agriculture is therefore of major importance in less developed and developing countries, but it also plays an important role in Europe and EOCD countries (HLPE 2013).

Farmers, or *peasants*, as they often name themselves, have a mainly economic relationship with the land they have traditionally used, since they depend on it for their families' subsistence. For peasants' communities, therefore, land is implicitly connected with other resources and services, like water and genetic resources, and credit, technology and access to market. Many peasants' groups also emphasize a social significance of land, on which depends a particular culture and way of life (Gelbspan and Prioste 2013).

Pastoralists, fisherfolk and nomadic farmers equally depend on land access, and their position can be even more vulnerable compared to farmers, since their rights are more difficult to frame within a formalized titling scheme or in a modern property rights system. For indigenous peoples, land has a strong spiritual and cultural value, and their ancestral territory is a core element of identity (De Schutter 2010b). It must be noted that sometimes communities recognize themselves both as peasants, or pastoralists, and indigenous people; in general, although it can be difficult to clearly define the status of a group, what is certain is that all depend on land for their subsistence and for enjoying a decent standard of living. Women, finally, are the category that most relies on land for their own and their families' livelihoods, and the one that has most the difficulties in enjoying its rights.

Characteristics of new instruments regulating land governance and investments in agriculture: the soft law

One of the characteristics of globalization is the concentration of power in different poles and at different scales. If sixty years ago the national government was the higher level of policy decision affecting the grassroots, the actual scene is characterized by a multiplicity of actors (with different degrees of accountability, and, sometimes none) controlling a growing percentage of general affairs.

The emergence of the non-state actors' role in the international decision-making platforms has led to the creation of a new, multicentric governance for the design of policies concerning natural resources and agricultural issues. The major consequence of that is a progressive emergence of more blurred, elastic instruments, more likely to obtain the general consensus, thus permitting a faster reaction to the international community's concerns.

These instruments, defined as *soft law*, although not possessing the strict characteristic of recognized enforceability as commonly understood for law, depending on the circumstances may possess significant normative weight. Furthermore, their role in the development of international law can be critical according to an increasing number of authors, such as Rosalyn Higgins, who argues that individuals are international actors as well, and that existing international legal structures must be revaluated in order to accommodate this reality (Higgins 1994), or Dinah Shelton, who says that 'the growing complexity of the international legal system is reflected in the increasing variety of forms of commitment adopted to regulate state and non-state behaviour with regard to an ever-growing number of transnational problems' (Shelton 2000).

A quite restrictive but clear definition of *soft law* is found in a document of the CFS, which says that soft law is a term describing documents that:

> have been negotiated and/or endorsed through an intergovernmental process. They embody the consensus of a group of countries on certain principles or standards, and their good-faith commitment to their implementation. Although not legally binding per se, these instruments create a legitimate expectation that states will respect them: in practice, states are no longer in position to raise objections against the general orientation indicated in these documents. [Soft law influences] the way in which existing international obligations are interpreted and can lead to the development of legally binding instruments in the future. They can be used as a yardstick to determine correct behaviour.
>
> (CFS Open Ended Working Group 2013)

Protection of tenure rights and access to land in selected soft law documents

Following on from the above discussions, it is worth analyzing recently released soft law documents about the right to food and land tenure, and to examine their references to the right to access to land and related natural resources, in order to reconstruct the existing regulatory framework and possible evolutions of the subject. My focus will be on the Voluntary Guidelines for Progressive Realization of the Right to Adequate Food in the Context of National Food Security (hereafter VGRF) approved by the FAO Council in 2006, and the Voluntary Guidelines on the Responsible Governance of Tenure of Land, Fisheries and Forests (hereafter VGGTs), endorsed by the UN Committee of Food Security in 2012. The selection of those two documents, among the variety of initiatives that have blossomed in the last few years, is founded on a choice of some characteristics of the documents. Criteria that inspired the selection are:

- *A right-based approach*: documents make explicit reference to the international, regional and national obligations protecting human rights and the environment, and starting from existing rights, further elaborate principles able to enhance and develop them.
- *A multi-stakeholder initiative*: the document is negotiated by representatives of national governments, together with representatives of civil society and the private sector.
- *An official endorsement*: the documents are approved and formally endorsed within an official meeting of an international body.
- *A normative content*: the documents contain prescriptions and make reference to particular behaviour that targeted subjects should adopt in specific contexts.

Before analyzing the documents, it is worthwhile to take into consideration the 1999 General Comment no. 12 of the Committee on Economic, Social and

Cultural Rights on the Right to Adequate Food which, in paragraph 12, details the connection between the right to adequate food and access to land and other natural resources. In the document it is written that one of the meanings of *availability* of food implies 'the possibilities . . . for feeding oneself directly from productive land or other natural resources'.[12] Then, in paragraph 13, the General Comment, defining the meaning of food accessibility, states that '[s]ocially vulnerable groups such as *landless* persons and other particularly impoverished segments of the population may need attention through special programmes' (emphasis added). Also, '[a] particular vulnerability is that of many indigenous population groups whose access to their ancestral lands may be threatened.'[13] Despite the non-binding nature of General Comments, they constitute an authoritative interpretation of the right to adequate food, being a carefully considered and systematic analysis emanating from a body uniquely placed to offer an interpretation of the norms contained in the ICESCR (Narula 2006).

In the narrative of fundamental rights, the realization of a right comes through three progressive obligations: the obligation to *respect*, the obligation to *protect*, the obligation to *fulfil* (that is composed by the obligation to *facilitate* and the obligation to *provide*). This categorization of States' obligations was adopted by the former Special Rapporteur of the Right to Food, Olivier De Schutter, for indicating necessary steps to undertake in order to progressively realize the right to food. In the same way, this paragraph will try to adopt those categories to display the actions that States should assume in order to progressively conform their behaviour to existing binding norms as well as to the emerging international standards on land and related natural resources governance. For this purpose, the recommendations expressed by De Schutter in his 2014 report on land access and the right to food constitute a good base for the reconstruction of the obligations that a state should bear for gradually realizing the right to land and natural resources (De Schutter 2014).

The obligation to *respect* existing land access rights requires State parties not to take any measures that result in preventing such access. States should first of all recognize the existing land rights, therefore identifying existing customary tenure rights and informal uses of land. This principle is affirmed by the VGGTs through its general principle 1, where it is expressed that States should:

> Recognize and respect all legitimate tenure right holders and their rights. They should take reasonable measures to identify, record and respect legitimate tenure right holders and their rights, whether formally recorded or not; to refrain from infringement of tenure rights of others; and to meet the duties associated with tenure rights.

With reference to this obligation, the VGGTs dedicate the third part to the enunciation of what States should do in order to effectively respect tenure rights and access to land. Guideline 7 lists the safeguards that States should consider while recognizing and allocating tenure rights; Guideline 8 considers contexts in which land is owned by the State; Guideline 9 is dedicated to particular categories such as indigenous peoples and other communities with customary tenure systems; and finally, Guideline 10 provides guidance in cases where informal uses of land and other natural resources overlap with formal tenure.

The obligation to *protect* is found in General Principle 2 of the VGGTs, according to which states should '[s]afeguard legitimate tenure rights against threats and infringements. They should protect tenure right holders against the arbitrary loss of their tenure rights, including forced evictions that are inconsistent with their existing obligations under national and international law.'

Part four of the VGGTs is dedicated to the clarification of what States should do in order to protect the access to land and related natural resources: it addresses the governance of tenure of land, fisheries and forests when existing rights and associated duties are transferred or reallocated through voluntary and involuntary ways through markets, transactions in tenure rights as a result of investments, land consolidation and other readjustment approaches, restitution, redistributive reforms or expropriation.

In the VGRF, Guideline 8.1 states that 'States should respect and protect the rights of individuals with respect to resources such as land, water, forests, fisheries and livestock without any discrimination.' The obligation to *fulfil* is composed of two sets of different actions: to *facilitate* means that the State must proactively engage in activities intended to make land and related natural resources more concretely accessible to its people, without discrimination. According to General Principle 3 of the VGGTs, States should '*[p]romote and facilitate* the enjoyment of legitimate tenure rights. They should take active measures to promote and facilitate the full realization of tenure rights or the making of transactions with the rights, such as ensuring that services are accessible to all.'

The obligation to facilitate is also expressed in Guideline 8 of the VGRF, where it is affirmed that:

> States should facilitate sustainable, non-discriminatory and secure access and utilization of resources consistent with their national law and with international law and protect the assets that are important for people's livelihoods.
>
> (Guideline 8.1)

> States should take measures to promote and protect the security of land tenure, especially with respect to women, and poor and disadvantaged segments of society, through legislation that protects the full and equal right to own land and other property, including the right to inherit.
>
> (Guideline 8.10)

The obligation to *provide* is stronger, and it indicates the action that the State should undertake in order to effectively allow access to land and related natural resources to people who, for reasons beyond their will, are unable to purchase it by other means. If in the VGGTs this set of actions is expressed in softer terms, the VGRF is much clearer:

> Where necessary and appropriate, States should carry out land reforms and other policy reforms consistent with their human rights obligations and in accordance with the rule of law in order to secure efficient and equitable access to land and to strengthen pro-poor growth.
>
> (Guideline 8.1)

> As appropriate, States should consider establishing legal and other policy mechanisms, consistent with their international human rights obligations and in accordance with the rule of law, that advance land reform to enhance access for the poor and women.
>
> (Guideline 8.10)

The VGGTs, on the contrary, in paragraph 15 talk about redistributive reforms and say that:

> In the national context and in accordance with national law and legislation, redistributive reforms *may* be considered for social, economic and environmental reasons, among others, where a high degree of ownership concentration is combined with a significant level of rural poverty attributable to lack of access to land, fisheries and forests respecting . . . the rights of all legitimate tenure holders. Redistributive reforms should guarantee equal access of men and women to land, fisheries and forests.

Both documents reserve special consideration to particular groups as indigenous peoples, peasant communities and minorities, and make reference to the need to put forward a sustainable use of land and related natural resources, therefore helping in framing a holistic picture of the several meanings of 'land' for different categories of peoples and for the essential functions that land and related natural resources are essential to fulfil.

Conclusion and future perspectives

What emerges from the examination of the extracts of the two documents is that both sets of voluntary guidelines follow the strategy of categorizing progressive actions for the implementation of rights while describing the principles regulating access to natural resources and, in particular, land. This structure leads to the consideration that, despite the manifest voluntary nature of the guidelines, there is an agreement within the international community on the fact that the issue of land access is sensible to the point that it should be considered along the same lines as a fundamental right.

If this hypothesis is true, it opens the floor to further research. As set out above, this time for land access attention is not the first. If the topic was considered compelling for the wellbeing of all the world's people in the 1970s, now the issue has reached a tipping point for multiple reasons: population growth, climate change, natural resource depletion and increasing interests for different purposes on land make a sustainable and comprehensive regulation of the matter at the international level of the utmost importance.

To ensure that the words of these documents are put into action, a set of issues deserves further research and policy attention. First of all, there should be an honest reconstruction of what the sources of international law are nowadays and what their value and effects are, in order to frame the efficacy and legitimacy of soft law instruments as currently utilized. Secondly, there should be, at the international governance level, the will to create a concrete coordination among several branches of existing international law (i.e. among international investment law, trade law, environmental law and human rights law) so as to avoid the fact that developments by one side are not undermined by contrasting policies from other sides. Thirdly, policy implementation on different governance levels should be assured and facilitated, especially with regard to effectively recognizing and operatizing collective rights and benefit sharing; community participation in decision making processes; systematization of norms and competing interests. Finally, but by no means of lesser importance, monitoring and accountability mechanisms should be implemented to test regulations' efficacy and to challenge the responsibilities of all actors involved.

Notes

1 United Nations Declaration on Human Environment, Stockholm 1972.
2 Basel Convention on the Control of Transboundary Movements of Hazardous Wastes and their Disposal, adopted on 22 March 1989 and entered into force on the ninetieth day after the date of deposit of the twentieth instrument of ratification, acceptance, formal confirmation, approval or accession by a country to the Convention, 5 May 1992.
3 United Nations Convention to Combat Desertification in Those Countries Experiencing Serious Drought and/or Desertification, Particularly in Africa (UNCCD), adopted in Paris, France on 17 June 1994 and entered into force in December 1996.
4 UN General Assembly Resolution no. 1803 (XVII) Permament Sovereignty over Natural Resources. 14 December 1962.
5 ILO Covenant no. 169 on Indigenous and Tribal People's Rights.
6 UN General Assembly resolution 2106 (XX) – entered into force 4 January 1969. Article 5d(v) affirms everyone's 'right to own property alone as well as in association with others'.
7 UN General Assembly resolution 34/180 of 18 December 1979. Entered into force on 3 September 1981. Article 16(h): 'the same right for both spouses in respect of the ownership . . . of property'. Interestingly, women's right to own property is affirmed only with reference to marriage rights, or agricultural land systems.
8 International Covenant on Economic, Social and Cultural Rights, General Assembly Resolution 2200A (XXI) of 16 December 1966, entry into force 3 January 1976 (emphasis added).
9 UN Doc. E/CONF. 65/20, 1974.
10 Part II of the Declaration, entitled 'Access to Land, Water and other Natural Resources' addresses the issue in seven paragraphs: a. reorganization of land tenure, b. tenancy reform and regulation of rural wages, c. regulation of changes in customary tenure,

d. land consolidation, promotion of group farming cooperatives and collective and state farms, e. community control over natural resources, f. settlement of unoccupied public lands, g. reduction of inter-regional and inter-community inequalities.

11 In the World Bank Group (2011) report we read: 'The largest donors have been the World Bank, the European Union, Japan, and the United States; on a regional basis, the top donors have been the African Development Bank and the Asian Development Bank. China, India, and major foundations (such as the Bill and Melinda Gates Foundation) are increasingly important sources of funding for developing countries, particularly in sub-Saharan Africa. The Republic of Korea, Mexico, Saudi Arabia, and Turkey have also begun to provide development assistance to agriculture'.

12 UN ESCR Committee General Comment No. 12, The Right to Adequate Food, 1999, UN Doc. E/C.12/1999/11.

13 *Ibid.*

References

Brunori, M. (2013) Banca Mondiale, interventi in agricoltura e diritti umani. Rivista di Diritto Agrario XCII fasc. 2. Available at www.academia.edu/3443184/Banca_Mondiale_Interventi_in_Agricoltura_e_Diritti_Umani

CFS Open Ended Working Group (2013) Consultancy Output 1: Summary of International Initiatives that Provide Guidance on Responsible Investment: Key Characteristics. CFS Open Ended Working Group on Principles for Responsible Agricultural Investments which Enhance Food Security and Nutrition. 29 January.

De Schutter, O. (2006) *International Human Rights Law: Cases, Materials, Commentary*. Cambridge: Cambridge University Press.

—— (2010a) The Emerging Human Right to Land. *International Community Law Review* 12: 303–34.

—— (2010b) *Access to Land and the Right to Food*. Report of the Special Rapporteur on the Right to Food presented at the 65th General Assembly of the United Nations. Doc. A/65/281. 21 October. New York: UN.

—— (2014) *Report of the Special Rapporteur on the Right to Food*. 24 January. Doc. A/HCR/25/57. New York: UN.

de Soto, H. (2000) *The Mystery of Capital: Why Capitalism Triumphs in the West and Fails Everywhere Else*. New York: Basic Books.

Gelbspan, T. and F. Prioste (2013) *Land and the Struggle for Social Justice: Social Movement Strategies to Secure Human Rights*. Curitiba: Tierra de Direitos.

Higgins, R. (1994) *Problems and Process: International Law and How We Use It*. Oxford: Clarendon Press.

HLPE (2013) *Investing in Smallholder Agriculture for Food Security*. Rome: High Level Panel of Experts on Food Security and Nutrition of the Committee on World Food Security.

Marauhn, T. (2010) Land Tenure and Good Governance from the Perspective of International Law. Opening address at the Colloquium 'Good Governance in Land Tenure', Potchefstroom, 22–23 April. *Potchefstroom Electronic Law Journal* 14(3): 5–23.

Mazoyer, M. and L. Roudart (2006) *A History of World Agriculture : From the Neolithic Age to the Current Crisis*. London: Earthscan.

Narula, S. (2006) The Right to Food: Holding Global Actors Accountable Under International Law. *Columbia Journal of Transnational Law* 44: 691.

Shelton, D. (ed.) (2000) *Commitment and Compliance: The Role of Non-Binding Norms in the International Legal System*. Oxford: Oxford University Press.

World Bank Group (2011) *Growth and Productivity in Agriculture and Agribusiness*. Washington, DC: Independent Evaluation Group, World Bank Group.

Part II

The right to water and to food – climate change

6 The right to water

The intersection between international and constitutional law

Antonio D'Aloia

Introductory notes about right to water

In recent years, the right to water has become one of the most complicated and fascinating paradigms of human rights discourse. It is (now) a very reliable forecast of the contradictions and the hazards endemic to economic globalization.

It is a multiform right, as is its content. A novel by the Italian writer Andrea Camilleri is titled *La forma dell'acqua* – 'The shape of water'. In this story, one person asks another 'What is the shape of water?' The answer: 'But water has no shape! . . . It takes the shape that is given to it.'

And this is exactly the point: even if we try to frame this 'word' from a legal point of view, with the implications and the challenges that it raises to the legal reflection. As a scholar has written on the right to water recently:

> in the absolute simplicity of the fundamental element, however it is hidden the deep complexities of its legal framework. . . . Fluid and elusive, the water escapes from easy legal definitions. It is not only a good, it is not a right like others, it is not just a public service. Maybe it is all these three things, or perhaps even more so.
>
> (Briganti 2012: 11)

At the same time, the meanings of the right to water and the reflection of its qualification as a fundamental right change in the different legal orders.

In some countries, water is defined as a sort of 'basic' right, and adequate access to water resources becomes a problem of survival, a watershed among a minimally dignified life, the total abandonment to an extreme poverty and the total lack of minimum conditions of existence. As Thielborger (2013: 1) put it, 'water is life. Lack of it leads to illness, misery, death.'

Some data may help us to appreciate the dramatic urgency of the problem, which has truly reached a tipping point. One person out of six does not reach the minimum standards set out by the UN of 20–50 litres of fresh water per capita daily, which is necessary to ensure the basic needs related to nutrition and hygiene. Furthermore, the Millennium Development Goals target to halve, by 2015, the proportion of the population without sustainable access to safe drinking water and basic sanitation, will be realized only partly.

On the other side, in the countries where fresh water availability is (still) far from a situation of extreme deprivation, the right to water has become a new emblem of the traditional conflicts between (i) privatization (and, in that case, the risks that it implies when it is applied to goods, such as water, that also have an economic dimension, but are essential for life and dignity of people) and (ii) the crucial redistributive role and the function of realizing equality both of the public authorities in order to protect the welfare and the social cohesion.

In 2000, the World Water Forum of The Hague resulted in a document in which water was qualified as an economic good of industrial relevance. This was an unwieldy and dangerous definition, because it placed water in the system of international trade, within the WTO model (Garofalo 2010: 16). Thus water becomes 'the object of rivalry between individuals, local communities, States, for alternative and competing uses, and the result is the exclusion of the less strong ones, the less skilled ones, the less competitive ones' (Petrella 2006: 3).

In Italy, for example, there was a referendum in 2011 in which the main result in political and media terms was the assertion of the principle of public water, even as a prototype of a new category of goods, the common goods.

This is a controversial category, one that is facing different translations of the concept; it is one that disrupts the dichotomy between the two traditional forms of property (public or private). As has been said by one of the most important theorists of this new movement, 'not another form of property, but the opposite of property' and individualism (Rodotà 2012: 81–2). Among the many definitions that have been given of the 'commons', particularly enlightening is one that defines the common goods (including rivers, streams, their sources, lakes and other waters) as 'those things that express functional utility to the exercise of the fundamental rights and to the free development of a person' (*ibid.*); for this reason, the common goods 'must be preserved subtracting them to the destructive logic of short period projecting their protection in the world farther, inhabited by future generations' (*ibid.*).

However, the gap between the different areas of the world is decreasing, and unfortunately not in a positive manner. Water resources are diminishing everywhere, and the scarcity of potable water available to human beings amounts to 0.001 per cent of the total water resources of the planet. This may lead us to agree with Samuel Coleridge, in the 'Rime of the Ancient Mariner', that there is 'Water, water, everywhere, / Nor any drop to drink'. It is a growing problem, one that also involves the more successful and developed countries. According to some studies, for the years ahead the expected scenario will be much worse than the current situation. The Organisation for Economic and Co-operation and Development (OECD) estimates that by 2050, the world's demand for water will grow by 55 per cent and the competition between water users and nations demanding water resources is predicted to escalate radically.

Many overlapping causes have an impact on this global water crisis, such as the increase of population, the explosion in consumption (related, in part, to the spread of welfare and development and the consequent changes in life-style, eating habits and hygiene), pollution and effects of climate change, the increasing urbanization and expansion of economic activities, and the production of biofuels.

In other words, we are faced with a double movement, which has a negative impact in these two dimensions on the availability of water resources: from one side, there are factors that reduce the amount of available water; on the other side, many factors increase the water demand.

In this context it is inevitable that the legal system deploys its more important and, at the same time, fragile resources – in other words, the instruments related to human rights. The right to water is therefore the best response to the issues raised above.

The right to water and the escalation of international water law

The path that led to the acknowledgment of a fundamental right to water has been developed on the two levels of international law and constitutional law – two separate tracks that have resulted in some partial convergence and integration.

Historically, inland waters were not regulated by international law. They constituted a reserved domain of states, tightly enclosed inside the fence that protected national sovereignty (Tanzi 2012: 517). What has happened to change this state of affairs? First of all, the scarcity of water, and the fact of it being a more and more 'disputed' good (Staiano 2011: 4), along with the shame of such extreme inequality in the availability and access to water resources (between countries and within a single country), has somehow crossed national borders: it has increasingly become a global problem of humanity, just another element of the complex mosaic of human rights.

From a different point of view, even the location of water defines an extreme asymmetric framework, which inevitably requires international rules, or a multi-level governance system, to overcome the different national interests. Thirteen states have almost 65 per cent of the global water resources, including three of the so-called BRICS countries: Brazil, Russia and China alone have nearly 30 per cent of water resources, whereas India, despite having 3.5 per cent of the world's water resources, is one of the states that risks water stress because of the explosion of population and the deteriorating quality of the water. Moreover, the fifth member of this new and controversial club of the economic globalization (South Africa) suffers the greatest water scarcity (Federico 2012: 566). As Jeffrey (2006: 11) wrote, 'water is international, national, regional and local, having geographical mobility'.

In that way, water gained consideration in the evolution of international law, to start on the basis of the principles of rational and equitable balance of interests in the international community (Tanzi 2012: 518), and in the meantime, especially after the Second World War, the international and the constitutional law have begun to move towards each other in a two-way process of mutual influence, particularly in the field of rights.

Constitutional law has opened its horizons to the law 'beyond the state'. It has agreed to combine and incorporate its principles with those resulting from international law, forfeiting elements of sovereignty in order to reach the wider goals of peace and justice between nations, for the protection of human rights; at the same time, international law, assuming the character of a constitutional law, has gained greater strength and ability to constrain the decisions of national legal systems.

This process has not had linear or unique developments. In the case of the right to water, the legal definition of it as a fundamental human right (in other words, inseparable from other rights) is a recent achievement, and this may explain the uncertainties and the fragility that still characterize its effective implementation.

This formula is used in UN Resolution 64/292 of 28 July 2010 ('The Human Right to Water and Sanitation'), which 'Recognizes the right to safe and clean drinking water and sanitation as a human right that is essential for the full enjoyment of life and all human rights'. Previously, the expression 'right to water' has been found also in the Convention on the Elimination of All Forms of Discrimination against Women (1979), and in the Convention on the Rights of the Child (1989).

Moreover, as early as 2002, the UN Committee on Economic, Social and Cultural Rights officially recognized water as a 'limited natural resource, a public good and, above all, the human right to water entitles to sufficient, safe, acceptable, physically accessible and affordable water for personal and domestic uses'.

Also significant is the UNECE Protocol of London of 1999, which entered into force in 2005. This Protocol on the interaction between water and health is considered to be an example of how the instruments of international law, even when they are not originally self-executing or do not confer rights directly, may trigger mechanisms that affect behaviours of states to generate progressively justiciable obligations (Tanzi 2012: 521ff.).

UN Resolution 64/292 stresses repeatedly that water for drinking and hygiene, as well as being an 'autonomous' right for every man, is 'derived from the right to an adequate standard of living and inextricably related to the right to the highest attainable standard of physical and mental health, as well as the right to life and human dignity'. In this sense, we can highlight that regardless of the explicit statement of the access to the water as a human right, right to water already exists as an implicit right in other fundamental rights, and it takes from them its primary relevance. This is not necessarily a reduction of the right to water: indeed, the nature of primary rights such as life, health and dignity have to be included in the legal value of water and its protection instruments.

For example, in the Gabčíkovo-Nagymaros decision of 1997, concerning the environmental impact of the construction of a system of dams on the river Danube, a controversy between Hungary and Slovakia, the International Court of Justice (ICJ), while still not recognizing a right to water, performed some general considerations on the environment that would have subsequently become the basis of the qualification of water as a fundamental right. According to the ICJ, the 'environment is not an abstraction but represents the living space, the quality of life and the very health of human beings, including generations unborn'. Even more significantly, in the dissenting opinion of Justice Weeramantry, it was underlined that 'damage to the environment can impair and undermine all the human rights spoken of in the Universal Declaration and other human rights instruments', and it recalled the need for international law in the environmental sector to intensify attention 'to the global concerns of humanity as a whole'.

In more recent years, a lot of international bodies, judicial or quasi-judicial, have recognized the essential nature of water as a prerequisite for the satisfaction of the rights to health and a dignified life. The most significant cases have occurred in Latin America, where the International Center for Human Rights (ICHR) has condemned Paraguay twice for violating the right to life of the Yakye Axa and the Sawhoyamaxa indigenous communities, who claimed the return of their ancestral lands. In the meantime they had been placed in temporary settlements without basic minimum services, including clean water and toilets. The court declared that 'special detriment to the right to health, and closely tied to this, detriment to the right to food and access to clean water, have a major impact on the right to a decent existence'.

These two decisions have imposed on the state immediate, regularly and permanently, the obligation to adopt measures to supply sufficient drinking water for consumption and personal hygiene to the members of the community.

In Africa, both the African Commission on Human and People's Rights and the Community Court of Justice of ECOWAS have adopted important decisions on environmental protection (and on water) in certain cases involving the exploitation and the pollution by private enterprises of the area of the Niger Delta. In these decisions, there are remarkable general statements on the legal significance of the collective environmental rights as 'essential elements of human rights in Africa' (Pineschi 2014). Moreover, we have to remember the removal of the protected human rights in the international arena by the devaluation made by the national Constitution of Nigeria, which considers economic, social and cultural rights, in general, as mere political directives and not justiciable rights.

In addition, the Community Court of Justice of ECOWAS distinguishes between rights whose implementation depends on the economic resources of the State, and those for which 'the only obligation required from the State to satisfy such rights is the exercise of its authority to enforce the law that recognizes such rights and prevent powerful entities from precluding the most vulnerable from enjoying the right granted to them'. In the latter situation the protection of the right, and the obligation of the state to ensure this protection, should have the fullest possible implementation.

Even the European Court of Human Rights (ECHR), despite there being no specific statements on the protection of right to water in the Convention, has recognized (in the case of *Sufi and Elmi v. the UK*) that the lack of access to fresh water and basic health services may be relevant in the proceedings for the expulsion of foreign citizens (in this case, coming from Somalia) to evaluate the risk of being subjected, once back to their homeland, to inhuman and degrading treatments (Staiano 2013: 14).

In short, these fragments of an international right to water, still incoherent and incomplete (Pineschi 2014: 131), show that opinion is gradually consolidating towards the principle that water belongs to the core of human rights, along with life, dignity and health. It is a condition of effectiveness of these other rights, and there are minimum standards that have to be guaranteed.

In my opinion, it is imperative that this movement (composed of legal and cultural issues, and social struggles) manages to eventually reach the form of a directly binding international treaty. In this way the right to water would follow the same evolution as other human rights, to which the right to water is intrinsically linked.

In fact, when we discuss life, dignity and health, we are in the area of *jus cogens*. It is only a matter of time; it is a process that may have a gradual trend, and may meet some resistance, but, in the end, the outcome is written in the very substance of the right to water.

Water is a common concern. It is not necessary to imagine an infringement of the principle of sovereignty of the states or a global governance of water, but merely to introduce more binding elements of solidarity and cooperation. In other words, a 'responsible' sovereignty, which puts states in a new role: that of trustees of humanity (Benvenisti 2013; Pineschi 2014: 148).

Convergences of constitutional law on the right to water

The circuits of international law and constitutional law on the right to water have gradually reached common results, which certainly enhance and strengthen, in a reciprocal perspective, the two dimensions of protection (Thielborger 2013: 123, 129–30). In the most recent aspects of constitutionalism, the right to water, or less directly the duty of states and public authorities to ensure conditions of effective and safe access to drinking water, appears in many fundamental charters of human rights.

South America is probably the most interesting example, and in this context Bolivia and Ecuador may be considered as typical cases (Mezzetti 2012: 553ff.). Article 16 of the Bolivian Constitution of 2009 states that 'Every person has the right to water and food. The State has the obligation to guarantee food security, by means of healthy, adequate and sufficient food for the entire population.' This principle is followed by a series of other principles and provisions for implementation, according to which:

> Water constitutes a fundamental right for life, within the framework of the sovereignty of the people and its development.
>
> (Articles 373/1 and 376/1)

> Water resources in all their states, surface and subterraneous, constitute finite, vulnerable, strategic resources, and serve a social, cultural and environmental function. These resources cannot be the object of private appropriation and

they, as well as water services, shall not be given as concessions and are subject to a system of licensing, registration and authorization pursuant to the law.

(Article 373/2)

The State shall protect and guarantee the priority use of water for life.

(Article 374/1)

The State shall recognize, respect and protect the uses and customs of the community, of its local authorities and the rural native indigenous organizations over the right, management and administration of sustainable water.

The fossil, glacial, wetland, subterraneous, mineral, medicinal and other waters are priorities for the State, which must guarantee its conservation, protection, preservation, restoration, sustainable use and complete management; they are inalienable, not attachable and cannot be limited.

The water is an essential part of the people's sovereignty, and its development.

(Articles 373/1 and 376/1)

Equally binding and strong are the rules contained in the Constitution of Ecuador (2008). Moreover, the recognition of the primary importance of water as a good and as a right of everyone in the Bolivian and Ecuadorian Constitutions is part of a framework of constitutional values that seem to outline a new way of constitutionalism, which is particularly interesting from the perspective of the protection of the environment and of a good such as water.

The pillars of this new construction are nature – harmony with nature, and the respect of what the Andean tradition and the Constitution call 'Pachamama', which can be identified with the world, as a 'global' entity, which is everywhere, in all places and at all times (Zaffaroni 2012: 422ff.).

The ideology of 'Pachamama' puts the emphasis on duties, on the responsibilities of everyone towards others and nature. The *buen vivir* is something more and different than our welfare and the same concept of sustainable development, even if this has a partially similar content: it is a new form of citizenship that has enormous implications on the political and economic level, and on the same way of thinking about rights, beginning with the identification of 'nature' as a legal entity. It opens a new chapter in the history of law (*ibid.*: 433), the implications of which are not easy to predict.

Africa, like Latin America, is a very lively workshop of this new strategy to a constitutional strengthening of the right to water. Even here there is a 'monster' case: that of South Africa, whose Constitution (Article 27) confirms that everyone has the right to access to sufficient food and water, while another constitutional provision (Article 24) protects the right to safe environment, nowadays and for future generations (Federico 2012: 572).

In that country, a famous court case concerned the right to water, which was decided ultimately by the Constitutional Court (*Mazibuko v. Johannesburg Water Ltd*).

The basis of the dispute was the constitutionality of the law with reference to a system under which, in the poorest areas of the city, pre-paid water meters are installed, with the automatic interruption of service when credit is exhausted.

The premise of the court is that the determination of the minimum core in the satisfaction of social rights (such as water) belongs to the political circuit between government and parliament in the context of a progressive and reasonable implementation of these rights, and cannot be determined by the judge. On this specific point, the Constitutional Judge states that the system of pre-paid water meters is 'unlawful' and incompatible with Article 27 of the Constitution, and also that it constitutes a discrimination against those who live in the poorest neighbourhoods. In fact, discrimination arises directly from the fact that access to water is differentiated because of economic and social conditions, creating an unfair differentiation between the inhabitants of the same community by reason of the area of residence (Federico 2012: 580).

Many other African countries have provisions in their constitutions that recognize water (and sanitation) as a human right, and oblige the state to facilitate equal access to clean and safe water (e.g. Gambia, Mauritania, Uganda). Finally, Tunisia has also included in its new Constitution, adopted on 26 January 2014, an article (Article 44) which states that 'the right to water shall be guaranteed. Conservation and rational use of water shall be a duty of the State and society'.

Nature, contents and implications of right to water

The right to water has a constitutional element, independent of its formal recognition within the constitutions. In this regard it is similar to the right to life, which is always included in the discourse about the right to water. Many constitutions have no explicit formulation for the protection of the right to life; nevertheless, the right to life is the assumption that makes every other right conceivable.

The right to water is also a social right, according to two different (but complementary) perspectives. First, it is a social right because it requires infrastructure, services (transport and distribution) and means to make the right (and the good to which it relates) accessible and usable in practice. On a different perspective, the 'social' nature of the right to water appears related to the fact that this is a right which creates a bond, sharing, solidarity and responsibility. In other words, anyone who wants to have access to water knows that his right should be exercised in ways that do not prevent other people from doing the same thing, but also knows that this right, the measure of its satisfaction, is the boundary among different conceptions of the world.

There is a moral sense in these rights that relates to the basic conditions of human survival (environment, peace, sustainable development, future generations, water). Along with the individual interest there is the perspective towards a fairer order, the point of emergence of community needs.

There is also a collective element, and this definition may take several forms, not without potential contradiction. In this context, the availability of sufficient and healthy water resources (as well as other natural resources) is linked to international law, and to self-determination of peoples to decide to freely pursue their economic, social and cultural development (Zolo 2005; Bobbio 1990). On the other hand, however, the sovereignty of states in the exploitation of natural resources in their territory could impede the process of evolution towards forms of enhanced cooperation among states (especially with respect to an asymmetrically distributed resource, such as water).

Policies concerning water should not only be local, but should be aware of the interest of humanity; they should tend to benefits that are shared and equally distributed among all nations, and incorporate positive obligations of cooperation and of resource sharing. This probably requires an international authority of regulation with effective powers.

The international obligations may be subdivided into duties to respect, protect and fulfil. The duty to respect means, above all, not to interrupt an existing water supply (Thielborger 2013: 179). The duty to protect may be included in the interventions replacing the failure of third state investments, which operate in another state for the provision of water. This would be an important tool against some risks of water privatization.

Most important and controversial, finally, is the obligation to fulfil the human right to water; the heart of the debate, according to Thielborger (*ibid.*: 182). This obligation may be implemented in several ways, including:

* to ensure the exchange of expertise in the field of water and to support further research on water management;
* to insert in agreements on trade liberalization and water privatizations rules that are able to avoid the capacity of countries to fulfil the right to water being affected; and
* to promote, in their capacity as a member of international organizations and international financial institutions, the development of lending policies, credit agreements and other institutional measures that are consistent with the right to water (*ibid.*: 183–4).

The collective nature of right to water may emerge even within each legal order, like a right of each community (ethnic, linguistic, cultural minorities) and the state to which it belongs. This right should find corresponding instruments of collective action (class actions, popular actions, civic organizations), protecting at the same time individual interests and collective interests, because they are essential to the life of the community (Iannello 2013: 147–8). We have seen examples in the cases of indigenous communities such as *Yakye Axa and Sawhoyamaxa v. Paraguay* (before the ICHR), and in the cases resolved by the Argentine jurisprudence in favour of Menores Paynemil and Valentina Norte communities.

Water is typically an intergenerational right (or interest), which has to be protected and guaranteed. Often the reference to water is associated with the assertion that the use of water resources has to be rational and sustainable, according to the principle of solidarity and responsibility towards unborn generations. In practice, what this means is to say that water is an intergenerational good (or right). Certainly, new categories require new solutions. Prohibitions, penalties and obligations unilaterally imposed may not be sufficient or fully effective.

A 'law for the future' needs participation, social awareness and incentive-based rules that encourage the sustainable use of water and of natural resources in general. Above all, we should find a way to insert, in the mechanisms of the governance nowadays prevailing, the topic of the future as something not abstract or undefined, but essential to problems belonging to humanity and future generations. In this sense, there are interesting proposals and experiments concerning court proceedings (admissibility of actions of associations representing the interests of children or future generations, as in the case Minors Oposa decided by the Supreme Court of the Philippines in 1993), the establishment of independent bodies and authorities for the protection of the interests of those who still have no voice, and the inclusion in the law-making and administrative processes of a test about intergenerational impacts.

Water between rights discourse and policies: problems of privatization and perspectives of the subsidiarity

Recognizing that water is a right, however fundamental, is very important, but is not sufficient to solve the problems and challenges that we face. A complex and layered right such as the right to water needs policies that may grasp the multiplicity of its connections with other topics, such as nutrition, energy levels and the structure of the economy of a country. For this, we have to underline that the challenge is one that touches the water companies in their complex structure, and which cannot be addressed only on the institutional or regulatory side. Regulatory and policies measures need to be supported and enhanced by a social culture that takes upon itself the responsibility of a change in the way we conceive of living the relationship between man, society and nature.

In a decision of the High Court of Kerala State in India, it is stated that ground-water is a 'national wealth' that belongs to 'the entire society', and more generally that 'certain resources like air, sea waters and the forests have such a great importance to the people as a whole that it would be wholly unjustified to make them subject of private ownership'.

Moreover, the regulatory landscape is not an even one. There are countries where public management is established as a principle of legislation (Switzerland, Netherlands, Belgium), while many other countries adopt mixed systems, or with a prevalence of privatized management (Germany, France, England, for example).

On the one hand, it is true that many of the struggles that led to the rediscovery and revival of the issue of public water and the fundamental right of the water were a reaction to the failures and dangers of a 'wild' privatization that had produced a rise in prices and rates, likely to jeopardize the possibility of a part of the population to access the resource (typified by the movement of Cochabamba in Bolivia).

Similarly, the unacceptable face of the 'private' in the water sector also emerges when analysing many of the service contracts between states and private companies, which often provide excessive run times, substantial freedom to set rates in relation to the economic recovery of the investments made, and massive penalties (especially unbearable for economically weak countries) in the event of the withdrawal of these contracts.

Litigation on these contracts (often assigned, at least in the first instance, to arbitration established by international treaties, such as the International Centre for the Settlement of Investment Disputes) is a very important test to understand how there may be 'serious' and effective progressive implementation of the right to water. It may be put on the same level as the obligation of a state to protect human rights and to comply with foreign investment. The cases of *Republic of Argentina v. Suez, Azurix Corp. v. Republic of Argentina* and *Aguas del Tunari v. Republic of Bolivia* are very significant in identifying the difficulties that a nation can encounter in restoring fairer prices unilaterally, or in the termination of contracts considered excessively favourable to private corporations (Tanzi 2012: 530–32).

On the other hand, however, it would be wrong, and merely 'ideological', to say that 'the public' is always good, and forget many cases of ineffective management of water resources by public companies. On this basis, despite the risks that have been reported, and then social reactions and policies that result, it is possible that the management of the distribution of water resources could be entrusted to private entities, ensuring, however, on the public level, that rates remain just and sustainable, that they are graded in relation to the use of water resources, and that these items are properly controlled by public authority and participatory bodies of users. In other words, as has been effectively argued, the fact that water is a public good 'cannot fail to produce a result on the ways of management, regardless of the operator, public or private' (Staiano 2011: 21).

The decisive point is the ability of public authorities (state and local) to define the rules of the game in which the fundamental decisions on rates and how to provide the service can be led back to a sphere of social utility, consistent with the constitutional status of the right to water (Frosini 2011). In fact, for some risks, such as increased tariffs, non-performance of the obligations of the development and maintenance of the water network, an attitude of inattention to the social importance of access to water and to the condition of the most marginal could be managed and controlled through strict regulation of service contracts, or providing for a subsidiary state liability.

One of the great principles of democratic constitutionalism is the principle of subsidiarity, which may have a 'horizontal' meaning (i.e. on the participation of citizens and associations of citizens, in the management or control on the management of the commons). The road of subsidiarity is not easy, and requires a sense of responsibility, civic culture, and awareness that water has played a crucial role in the maintenance and revitalization of constitutionalism of rights (and duties). Subsidiarity as participation may produce important benefits, such as better-informed decisions through (local) stakeholder input; education of the lay public; enhanced legitimacy of decisions; and thus more widely accepted decisions and more effective delivery.

In this process the definition of water as a 'common good' could really materialize, generating what the Overarching Conclusion of World Water Week 2013 in Stockholm called the 'water esperanto': a new and shared language for a more sustainable world.

References

Benvenisti, E. (2013) Sovereigns as Trustees of Humanity: On the Accountability of States to Foreign Stakeholders. *American Journal of International Law* 107: 295–333.

Bobbio, N. (1990) *L'età dei diritti*. Turin: Einaudi.

Briganti, R. (2012) *Il diritto all'acqua tra tutela dei beni comuni e governo dei servizi pubblici*. Napoli: ESI.

Camilleri, A. (1994) *La forma dell'acqua*. Palermo: Sellerio.

Dogliani, M. (2010) (Neo)costituzionalismo: un'altra rinascita del diritto naturale? Alla ricerca di un ponte tra neocostituzionalismo e positivismo metodologico. Available at www.costituzionalismo.it/articoli/357.

Federico, V. (2012) 'Ogni persona ha diritto di accesso all'acqua': Acqua e diritti fondamentali in Sudafrica. *Diritto pubblico comparato ed europeo* 2012(2): 566.

Frosini, T. E. (2011) *La lotta per i diritti. Le ragioni del costituzionalismo*. Napoli: ESI.

Garofalo, L. (2010) Osservazioni sul diritto all'acqua nell'ordinamento internazionale. *Analisi Giuridica dell'Economia* 1.

Iannello, C. (2013) *Il diritto all'acqua. L'appartenenza collettiva della risorsa idrica*. Napoli: La scuola di Pitagora Ed.

Jeffrey, P. (2006) The Human Dimensions of IWRM: Interfaces between Knowledge and Ambitions. In P. Hlavinek (ed.), *Integrated Urban Water Resources Management*. Berlin: Springer.

Mezzetti, L. (2012) Diritto all'acqua negli ordinamenti dei Paesi latinoamericani: evoluzioni recenti e prospettive. *Diritto pubblico comparato ed europeo* 2012(2): 553.

Petrella, R. (2006) Diritto all'acqua per tutti e beni comuni mondiali: giustizia e solidarietà. *Cooperazione Mediterranea (Rivista Quadrimestrale dell'Isprom)* 3.

Pineschi, L. (2014) Un'evoluzione imperfetta nella tutela del diritto a un ambiente soddisfacente: la sentenza della Corte di giustizia dell'ECOWAS sul caso SERAP c. Nigeria. *Diritti Umani e Diritto Internazionale* 2014(8).

Rodotà, S. (2012) *Il diritto di avere diritti*. Rome: Bari.

Staiano, S. (2011) Note sul diritto fondamentale all'acqua. Proprietà del bene, gestione del servizio, ideologie della privatizzazione. Available at www.federalismi.it/nv14/articolo-documento.cfm?Artid=17695&content−ote+sul+diritto+fondamentale+all%27 acqua.+Propriet%C3%A0+del+bene,+gestione+del+servizio,+ideologie+della+privatizzazione&content_author=Sandro+Staiano.

Staiano, F. (2013) La progressiva emersione di un diritto umano e fondamentale all'acqua in sistemi di diritto internazionale e costituzionale: principi generali e prospettive di implementazione. Available at http://www.federalismi.it/nv14/articolo-documento.cfm?Artid=21823&content=La+progressiva+emersione+di+un+diritto+umano+e+fondamentale+all%E2%80%99acqua+in+sistemi+di+diritto+internazionale+e+costituzionale:+principi+generali+e+prospettive+di+implementazione&content_author=Fulvia+Staiano.

Tanzi, A. (2012) Il tortuoso cammino del diritto internazionale delle acque tra interessi economici e ambientali. *Diritto pubblico comparato ed europeo* 2012(2). Available at www.

dpce.it/index.php/la-rivista/indici-2011-2014/1821-il-tortuoso-cammino-del-diritto-internazionale-delle-acque-tra-interessi-economici-e-ambientali.

Thielborger, P. (2013) *The Right(s) to Water: The Multi-Level Governance of a Unique Human Right.* Berlin: Springer.

Zaffaroni, E. R. (2012) Pachamama, Sumak Kawsay y Constituciones. *Diritto pubblico comparato ed europeo* 2012(2): 422.

Zolo, D. (2005) Il diritto all'acqua come diritto sociale e come diritto collettivo: Il caso palestinese. *Diritto pubblico comparato ed europeo* 2005(1): 125.

7   Law and the provision of water for megacities

Joseph W. Dellapenna

Introduction

In 1950, only New York City qualified as a megacity (i.e. a city or metropolitan area of more than 10,000,000; Chandler 1987). Today, there are at least 24, spread across five continents, with more on the way (Population Reference Bureau 2013). The largest is Tokyo, at nearly 35,000,000, followed by Guangzhou with nearly 32,000,000. New York City comes in at number 10 today, with a population a bit under 22,000,000. New York isn't even the largest megacity in North America anymore (that honour goes to Mexico City at nearly 24,000,000). What these megacities have in common are staggering problems that often appear insoluble (Liotta and Miskel 2012). That should hardly be a surprise. Jane Jacobs taught us 45 years ago that at any given time the largest cities in the world are impractical and inefficient because we have not yet learned how to manage their needs (Jacobs 1969). As she pointed out, we found cities of 100,000 and then 1,000,000 in the past to be too large, but today we long for the relative ease of their problems while bemoaning the impossible difficulties of our even larger cities. Not the least of these problems is the need, often unmet, to provide water and sanitary services for their massive populations. This chapter addresses a small question related to the ability of megacities to provide water and sanitary services, namely the ability of governments or other management arrangements to reallocate water from current or traditional uses to meet the needs of the megacities and the extent to which various legal regimes facilitate or impede such reallocations.

The evolution of water law

Laws governing the allocation of water ('water laws') are found around the world in local customs and regulations, national legislation, regional agreements, and global treaties, together creating a complex legal governance framework for water (Dellapenna and Gupta 2009). The framework is a result of historical processes. Water laws are found in the earliest human civilizations. So central was the need to regulate water in these early civilizations that Karl von Wittfogel concluded that this need drove the emergence of basin-wide or other hydraulically-focused empires in early civilizations (Wittfogel 1981).

Through the centuries, water laws developed in a context reflecting the history, geography, and political systems of the countries concerned. Early and modern water laws exhibit certain recurring patterns. Some of these are purely cultural, reflecting the predominant forms of social structure at a time and place. Other features reflect the nature of the resource and of patterns of use. Thus the right to use water is variously granted to owners of riparian land (land contiguous to a water source) or because of temporal priority in using the water (first in time, first in right) (Scott and Coustalin 1995). The riparian approach generally required a sharing of the water, while the priority approach often did not. Some cultures would mix the two principles, while others gave preferences to particular types of use (e.g. irrigation versus municipal uses). And from the beginning, the laws addressed questions of pollution as well as the allocation to particular uses, such as prohibitions on allowing cattle to defecate in flowing water. Water laws tended to be most developed in arid or semi-arid regions (Grossfeld 1984).

The nature of water resources and the nature of the uses of the resource to some extent provide a measure of unity to patterns of water law, and to the continuing and sometimes intense debate about which legal approach is best (Dellapenna 2008a; Trelease 1974). The purely social, or jurisprudential, features of water laws create a possibility of water laws that do not simply reflect the nature of the resource or its uses. These various influences can co-exist, often resulting in a process that Francis Cleaver has termed 'bricolage' (Cleaver 2012). By bricolage, Cleaver means an uneven blending of old practices and norms with new practices and norms. Institutional and legal bricolage involves a constant renegotiation of norms and reinvention of traditions. The result today is almost 200 different national water law systems, each with country-specific characteristics. These systems are composed of overlapping and contradictory elements derived from residual indigenous laws, water laws imposed by colonial regimes or imported from 'more advanced' systems, and attempts at water law reform deriving from international legal standards or the prevalent thinking of epistemic communities (Dellapenna and Gupta 2009).

This leaves multiple systems of water law competing for application within a country as well as between countries. The resulting pluralism could be seen as positive, recognizing interests that cannot be aggregated in universalist approaches (Krisch 2006), or as negative, fragmenting interests and policies and breaking down legal structures. Recent efforts to integrate different regulations into a comprehensive water code sometimes succeed for better or worse (Kotov 2009; Laster and Livney 2009). In other cases, they founder on the resistance of those who are committed to earlier regimes (Cullet and Gupta 2009; Farias 2009; Nilsson and Nyanchaga 2009).

Contemporary patterns of water law at the local or national level and their impact on the provision of water for megacities

Despite the highly varied origins and evolution of local and national water laws, the nearly 200 national water legal systems define the right to use water according to only a few possibilities (Gupta and Dellapenna 2009). The right to use water might be defined in terms of the relationship of the use to the water source:

1 based on the location of the use (a riparian connection);
2 the timing of the use (a temporal or seasonal priority system); or
3 the nature of the use (preferences for the most socially important uses).

Rights to use water are often characterized as a kind of property, which allows a different typology:

1 common property (the resource is used freely by those with lawful access, without collective decision-making);
2 private property (defined water rights are allocated to particular users with considerable control over 'their' water); or
3 community or public property (water is managed jointly by those entitled to share the resource) (Ostrom 1990; Dellapenna 2010).

Each type of property right must recognize to some extent the public nature of water resources, and therefore even in the most thoroughly privatized water property regime there will be regulations to:

1 enforce the property or water right regime;
2 protect the resource from pollution or degradation; and
3 promote or preclude markets.

A decision to conceive of these various approaches as property rights introduces a certain rigidity that can make it difficult, or at least expensive, to provide water for megacities. Yet the correspondence between the forms of water law to the several basic models of property rights allows prediction of whether a form is adaptable to changing circumstances, or whether an entirely new form of water law must be substituted when circumstances of water demand or supply change dramatically (as they have with the rise of megacities). Treating water as common property leads into a free-for-all that results in the tragedy of the commons (Hardin 1968): As soon as water becomes scarce relative to demand in a particular region, unregulated individual decision-making accelerates the destruction of the resource (Dellapenna 2000; Sinden 2007). A private property model for water, however, freezes patterns of use, at least at the large scale, rather than opening up the possibility for markets to reallocate the huge amounts of water necessary to meet the needs of megacities (Dellapenna 2008a). Markets fail because of the need to protect third-party rights if society is genuinely going to protect private rights to use water (Dellapenna 2000). Because of the utter failure of true markets (Dellapenna 2013a: §§6.01(b)–6.01(b)(3)), the admittedly imperfect public property model serves as the best available (Dellapenna 2013a: §9.03(a)(5)(D)). Various economic incentives – including fees, taxes and 'water banks' – can be good management tools, but they shouldn't be confused with true markets (Seroa da Motta *et al.* 2005; Stern 2006; Wichelns 2006) The real problem is how to transition from an older and less flexible model to the public property approach without excessive economic and social costs (Dellapenna 2013b).

The evolution of water law at the international level

None of the 24 megacities that already exist straddle international boundaries, with the possible exception of Buenos Aires (if one includes Montevideo, although probably the officials of neither city would endorse that idea). Many present megacities depend on water sources that do straddle such boundaries. Therefore, in considering the role that law can play in facilitating or impeding the provision of water to megacities, regional and global international law must play a role along with national and local law. Although international water agreements go back at least 800 years, true international water law only developed in the last two centuries. International law in general provides a framework, with rules for treaty making and interpretation and means for dispute resolution. International law empowers international actors by legitimating their claims, but it also limits the claims they are allowed to make (Dellapenna 2008b). International water law today is found in numerous treaties, including the UN Convention on the Law of Non-Navigational Uses of International Watercourses (hereafter the 'UN Watercourses Convention'), which entered into force in August 2014, and in a body of customary international law, the reach of which is broader than the convention both in the sense of covering more issues and in the sense of extending to more countries.

Customary international law develops through states making claims and counterclaims until they agree on what the law requires (Danilenko 1993). Identifying customary law is informal and challenging. Customary international water law evolved largely through water treaties, beginning in the late eighteenth century. The treaties focused first on freeing navigation, then (because of the industrial revolution in the nineteenth century) on water allocation, and finally on cooperative or joint management regimes in the twentieth centuries (Dellapenna 1994). Contemporary customary international water law resembles the common principles underlying national water laws, including recognizing rights in riparian or aquifer states, considering temporal priority to a limited extent, and emphasizing the nature of and need for particular uses. These principles often take on different colorations when applied to an incompletely organized community of states.

Bilateral or regional international water agreements provide sources of law for participating states as well as for inferring a developing customary international law. There are literally hundreds of such international water agreements which together give rise to a body of international customary law that sets basic standards even for water resources not covered by an international agreement (Dellapenna 2001). What today is effectively the codification of customary international water law is found in the UN Watercourses Convention (International

Court of Justice 1997). Customary international water law primarily includes three principles:

1   limited territorial sovereignty over national waters (requiring states to consider the needs of other riparian states; UN Watercourses Convention 1997, arts. 5, 6);
2   the no-harm principle (derived from the Roman law maxim, *sic utero tuo ut alineium non laedes* – 'Use not your property so as to injure the property of another') (UN Watercourses Convention 1997, art. 7; Dellapenna 1996); and
3   the obligation to settle disputes peacefully (UN Watercourses Convention 1997, arts. 30–33).

Some states also claim historic rights (i.e. the right to use the quantity of water they have been using for a significant period of time; Brunnée and Toope 2002). Limited territorial sovereignty, expressed as the principle of equitable utilization – that is, the need to share international waters equitably (fairly) – is the dominant principle (UN Watercourses Convention 1997, arts. 5, 6; Dellapenna 2001).

The UN Watercourses Convention, however, only provides a limited framework for structuring negotiations. Although it includes environmental values and some modern ideas about water governance, arguably it was out of date when it was approved by the General Assembly (1997) for it scarcely refers to legal developments in the environmental, human rights and investment arenas. The UN Economic Commission for Europe has provided an alternative template in the Convention on the Protection and Use of Transboundary Watercourses and International Lakes (or 'Helsinki Convention'; Economic Commission for Europe 1992). This convention became open for signature by any country in the world in 2014. It focuses much more on the protection of water and on building institutions than does the UN Watercourses Convention, but is so detailed that it could prove far too rigid for use outside of Europe.

The most recent effort at a comprehensive codification of customary international water law is the Berlin Rules on Water Resources, approved by a unanimous vote by the plenary meeting of the International Law Association in Berlin in 2004 (International Law Association 2004). The Berlin Rules are grounded in existing law interpreted in light of evolving changes in global water law. They integrate insights from environmental, humanitarian, human rights and resource law. Where international law (particularly its rules on the environment and on public participation) justifies it, these comprehensive rules cover all national and international fresh waters and related resources (the aquatic environment) and thereby penetrate national jurisdiction. In addition to closely tracking the rules in the UN Watercourses Convention, the Berlin Rules include the principles of public participation, the obligation to use best efforts to achieve conjunctive and integrated management of waters, and the duties to achieve sustainability and minimize environmental harm. The rules identify the rights and duties of states and persons, the need for environmental impact assessments, and rules relating to extreme situations including accidents, floods and droughts. And the Berlin Rules included the first attempt at a com prehensive codification of the customary international law applicable to ground-water (International Law Association 2004: ch. 8). The subsequent attempt by the International Law Commission (a UN agency) to codify the customary international law of ground-water (International Law Commission 2008) was unable to gain approval by the UN General Assembly because of the evident deficiencies of the rules they produced, rooted in its decision to focus on aquifers rather than on waters (UN General Assembly 2008; Dellapenna 2011; McCaffrey 2009).

Conclusion

Despite talk of 'water wars', water resources tend not to be a key reason for conflict (Kalpakkian 2004). Instead, at the national, regional, and international levels, water law has served to mediate conflict and resolve disputes. Yet after 4,000 years, water law remains tied to old models that, at least in general terms, can be traced back to the earliest extant historical records. Today, many challenges exist worldwide to water management and to water law, including especially the problem of providing water for the exploding number of megacities.

Communities at all levels face global water problems such as access, sanitation, pollution, ecosystem destruction and changing flow regimes as a result of dams, other human activities, and the increasingly disrupted climate. Governance systems themselves are in a state of flux (Gupta 2011). Some might see law – local, national and international – as an impediment to coping adequately with the water needs of the coming century (Brandes and Nowlan 2009). Conceiving of rights to use water as property rights in itself introduces a kind of rigidity that can make it more difficult to introduce change into the legal structure of water use. All of this suggests that law or legal regimes will be impediments to meeting the water and sanitation needs of megacities.

Because issues of water governance become very technical, technocratic solutions may lead to growing formal and informal administrative law and governance in the water field, some of which might be adopted through international cooperation processes but without necessarily a formal international legal agreement or even consensus. History shows, however, that water law is able, if slowly, to rise to the challenge of change. Water law is moving forward through regional agreements, administrative frameworks, and joint water management bodies at all levels of governance from the community up to the global level. Legal systems, however slow their development, have the authority of history behind them and may ultimately provide the vehicle for problem solving and conflict resolution in the twenty-first century. Water law will figure prominently as water management systems and social justice processes struggle to cope with tomorrow's needs, including the needs of megacities.

References

Brandes, O. M., and L. Nowlan (2009) Wading into Uncertain Waters: Using Markets to Transfer Water Rights in Canada – Possibilities and Pitfalls. *Journal of Environmental Law and Practice* 19: 267–87.

Brunnée, J., and S. J. Toope (2002) The Changing Nile Basin Regime: Does Law Matter? *Harvard International Law Journal* 43: 105–59.

Chandler, T. (1987) *4,000 Years of Urban Growth: An Historical Census.* Lewiston, NY: Edwin Mellen Press.

Cleaver, F. (2012) *Development through Bricolage: Rethinking Institutions for Natural Resources Management.* Abingdon: Routledge.

Cullet, P., and J. Gupta (2009) India: The Evolution of Water Law and Policy. In J. W. Dellapenna and J. Gupta (eds), *The Evolution of the Law and Politics of Water,* 157–74. London: Springer.

Danilenko, G. M. (1993). *Law-Making in the International Community.* Leiden: Martinus Nijhoff.

Dellapenna, J. W. (1994) Treaties as Instruments for Managing Internationally Shared Water Resources: Restricted Sovereignty vs. Community of Property. *Case Western Reserve Journal of International and Comparative Law* 26: 27–6.

—— (1996) Rivers as Legal Structures: The Examples of the Jordan and the Nile. *Natural Resources Journal* 36: 217–50.

—— (2000) The Importance of Getting Names Right: The Myth of Markets for Water. *William and Mary Environmental Law and Policy Review* 25: 317–77.

—— (2001) The Customary International Law of Transboundary Fresh Waters. *International Journal of Global Environmental Issues* 1: 264–05.

—— (2008a) Climate Disruption, the Washington Consensus, and Water Law Reform. *Temple Law Review* 81: 383–432.

—— (2008b) International Water Law in a Climate of Disruption. *Michigan State Journal of International Law* 17: 43–94.

—— (2010) Global Climate Disruption and Water Law Reform. *Widener Law Review* 15: 409–45.

—— (2011) The Customary Law Applicable to Internationally Shared Groundwater. *Water International* 36: 584–94.

—— (2013a) Riparianism. In A. Kelly (ed.) *Waters and Water Rights,* vol. 1, chs. 6–9. LexisNexis.

—— (2013b) The Rise and Demise of the Absolute Dominion Doctrine for Groundwater. *University of Arkansas-Little Rock Law Review* 35: 291–355.

Dellapenna, J. W., and J. Gupta (2009) *The Evolution of the Law and Politics of Water.* London: Springer.

Economic Commission for Europe (1992) Convention on the Protection and Use of Transboundary Watercourses and International Lakes. Available at www.unece.org/fileadmin/DAM/env/documents/2013/wat/ECE_MP.WAT_41.pdf.

Farias, P. J. L. (2009) Brazil: The Evolution of the Law and Politics of Water. In J. W. Dellapenna and J. Gupta (eds), *The Evolution of the Law and Politics of Water,* 69–86. London: Springer.

Grossfeld, B. (1984) Geography and Law. *Michigan Law Review* 82: 1510–19.

Gupta, J. (2011) Developing Countries: Trapped in the Web of Sustainable Development Governance: Performance, Legal Effects and Legitimacy. In O. Dilling, M. Herber, and G. Winter (eds), *Transnational Administrative Rule-Making: Performance, Legal Effects and Legitimacy,* 305–30. Oxford: Hart Publishing.

Gupta, J., and J. W. Dellapenna (2009) The Challenge for the Twenty-First Century: A Critical Approach. In J. W. Dellapenna and J. Gupta (eds.), *The Evolution of the Law and Politics of Water*, 391–410. London: Springer.

Hardin, G. (1968) The Tragedy of the Commons. *Science* 162: 1243–48.

International Court of Justice (1997) *Case of the Gabçikovo-Nagymaros Dam (Hungry/Slovakia)*. Judgment of 25 September 1997, 1997 ICJ no. 92.

International Law Association (2004) Berlin Rules on Water Resources. In *Report of the Seventy-First Conference*, 334–421 (Berlin). London: International Law Association.

International Law Commission (2008) The Law of Transboundary Aquifers. UN Doc. A/CN.4/L.724 (second reading).

Jacobs, J. (1969) *The Economy of Cities*. New York: Random House.

Kalpakkian, J. (2004) *Identity, Conflict and Cooperation in International River Systems*. Aldershot: Ashgate.

Kotov, V. (2009) Russia: Historical Dimensions of Water Management. In J. W. Dellapenna and J. Gupta (eds), *The Evolution of the Law and Politics of Water*, 139–55. London: Springer.

Krisch, N. (2006) The Pluralism of Global Administrative Law. *European Journal of International Law* 17: 247–78.

Laster, R. and D. Livney (2009) Israel: The Evolution of Water Law and Policy. In J. W. Dellapenna and J. Gupta (eds), *The Evolution of the Law and Politics of Water*, 121–37. London: Springer.

Liotta, P. H., and J. F. Miskel (2012) *The Real Population Bomb: Megacities, Global Security and the Map of the Future*. Dulles, VA: Potomac Books.

McCaffrey, S. C. (2009) The International Law Commission Adopts Draft Articles on Transboundary Aquifers. *American Journal of International Law* 103: 272–93.

Nilsson, D., and E. N. Nyanchaga (2009) East Africa. In J. W. Dellapenna and J. Gupta (eds), *The Evolution of the Law and Politics of Water*, 105–20. London: Springer.

Ostrom, E. (1990) *Governing the Commons: The Evolution of Institutions for Collective Action*. Cambridge, UK: Cambridge University Press.

Population Reference Bureau (2013) World Population Data Sheet. Available at www.prb.org/pdf13/2013-population-data-sheet_eng.pdf.

Scott, A., and G. Coustalin (1995) The Evolution of Water Rights. *Natural Resources Journal* 35: 821–979.

Seroa da Motta, R., A. Thomas, L. Saade Hazin, J. G. Féres, C. Nauges and A. Saade Hazin (2005) *Economic Instruments for Water Management: The Cases of France, Mexico and Brazil*. Northampton, MA: Edward Elgar.

Sinden, A. (2007) The Tragedy of the Commons and the Myth of the Private Property Solution. *University of Colorado Law Review* 78: 533–612.

Stern, S. (2006) Encouraging Conservation on Private Lands: A Behavioral Analysis of Financial Incentives. *Arizona Law Review* 48: 541–83.

Trelease, F. J. (1974) The Model Water Code, the Wise Administrator, and the Goddam Bureaucrat. *Natural Resources Journal* 14: 207–29.

UN General Assembly (1997) Convention on the Law of Non-Navigable Uses of International Watercourses. GA Res. 51/229, 21 May 1997, UN Doc. A/51/49. Available at http://untreaty.un.org/ilc/texts/instruments/english/conventions/8_3_1997.pdf.

UN General Assembly (2008) The Law of Transboundary Aquifers. GA Res. A/Res/63/124, 11 Dec. 2008. Available at http://daccess-dds-ny.un.org/doc/UNDOC/GEN/N08/478/23/PDF/N0847823.pdf?OpenElement.

Wichelns, D. (2006) Economic Incentives Encourage Farmers to Improve Water Management in California. *Water Policy* 8: 269–85.

Wittfogel, K. A. (1981) *Oriental Despotism: A Comparative Study of Total Power* (reprint ed.). New York: Vintage Books.

8 A critique of subsidies for industrial livestock production in the European Union and the United States

Constanze Frank-Oster

Industrial livestock production poses a threat to food justice in many ways.[1] It causes environmental problems such as deforestation, land degradation, loss of biodiversity, acidification and pollution of rivers and lakes. It is one of the biggest emitters of greenhouse gases, making the sector an important contributor to climate change. Other, more immediate issues concern negative impacts on public health, problematic working conditions in slaughterhouses and feeding operations as well as animal suffering.

This chapter will analyse the ethical implications of industrial livestock production. It will further address the question as to whether livestock subsidies can be justified against this background. American as well as European subsidies shall be examined.

Human beings are dependent on predictable and useable environmental services. Food production, for example, requires soils with a healthy sulphur content, predictable weather patterns and the availability of water. This shows that moral duties towards human beings necessarily include moral duties regarding nature. It is not essential, however, to accept the non-instrumental value of nature or moral duties to nature in order to criticize the current practice of industrial livestock production.

The industry is responsible for the emission of large amounts of sulphur, which are needed in order to fertilize the soils for feed production. This causes the soil to be oversupplied with sulphur, which leads to a decrease in the productivity of forests and pastures as well as to acidification of rivers and lakes. Today, more than 90 per cent of western European ecosystems are threatened by sulphur pollution (Steinfeld *et al.* 2006). Clearly, sulphur is not only used in feed production but also for the production of plant food for human beings. Nevertheless, livestock production accounts for the majority of sulphur pollution. In 2012 corn and soy, the classic feed crops, were cultivated on 173 million acres in the USA (NASS 2013a, 2013b), whereas wheat, the crop most widely grown for human consumption, only accounted for 49 million acres (NASS 2013c), rice for 2.6 million acres (NASS 2013d) and tomatoes for 94,700 acres (NASS 2013e). Also, feedcrops such as corn have an especially high demand for fertilisers (Magdoff 1991). Regardless of this, the production of 1 kg of animal protein uses 6 kg in plant protein (Pimentel and Pimentel 2003) because grains are being fed to the animals of which the animals use the biggest part to cover their everyday need of calories. Only one sixth of the protein that has been fed is actually available after slaughter. This helps to explain why livestock production has a greater sulphur use than the production of plant food. Industrial livestock production also requires a great amount of water. A plant-based diet uses 0.5 m^3 water per 1000 kcal, whereas a diet with 20 per cent animal protein uses 4 m^3 per 1000 kcal (Deutsch *et al.* 2010). It is important to not only consider the use of blue water (i.e. the water that is available in liquid form in lakes and rivers) but also green water (i.e. water that is bound in roots of the soil). In order to get 1 kg of plant protein it is only necessary to irrigate the plants that are needed for this. In order to get 1 kg of animal protein, six times that amount of blue water is needed for feed irrigation. Pasture farming furthermore accounts for a large share of green water usage. As grass and hay have a much lower energy density than corn and soy, much larger areas are needed in order to produce the same amount of calories. Also, a large amount of water is lost to the atmosphere through evapotranspiration[2] due to grazing (Deutsch *et al.* 2010). Furthermore, livestock production changes the distribution of water and the storage capacity of soils by working surfaces (*ibid.*).[3] After all, no kind of agricultural usage accounts for a higher demand for surfaces. Subsequent changes in the water cycle have significant impacts on ecosystems and environmental services. This directly influences human beings (e.g. through desertification of the Amazon or through changes in African and Asian rain season patterns; *ibid.*). Many environmental problems that arise because of industrial livestock production are caused by the fact that the amount of manure it produces by far exceeds the amount that can safely be used for fertilising fields. This leads to soil pollution. Moreover, manure storage causes a great risk of pollution of close-by soils, waters and the air (Burton and Turner 2003; Menzi *et al.* 2010). It also immediately threatens human health through contamination of water and the air with microorganisms and parasites (Steinfeld *et al.* 2006). Pasture management,[4] which has been practised in a sustainable way for the last centuries, today threatens ecosystems and therefore human beings. Due to its intensification it now leads to land degradation, which affects about one third of Asian, half of Latin American and two thirds of African soils (*ibid.*). Furthermore, industrial livestock production contributes to the deforestation of large parts of the rain forest, which not only threatens biodiversity, but also contributes to climate change (*ibid.*). The intensification of fishing and aquacultures also lead to many environmental problems (Pillay 2004; Stotz 2000; Naylor *et al.* 2000). The large amounts of fish that are being fished lead to overfishing, which threatens marine ecosystems. Aquacultures add to this problem because they demand huge quantities of wild fish for feed.

The industry's contribution to climate change is of special significance. According to the Food and Agriculture Organization of the United Nations (FAO) it accounts for 18 per cent of worldwide anthropogenic greenhouse gas emissions (Steinfeld *et al.* 2006). The WorldWatch Institute, which accuses the FAO of highly underestimating emissions caused by the use of land and by the animals' digestive systems, estimates that the industry is responsible for as much as 51 per cent of anthropogenic greenhouse gas emissions (Goodland and Anhang 2009). A significant change of climate undoubtedly poses a threat to human welfare, especially but not only in those countries that do not have the financial and technological means for adaptation (*ibid.*). Probable effects include extreme weather events, such as storms, tsunamis, droughts and floods (also caused by the rise of the sea level), as well as severe influences on agriculture, which could threaten food security in countries where the supply of food is not problematic at the moment and even further exacerbate the situation in other regions of the world (*ibid.*). An especially drastic depiction of the impacts' extent is provided by the Pacific Access Category, an agreement between Tuvalu and New Zealand. It obliges New Zealand to each year accept 75 climate refugees from the pacific island, which is going to go under water due to climate change (Ralston *et al.* 2004).

All of these impacts pose a threat to human welfare and livelihood for the sake of pure aesthetic preferences and therefore violate moral duties towards human beings living in the present and in the future. Eating the large amounts of animal products that we eat at the moment is not a nutritional necessity at all. Rather, we choose to eat milk, meat and eggs every day because we like the way they taste. Compared to a being's interest in not suffering (as in not having to abandon one's land in order to survive, having enough food to not go hungry and not having to face drastic weather events such as tsunamis, floods and droughts), the aesthetic preference for animal products doesn't seem very important. According to the pathocentric view, which I accept, a being's interest in not suffering is its most basic interest, which cannot be overridden by mere taste preferences. Which moral duties does this view entail? Does this mean that we are responsible for every living being's contentment? Are we morally obliged to make everybody's life good? This is a very popular rhetorical question when facing the threat of having to change one's behaviour. In asking it, we try to make the point that as we are simply not able to ensure everybody's happiness, we do not have to change our own destructive behaviour (e.g. eating large amounts of animal products), as changing it would still not make us meet our moral duties. On the contrary, we are not responsible for everybody's contentment, because responsibility for something entails that one's actions are causally linked to it. As my decision to eat meat on a daily basis is not causally linked to the exploitation of children in Indian sweat shops, I am therefore not responsible for it. I nevertheless am responsible for the direct causes of my consumption. Also, it is not necessary to embrace positive duties, but rather negative duties are sufficient to show that increasing the consumption of animal products by subsidizing industrial livestock production is not acceptable. Even if we do not assume that there is a moral duty to increase any living being's contentment, we still shouldn't consume as many animal products, as there are at least moral duties not to harm anybody for profane reasons such as taste preferences. The pathocentric view presupposes the ability to suffer in order to ascribe moral duties, which means that there can be moral duties neither towards plants nor ecosystems or whole species. It also means that we do have moral duties towards most animals. Species doesn't matter when weighing a being's interest in its own welfare (Singer 1997). What also does not matter is whether the being exists now or will exist in the future. We can be relatively sure that there will be human beings in the near future that will probably share our interest in not suffering (Birnbacher 1988). Nevertheless, we act as if future people matter significantly less than people living today, when considering the environmental damage we cause mentioned in the section above. First of all, there are no good reasons to believe this. And second of all, it is highly counterintuitive, as most of us do not think that our grandchildren should have to suffer from hunger or extreme weather events just because we like to eat meat every day.

In addition to environmental damage, industrial livestock production also threatens public health in several ways. Not only do workers have to face dangerous working conditions in slaughterhouses and feeding operations, which jeopardize their mental and physical health, the industry also affects consumers' health and that of people living near production facilities. It furthermore threatens the health of every human being regardless of their consumption of animal products or proximity to production facilities. The next paragraph will focus on these public health impacts.

The combination of long working hours, physically exhausting work and the necessity to work with fast and risky equipment such as knives make the work in slaughterhouses very dangerous. According to a study by the University of Arkansas, the number of injuries reported among 100 full-time employees per year is 9.8. Employees in production are injured approximately twice as often. A disturbing 51 per cent of slaughterhouse workers are injured every year (Worrall 2004). Furthermore, slaughterhouse workers have a significantly higher risk of psychological damage than workers in other fields. Compared to employees of supermarkets and offices, slaughterhouse workers show higher rates of anger and aggression, anxiety problems, phobias and psychoses (Emhan *et al.* 2012). They are also, just like soldiers and executioners, considered a high risk group for a special case of post-traumatic stress disorder, which is caused by the patients' own participation in the disturbing event (MacNair 2002). Employees of animal feeding operations also face drastic health risks, for example due to high concentrations of hydrogen sulphide, ammonia (which is also a precursor of particulate matter), carbon dioxide and methane as well as endotoxins (MacNair 2002). Ammonia concentrations above 7.5 ppm are generally considered health threatening (Donham 1995). In German veal barns, ammonia concentrations of up to 20 ppm are considered legal, which should show that the strain on employees is severe (Verordnung zum Schutz landwirtschaftlicher Nutztiere, 2001). While 5.7 per cent of all employees in non-agricultural jobs and 7.2 per cent of all employees in grain-producing agriculture suffer from chronic bronchitis, the number is 15.3 per cent in industrial livestock production (Zejda 1993). Long-time contact with hydrogen sulphide leads to neurological and cardiological problems, while the frequent inhalation of particulate matter causes damage of the respiratory system and cardiological problems such as arrhythmias and heart attacks (ASTDR 2013; EPA 2013). In addition, employees in industrial livestock production very often suffer from musculo-skeletal disorders (Kolstrup 2008).

People living near livestock production facilities are very likely to suffer from diseases of the respiratory system because of high ammonia concentrations in the air (Schiffmann 1998; Thu *et al.* 1997). As a lot of animals are kept in a small space, high ammonia concentrations caused by the animals' manure emerge. Ammonia is a precursor of fine particulate matter, which is probably the most important environment-related public health threat in US livestock production, is a major emitter of ammonia (Shih *et al.* 2006). Public health is also threatened by water polluted with pathogens (*ibid.*). In order to control pathogens in industrial livestock production systems, antibiotics are being used on a prophylactic basis. This has even more severe consequences for human health as it leads to an increase in pathogens that are resistant to antibiotics. In the EU, external costs caused by resistant pathogens are estimated to be 1.9 billion US dollars (European Parliament 2011). In the US, costs exceed 30 billion dollars per year. These externalities consist of the costs of hospital care, physician services, drugs, nursing homes and labour losses. Some 88 per cent of the costs are caused by untimely deaths, which shows that there is more to consider than economic factors (Phelps 1989). Antibiotics are also used to promote the animals' growth in the US; this has been banned in the EU since 2006 (Regulation (EG) Nr. 1831/2003). Despite these efforts, pathogens from industrial livestock still have an immense impact: 60 per cent of all pathogens are of zoonotic origin and it is estimated that 75 per cent of all reported new pathogens are being transmitted by animals (Todd 1997). Furthermore, there are severe health risks that emerge from the use of pesticides in feed production. In 2002, 58,000 cases of pesticide poisoning were recorded in the US. Pesticides also cause eye injuries, diseases of the respiratory system, hormonal imbalances, birth defects, nerve damage and cancer (Tegtmeier and Duffy 2004). Again, pesticides are obviously also used in the production of plant food, but livestock production has an exponentially higher demand, as was explained in a previous section. Industrial livestock production also causes health damage via contaminated animal products. According to the WHO, hundreds of millions of people are affected by food-borne diseases, with animal products being at the top of the list (Todd 1997). Diseases caused by salmonella, for example, lead to costs of 2.5 billion dollars per year in the US alone (Gurian-Sherman 2008). And even in cases where animal products are not contaminated they cause health issues, as the consumption of animal products in industrialised countries clearly surpass healthy quantities (Hilbig *et al.* 2011; US Department of Agriculture 2006). This leads to, among others, diseases of the cardiovascular system (Popkin and Du 2003), diabetes (Melnik 2009) and obesity (Wang and Beydoun 2009; Fung *et al.* 2001).

The consequences for public health mentioned above clearly violate the moral duty not to harm human beings for secondary reasons in a very direct way.

As the pathocentric ground of morals states that suffering is morally relevant irrespective of the suffering being's species, industrial livestock's impact on the welfare of animals is of interest as well. I will nevertheless refrain from depicting housing conditions, as they should be generally known. It should only be mentioned that the European Food Safety Authority (EFSA), which is not primarily concerned with animal welfare issues, deems the housing conditions of livestock in industrial production systems in the EU to be a threat to animal welfare (EFSA 2004, 2005, 2006, 2007). As animal welfare laws are comparably strict in the EU, it is sound to assume that conditions are even worse in other regions of the world, such as the United States.

I conclude that the current practice of industrial livestock production violates very basic moral duties towards human beings living now and in the future as well as towards animals capable of suffering. Moreover, all of these problems are very likely to worsen during the next decades due to the drastic population growth predicted for the near future (Steinfeld *et al.* 2006). An increase in population is going to increase the demand for cheap animal products, which will be met with higher productivity. This will lead to an exponential aggravation of the problems mentioned above.

Against this background the question arises in which way the large subsidies that are being granted to this industry can be justified. The livestock sector is one of the most heavily subsidized sectors in the world (Steinfeld *et al.* 2006). Between 1995 and 2011, livestock subsidies in the USA totalled 3.7 billion US dollars per year (EWG 2012). Intensive meat production alone was subsidized with 1.3 billion dollars per year in 2008 and 2009 (Benning and de Andrade 2011). In the EU, subsidies make up 70 per cent of a beef farmer's income, in the US they make up 40 per cent of a dairy farmer's income (De Haan *et al.* 2010). What catches the eye is the fact that the payments privilege large systems compared to smaller businesses, although smaller ones are generally managed in a more sustainable way. In 2006, the 115 largest swine-feeding operations in the US, housing more than 50,000 swine, received 5.01 million dollars each, whereas smaller operations with up to 4,999 swine received 60,000 dollars each. If the payments for smaller operations were assessed using the same formula as for the payments for large operations, those for the large ones should not exceed 600,000 dollars (Gurian-Sherman 2008). This practice is also applied in the EU where Doux Poultry, for example, one of the world's five largest poultry producers, received 4.7 million euros (6.1 million US dollars) in 2008 and 2009 (Benning and de Andrade 2011). Moreover, feed subsidies disadvantage those farms that produce their own feed or practice pasture farming, because they graze their animals on pastures and therefore do not benefit from feed subsidies that are being paid for the production of feed crops such as corn or soy (Gurian-Sherman 2008). The environmentally sounder practice of pasture farming therefore entails financial disadvantages compared to the more destructive practice of feeding the animals grains. Direct subsidies through the Environmental Quality Incentives Program, were originally intended to help small- and medium-sized farms face environmental challenges. After a dramatic change of the regulations, today the program advantages the largest intensive operations. Maximum payments, which were at 50,000 dollars, were changed to 450,000 dollars.

Furthermore, the regulations now state that 60 per cent of EQIP payments have to go to livestock production. This change was allegedly made in order to subsidize those operations in which the biggest environmental benefits can be expected (NRCS 2003). A different interpretation of this reorientation suggests that the biggest financial support is provided for those operations that cause the most severe environmental problems. This not only seems to undermine any incentives for acting in environmentally sound ways, it also provides a strong incentive for establishing new systems of intensive livestock production. The EU also grants subsidies that are linked to environmental protection. Although this 'second pillar' of financial support, which focuses on local marketing, organic production as well as protection of the environment and animal welfare, as opposed to the EQIP payments, does form an incentive for environmentally sound conduct, it is very small compared to those payments that are granted irrespectively of environmental protection and the like. Around 40 billion euros are given to farmers, advantaging those with the largest operations, which in most cases means those farms with the most destructive conduct. Only 10 billion euros are spent on the second pillar, forming an incentive for protection of the environment and animal welfare. This shows that, even though the EU provides an incentive for socially acceptable conduct, the incentive for the opposite is much more powerful (Benning and de Andrade 2011).

Subsidies are also granted for fisheries in order to reduce global overfishing. But according to an analysis of Poseidon Aquatic Resource Management, only 17 per cent of granted fisheries subsidies in the EU actually had a positive effect on the reduction of overfishing (Pew Environment Group 2010). The largest percentage, namely 54 per cent of subsidies had a neutral effect, whereas 29 per cent even worsened the problem (*ibid.*). Of US fisheries subsidies, 3 per cent have an ambiguous effect, 23 per cent have a negative effect and 74 per cent have a positive effect on overfishing (Sea Around Us Project 2013). It is important to note that apart from the payments that actually accomplish a positive effect by the scrapping of boats, investments in environmental protection programs and eco-labels, all the other subsidies for fisheries and industrial livestock production must be regarded as negative. As subsidies artificially lower the prices of animal products, they increase the demand, which is what leads to the problems mentioned above in the first place. Moreover, subsidizing US or European animal products leads to their competitive advantages on markets in developing countries. This causes price dumping and distortion of competition at the expense of peasant farmers (Benning and de Andrade 2011). As US and European products are very cheap due to subsidies, people in developing countries choose the imported goods over the locally produced ones, thereby denying local farmers and infrastructure their much needed monetary support.

All in all, the subsidies provided for industrial livestock production do not seem to be based on the principle "Public money for public goods". Rather, immense amounts of public money are being used to support and promote an industry the impacts of which violate moral duties towards human beings living today and in the future as well as of animals. My proposal to end subsidies for industrial livestock production is shared by the FAO (De Haan *et al.* 2010). A seemingly more feasible claim would be to link subsidies to standards of environmental protection, so that those operations which cause the most severe environmental damages will receive accordingly smaller payments (Benning and de Andrade 2011). Subsidies could also be linked to animal welfare[5] and public health standards (*ibid.*). Another proposal would be to reduce EQIP maximum payments so that they do not form an incentive for causing environmental problems anymore (Gurian-Sherman 2008). Also, it seems appropriate to make employment situations a condition for subsidies. It is not acceptable to financially support the creation of jobs that expose employees to significant health risks without adequate remuneration. Despite their current form and impact, livestock subsidies may even be used to provide an incentive for sustainable and socially acceptable conduct. They could, for example, create incentives for regional processing and distribution, thereby reducing livestock transports and greenhouse gas emissions as well as supporting local businesses (Benning and de Andrade 2011).

In March 2013 the European Parliament voted in favour of a change of the EU's Common Agricultural Policy towards more sustainable agricultural practices. The Parliament decided to make a certain percentage of direct sub-sidies dependent on measures promoting environmental services. It remains unclear, however, how large the percentage is going to be, which is why this change's impacts cannot yet be estimated. Furthermore, advantages for large operations are to be decreased by gradually reducing payments depending on operation sizes and by determining maximum payments (European Parliament 2011). However, the regulations include a passage stating that this should not lead to disproportionate disadvantages for large operations with many employees. This means that the operations that cause the most severe environmental damage and animal welfare problems will continue to receive the largest payments. All in all, the changes in the European Common Agriculture Policy can be seen as a step in the right direction. Unfortunately, they do not go far enough that they could effectively combat the systematic funding of practices that violate moral duties towards human beings and animals. Rather, EU subsidies in their new form will still support industrial livestock production to a large extent.

As long as subsidies for industrial livestock production still exist in the current form, it is up to the consumers to reduce their financial support for this industry.

Notes

1 Food justice means the fair distribution of benefits and risks of where, what, and how food is grown, produced, transported, distributed, accessed and eaten (Gottlieb and Joshi 2010).
2 Evapotranspiration accounts for the movement of water within a plant and the loss of water in the form of vapour through the leaves of a plant.
3 For example, ploughing or grubbing.
4 Grazing the animals on farmland.
5 This is already being done in the Netherlands; see Hart *et al.* (2011).

References

Benning, R., and C. de Andrade (2011) *Subventionen für die industrielle Fleischerzeugung in Deutschland. BUND-Recherche zur staatlichen Förderung der Schweine- und Geflügelproduktion in den Jahren 2008 und 2009*. Berlin: Bund für Umwelt und Naturschutz.

Birnbacher, D. (1988) *Verantwortung für zukünftige Generationen*. Stuttgart: Reclam.

Burton, C. H. and C. Turner (2003) *Manure Management: Treatment Strategies for Sustainable Agriculture*. Flitwick: Lister & Durling.

De Haan, C., P. Gerber and C. Opio (2010) Structural Change in the Livestock Sector. In Steinfeld, H., H. A. Mooney, F. Schneider and L. E. Neville (eds), *Livestock in a Changing Landscape: Drivers, Consequences and Responses*, volume I. Washington, DC: Island Press.

Deutsch, L., M. Falkenmark, L. Gordon, J. Rockström and C. Folke (2010) Water-Mediated Ecological Consequences of Intensification and Expansion of Livestock Production. In Steinfeld, H., H. A. Mooney, F. Schneider and L. E. Neville (eds), *Livestock in a Changing Landscape: Drivers, Consequences and Responses*, volume I. Washington, DC: Island Press.

Donham, K., K. Thu, R. Ziegenhorn, S. Reynolds, P. S. Torne, P. Subramanian, P. Whitten and J. Stookesberry (1997) A Control Study of the Physical and Mental Health of Residents Living Near a Large-Scale Swine Operation. *Journal of Agricultural Safety and Health*, 3(1): 13–22.

EFSA (2004) Opinion of the Scientific Panel on Animal Health and Welfare on a Request from the Commission Related to Welfare Aspects of the Castration of Piglets. *EFSA Journal* 91: 1–18.

—— (2005) Opinion of the Scientific Panel on Animal Health and Welfare on a Request from the Commission Related to Welfare Aspects of Various Systems of Keeping Laying Hens. *EFSA Journal* 197: 1–23.

—— (2006) Opinion on the risks of poor welfare in intensive calf farming systems. An update of the Scientific Veterinary Committee Report on the Welfare of Calves. *EFSA Journal* 366: 1–36.

—— (2007) Scientific Opinion of the Panel on Animal Health and Welfare on a Request from the European Commission on Animal Health and Welfare Aspects of Different Housing and Husbandry Systems for Adult Breeding Boars, Pregnant, Farrowing Sows and Unweaned Piglets. *EFSA Journal* 572: 1–13.

Emhan, A., A. S. Yildiz, Y. Bez and S. Kingir (2012) Psychological Symptom Profile of Butchers Working in Slaughterhouses and Retail Meat Packing Business: A Comparative Study. *Journal of the Faculty of Veterinary Medicine, University of Kafkas* 18(2): 319–22.

European Parliament (2011) Vorschlag für einen Beschluss des Europäischen Parlaments betreffend die Ausnahme von und das Mandat für interinstitutionelle Verhandlungen über den Vorschlag für eine Verordnung des Europäischen Parlaments und des Rates mit Vorschriften über Direktzahlungen an Inhaber landwirtschaftlicher Betriebe im Rahmen von Stützungsregelungen der Gemeinsamen Agrarpolitik (COM/2011). 0625 – C7-0336/2011 – 2013/2528(COD). Brussels: European Parliament.

EWG (2012) Farm Subsidy Database. 18 December. Available at http://farm.ewg.org/progdetail.php?fips=00000&progcode=livestock.

Fung, T. T., E. B. Rimm, D. Spiegelman, N. Rifai, G. H. Tofler, W. Willett and F. B. Hu (2001). Association between Dietary Patterns and Plasma Biomarkers of Obesity and Cardiovascular Disease Risk. *American Journal for Clinical Nutrition* 73(1): 61–7.

Goodland, R. and J. Anhang (2009) Livestock and Climate Change: What if the Key Actors in Climate Change are . . . Cows, Pigs, and Chickens? *WorldWatch Magazine* (November/December): 10–19.

Gottlieb, R. and A. Joshi (2010) *Food Justice*. Cambridge, MA: MIT Press.

Gurian-Sherman, D. (2008) *CAFOs Uncovered: The Untold Costs of Confined Animal Feeding Operations*. Cambridge: Union of Concerned Scientists Publications.

Hart, K., D. Baldock, P. Weingarten, B. Osterburg, A. Povellate, F. Vanni, C. Pirzio-Biroll and A. Boyes (2011) *What Tools for the European Agricultural Policy to Encourage the Provision of Public Goods?* Brussels: Directorate General for Internal Policies.

Hilbig, A., U. Alexy, C. Drossard and M. Kersting (2011) GRETA: Ernährung von Kleinkindern in Deutschland. (German Representative Study of Toddler Alimentation.) *Aktuelle Ernährungsmedizin* 36(4): 224–31.

Kolstrup, C. (2008) Work Environment and Health among Swedish Livestock Workers. Doctoral thesis, Swedish University of Agricultural Sciences, Alnarp, Sweden.

MacNair, R. M. (2002) *Perpetration-Induced Traumatic Stress: The Psychological Consequences of Killing.* Lincoln: Praeger.

Magdoff, F. (1991) Managing Nitrogen for Sustainable Corn Systems: Problems and Possibilities. *American Journal of Alternative Agriculture* 6(1): 3–8.

Melnik, B. (2009) Milchkonsum: Aggravationsfaktor der Akne und Promotor chronischer westlicher Zivilisationskrankheiten. *Journal der Deutschen Dermatologischen Gesellschaft* 7(4): 364–70.

Menzi, H., O. Oenema, C. Burton, O. Shipin, P. Gerber, T. Robinson and G. Franceschini (2010) Impacts of Intensive Livestock Production and Manure Management on the Environment. In Steinfeld, H., H. A. Mooney, F. Schneider and L. E. Neville (eds), *Livestock in a Changing Landscape: Drivers, Consequences and Responses*, volume I. Washington, DC: Island Press.

NASS (2013a) National Statistics for Corn. Available at www.nass.usda.gov/Statistics_by_Subject/result.php?093BD160-7A0F-3219-BA23-6FD584190AF9&sector=CROPS &group=FIELD%20CROPS&comm=CORN.

NASS (2013b) National Statistics for Soybeans. Available at www.nass.usda.gov/Statistics_by_Subject/result.php?65B8972C-0E05-3633-9600-499D951156DD&sector=CROPS&group=FIELD%20CROPS&comm=SOYBEANS.

NASS (2013c) National Statistics for Wheat. Available at www.nass.usda.gov/Statistics_by_Subject/result.php?AFE112BD-5125-34C1-86F5-2FF4DAF69809&sector=CROPS&group=FIELD%20CROPS&comm=WHEAT.

NASS (2013d) National Statistics for Rice. Available at www.nass.usda.gov/Statistics_by_Subject/result.php?3A819756-FEE1-3348-BAC7-C2021D3FCDE6&sector=CROPS &group=FIELD%20CROPS&comm=RICE.

NASS (2013e) National Statistics for Tomatoes. Available at www.nass.usda.gov/Statistics_by_Subject/result.php?A6AED7D0-8144-3C22-8498-F5E4E2589AEE&sector=CROPS&group=VEGETABLES&comm=TOMATOES.

Naylor, R. L., R. J. Goldburg, J. H. Primavera, N. Kautsky, M. C. M. Beveridge, J. Clay, C. Folke, J. Lubchenco, H. Mooney and M. Troell (2000) Effect of Aquaculture on World Fish Suppplies. *Nature* 405: 1017–24.

NRCS (2003) Environmental Quality Incentives Program. *Federal Register* 68(104): 32, 337–55.

Pew Environment Group (2010) *FIFG 2000-2006 Shadow Evaluation: Final Report.* Portmore: Poseidon Aquatic Resource Management.

Phelps, C. E. (1989) Bug/Drug Resistance. *Medical Care* 27: 194–203.

Pillay, T. V. R. (2004) *Aquaculture and the Environment*, 2nd edn. Oxford: Blackwell.

Pimentel, D. and M. Pimentel (2003) Sustainability of Meat-Based and Plant-Based Diets and the Environment. *American Journal of Clinical Nutrition* 78(3 Suppl.): 660S–63S.

Popkin, B. M. and S. Du (2003) Dynamics of the Nutrition Transition toward the Animal Foods Sector in China and its Implications: A Worried Perspective. *Journal of Nutrition* 133(11): 3893–906.

Ralston H., B. Hortmann and C. Holl (2004) *Climate Change Challenges Tuvalu.* Berlin: Bundesministerium für wirtschaftliche Zusammenarbeit und Entwicklung.

Schiffman, S. S. (1998) Livestock Odors: Implications for Human Health and Well-being. *Journal of Animal Science* 76: 1343–55.

Sea Around Us Project (2013) Fisheries, Ecosystems and Biodiversity: Fisheries Subsidies in the USA. Available at www.seaaroundus.org/Subsidy/default.aspx?GeoEntity ID=221.

Shih, J.-S., D. Burtraw, K. Palmer and J. Siikamäki (2006) *Air Emissions of Ammonia and Methane from Livestock Operations: Valuation and Policy Options.* Washington, DC: Resources for the Future.

Singer, P. (1997) Alle Tiere sind gleich. In Krebs, A. (ed.), *Naturethik: Grundtexte der gegenwärtigen tier- und ökoethischen Diskussion.* Frankfurt am Main: Suhrkamp.

Steinfeld, H., P. Gerber, T. Wassenaar, V. Castel, M. Rosales, Mauricio and C. de Haan (2006) *Livestock's Long Shadow: Environmental Issues and Options.* Rome: Food and Agriculture Organization of the United Nations.

Stotz, W. (2000) When Aquaculture Restores and Replaces an Overfished Stock, Is the Conservation of the Species Assured? The Case of the Scallop *Argopecten Purpuratus* in Northern Chile. *Aquaculture International* 8(2–3): 237–47.

Tegtmeier, E. M., and M. D. Duffy (2004) External Costs of Agricultural Production in the United States. *International Journal of Agricultural Sustainability* 2(1). Available at www.leopold.iastate.edu/pubs-and-papers/2004-01-external-costs-agricultural-production-united-states.

Thu, K., K. Donham, R. Ziegenhorn, S. Reynolds, P. S. Thorne, P. Subramanian, P. Whitten and J. Stookesberry (1997) A Control Study of the Physical and Mental Health of Residents Living Near a Large-Scale Swine Operation. *Journal of Agricultural Safety and Health* 3(1): 13–22.

Todd, E. C. (1997) Epidemology of foodborne diseases: a worldwide review. *World Health Statistics Quarterly* 50(1–2): 30–50.

US Department of Agriculture (2006) *Nutrient Content of the US Food Supply, 1909–2006.* Washington, DC: Center for Nutrition Policy and Promotion, US Department of Agriculture.

Verordung zum Schutz landwirtschaftlicher Nutztiere und anderer zur Erzeugung tierischer Produkte gehaltener Tiere bei ihrer Haltung (Tierschutz-Nutztierhaltungsverordnung-Tier-SchNutztV), 2001 (§6).

Wang, Y., and M. A. Beydoun (2009) Meat Consumption is Associated with Obesity and Central Obesity among US Adults. *International Journal of Obesity* 33: 621–8.

Worrall, M. (2004) Meatpacking Safety: Is OSHA Enforcement Adequate? *Drake Journal of Agricultural Law* 9(2): 301–23.

Zejda, J. E. (1993) Respiratory Health of Swine Producers: Focus on Young Workers. *Chest Journal* 103(3).

9 Promoting the ecological sustainability of climate change-related investments through the Holistic Impact Assessment

Massimiliano Montini

The climate change regulatory framework and the flexibility mechanisms

Introduction

The climate change challenge is certainly one of the top priorities of our times, as was recently confirmed by the outcomes of the September 2014 Climate Summit convened by the UN Secretary-General Ban Ki-moon. In the last two decades the growing awareness about the close connection between the increase in the concentration of anthropogenic greenhouse gases (GHG) emissions in the atmosphere and the modifications of the Earth's climate experienced all over the world has been aptly demonstrated by the five reports issued by the Intergovernmental Panel on Climate Change (IPCC).[1] This has shown the urgency of trying to devise more effective ways to promote and regulate global mitigation and adaptation initiatives.

The climate change regulatory framework

The climate change regulatory framework has been developed on the basis of the scientific findings of the IPCC Reports. However, despite the enormous advancements of climate change science in the last two decades, as witnessed by the findings of the most recent IPPC Reports, the main regulatory source for the promotion and coordination of the international efforts on climate change is still represented by the 1992 Framework Convention on Climate Change (UNFCCC), which was adopted following the publication of the first IPCC Report in 1990. Subsequently, the 1992 Convention was supplemented by the 1997 Kyoto Protocol, which contained the first binding targets for Annex I Parties to the UNFCCC for GHG mitigation purposes. The first commitment period envisaged by the Kyoto Protocol spanned from 2008 to 2012 and was quite successful in relative terms, although in absolute terms it contributed very little to a true global GHG mitigation. The first period has been followed by a second commitment period, covering 2013 to 2020, which, however, involves a more limited number of countries and rather shows that the Kyoto Protocol has meanwhile lost its propulsive role.

After many years of long lasting and quite disappointing international negotiations, the global community is now waiting for a new global agreement on climate change, which may hopefully be reached in the near future. In this sense, the harshest delusion was experienced at the 2009 Copenhagen Conference of the Parties (COP), which probably marked the lowest point for the cooperation efforts at the international level, showing the 'lost momentum' for the possibility to reach a real consensus at a global level on rules and actions needed to promote an effective fight against climate change.[2] However, new positive expectations have been raised after the 2011 Durban and 2012 Doha COPs, which identified the 2015 Paris COP as the possible right time for adopting a new global agreement on climate change for regulating the post-2020 period.[3] Despite the good intentions of several countries, the possibility to reach a *new protocol, another legal instrument or an agreed outcome with legal force* by 2015 COP, and to enhance the level of global mitigation ambitions, is still very uncertain. For sure, the persistent reluctance of the majority of the Parties to take on further GHG reduction commitments, even within the second commitment period of the Kyoto Protocol, foretells hard times for the near future.

The flexibility mechanisms

Notwithstanding the shortcomings of the Kyoto Protocol in terms of absolute emission reductions at a global level, it must be acknowledged that since its entry into force in 2005, it has influenced and promoted, both directly and indirectly, a wide array of international mitigation projects, often involving bilateral partnerships among Parties to the UNFCCC. This trend has been largely driven by the implementation of the so-called flexibility mechanisms devised by the Kyoto Protocol, which have promoted a large flow of investments aimed at reducing GHG emissions in a global perspective. The three flexibility mechanisms introduced by the Kyoto Protocol were two project-based instruments, namely the Clean Development Mechanism (CDM) and the Joint Implementation, and one market-based tool, the Emission Trading.

The most successful instrument, among the three mechanisms mentioned above, has so far been the CDM, which is defined in Article 12 of the Kyoto Protocol as a tool whose purpose is:

> to assist Parties not included in Annex I in achieving sustainable development and in contributing to the ultimate objective of the Convention, and to assist Parties included in Annex I in achieving compliance with their quantified emission limitation and reduction commitments under Article 3 of the Kyoto Protocol.

In brief, the CDM foresees project activities carried out by Annex I Parties into non-Annex I countries, which aim at earning Certified Emission Reductions

(CERs), each one being equivalent to one tonne of CO_2. The so obtained CERs are tradable and may be sold in the related carbon market, and ultimately may be used by industrialized countries to meet a part of their emission reduction targets under the Kyoto Protocol. To this effect, detailed rules regulating the CDM projects requirements, supervision, project cycle, entities and authorities involved, as well as CERs issuance, are spelled out mainly in the CDM modalities and procedures adopted by Decision 5/CMP.1. The available evidence shows that Annex I Parties made a great recourse to the CDM, which can be considered the most popular of the three flexible mechanisms. Indeed, the latest available data report a total of more than 8,694 CDM projects, for a total of 1,490,923,000 CERs issued.[4] The vast majority of CDM projects have been hosted by the Asia and Pacific Area (81.7%, out of which 55.7% were hosted by China and 29.2% by India), followed, at long distance, by Latin America (13.2%) and Africa (2.8%).[5]

Joint Implementation (JI) is also a project-based instrument, which partially differs from CDM in purpose and functioning and, most importantly, with regard to the Parties involved. JI is defined in Article 6 of the Kyoto Protocol as follows:

> For the purpose of meeting its commitments under Article 3, any Party included in Annex I may transfer to, or acquire from, any other such Party emission reduction units resulting from projects aimed at reducing anthropogenic emissions by sources or enhancing anthropogenic removals by sinks of greenhouse gases in any sector of the economy.

Like CDM, JI is also a project-based tool, which aims at generating carbon credits. However, contrary to the CDM, it allows the participation of Annex I Parties only and foresees the issuance of so-called Emission Reduction Units (ERUs), which likewise CERs generated under the CDM are equivalent to one tonne of CO_2. JI rules and procedures, covering project eligibility requirements, stages of approval, entities involved and ERUs issuance were adopted by Decision 2/CMP.2. The latest available data report a total of 751 JI projects, for a total of 849,859,000 ERUs issued. Ukraine is by far the leading host country with 321 JI projects, followed by the Russian Federation accounting for 182 projects.[6]

Finally, the only market-based mechanism foreseen by the Kyoto Protocol, namely the Emission Trading, also known as International Emission Trading (IET), is defined as follows by Article 17 of the Kyoto Protocol:

> Parties included in Annex B may participate in emissions trading for the purposes of fulfilling their commitments under Article 3. Any such trading shall be supplemental to domestic actions for the purpose of meeting quantified emission limitation and reduction commitments under that article.

In greater detail, the IET mechanism works as follows. Annex B Parties to the Kyoto Protocol (corresponding to the Annex I Parties to the UNFCCC) have been given reduction targets corresponding to specific levels of allowed emissions, or 'assigned amounts', over the 2008–2012 commitment period, which are divided into 'assigned amount units' (AAUs). The IET allows any Annex B party which has spare AAUs (i.e.: not used for compliance purposes) to sell them in the related carbon market to other Annex B countries that are exceeding their targets. Decision 11/CMP.1 provides modalities, rules and guidelines for IET under Article 17 of the Kyoto Protocol. So far, it is not quite clear whether the implementation of IET has been as successful as CDM and JI, since there are no available data published by either the UNFCCC or by other international organization on the number of AAUs traded.

In addition to the three flexibility mechanisms mentioned above, in the last few years the possibility to create a new market-based mechanism (NMM) has been envisaged in the framework of the international negotiations on the future climate change regime. In such a context, at the 2012 Doha meeting (COP-18), the COP requested the Subsidiary Body for Scientific and Technological Advice (SBSTA) to conduct a work programme to elaborate modalities and procedures for developing a new market-based mechanism, as previously defined at the 2011 Durban meeting (COP-17) and at the 2010 Cancun meeting (COP-16).[7] However, such modalities and procedures have not been defined yet, as the Parties are still negotiating on the characteristics of such a new market-based mechanism, which could possibly merge within a new single instrument some of the features presently recurring in both the project-based (CDM and JI) and in the market-based (ET) instruments presented above.

Climate change related investments and the rise of 'internal environmental' conflicts

The promotion of climate change related investments

Considering the three existing flexibility mechanisms and the prospective new market-based mechanism, it may be argued that a well-established climate change-related regulatory framework exists, which aims at promoting and overseeing the flow of low-carbon foreign direct investment at a global level. In fact, the operation of the project-based as well as the market-based mechanisms envisaged under the Kyoto Protocol and supplemented by the provisions of the various COP decisions all point to the direction of trying to facilitate climate change-related investments.

However, it should be highlighted here that a serious consequence arising from the implementation of such a 'facilitative' regulatory framework for the development of climate-related investment projects may consist in the emergence of a structural imbalance between the potentially conflicting interests relating on the one side to the investment promotion and on the other side to the protection of the environment.

The rise of 'internal environmental conflicts'

The structural imbalance mentioned above, which may derive from the application of the 'facilitative' regulatory framework aimed at the promotion of climate-related investment projects has not been subject yet to a careful and detailed analysis in the legal literature. However, there are many types of projects where the imbalance may arise. In such a context, one may think, for instance, of projects for energy production from unconventional fossil fuels or from renewable sources, as well as of carbon capture and storage facilities, which may negatively affect the land where they are going to be placed.

As far as energy production plants are concerned, some criticalities may arise involving both non-renewable and renewable sources of energy. As for non-renewable ones, an interesting example is represented by the so-called 'unconventional fossil fuels' which, due to their peculiar geological location, cannot be extracted with conventional drilling. Unconventional gas in particular, in the form of 'shale gas', is gaining growing attention at international level as a means to fight climate change. In fact, gas is the lowest CO_2 emitter if compared to the other fossil fuels.[8] However, this should be coupled with the following fact: methane, the main component of natural gas, 'when released into the atmosphere is 86 to 105 times as powerful as CO_2 at disrupting the climate over a 20-year period'.[9] Although compared with CO_2 methane emissions associated with the burning of fossil fuels are very small,[10] it should be taken into consideration that, as reported by Greenpeace, the IPCC calculated that methane is 34 times stronger as a heat-trapping gas than CO_2 over a 100-year time scale (according to EPA, the comparative impact of methane on climate change is over 20 times greater than CO_2 over a 100-year period),[11] meaning that its heat-trapping strength is nearly 40 per cent greater than the previously estimated 25 per cent.[12] Furthermore, the alleged unconventional gas advantage in terms of climate change mitigation should be carefully scrutinized and balanced with the possible (or likely) disadvantages in terms of its overall environmental impacts. In fact, the technique used to extract the shale-gas (fracking) has the potential to seriously interfere with an essential environmental resource, namely water, in two ways. First, such a technique consumes huge quantities of water (it has been estimated between 9,000 and 29,000 cubic meters of water per year for each gas well);[13] second, it affects the aquifers, by polluting them with the chemical additives used in the extraction process. Moreover, some concerns have been raised regarding the presumed implications of fracking for the overall geological stability of the concerned land. Lastly, and more surprisingly, the extraction of shale gas also entails fugitive methane emissions, which, to a certain extent, could counterbalance the low levels of GHG emissions usually associated with gas.

Also in the case of renewable sources, the construction of new plants may present some 'environmental criticalities' to be weighed against the 'climate change/energy advantages'. The most relevant criticalities in this sense may be represented by the aesthetic impacts on landscape and by the land consumption phenomenon.[14] In this sense, wind farms and ground-mounted photovoltaic plants are particularly relevant examples of projects where renewable energy interests may conflict with environmental protection ones. In fact, such types of energy plants may unarguably have a non-negligible impact on the land where they are placed, by altering the traditional landscape and by diverting land from its traditional use. In some circumstances, the construction of renewable energy plants in rural areas may also give rise to a hard competition between energy production and agriculture/food production. This is especially the case when such plants are being placed in areas of high landscape and naturalistic value.

Similar remarks could be made for carbon capture and storage (CCS) as well. Such a technique consists of capturing carbon dioxide emissions from coal- or gas-burning power plants and transferring the captured CO_2 as high-pressure liquid carbon dioxide to underground geologic formations. Due to its potentially relevant role in reducing CO_2 associated with the use of conventional fossil fuels, CCS is widely supported as an effective means to fight climate change, despite its limited application on a large scale to date. However, CCS has the potential to negatively impact on the land where the plant is located. In fact, as discussed more extensively elsewhere,[15] the complex CCS technology entails various risks, still partially unexplored, for both the environment and human health.[16] Furthermore, like fracking, the CCS technique, which involves the injection of large volumes of CO_2 into the brittle rocks, may represent a potential risk for geological stability.[17]

The new type of conflict between climate change/energy-related advantages on the one side and environmental risks on the other side which may occur in the above mentioned projects largely resembles the traditional opposition often arising between the specific interests related to investment promotion on the one side and to environmental protection on the other side. Such a traditional type of conflict has emerged in the last few years in many recent controversies, dealing with the interpretation and application of bilateral investment agreements (BITs) or of multilateral investment treaties, such as the NAFTA or the Energy Charter Treaty (ECT), which have normally been adjudicated by international arbitral tribunals set up under the ICSID Convention. The arbitral tribunals decisions which have settled such types of cases have given rise to a broad but often not very consistent case-law, which has been subject to criticisms by several international law scholars.[18]

In my view, if one compares the new type of conflict highlighted above with the traditional investment versus environment conflicts, a new specific feature emerges, which arises from the realization of climate change-related investments. Such a feature consists in the emergence of an 'internal environmental conflict', which is characterized by the contrast between two potentially legitimate environmental interests, the first one related to the fight against global climate change and the second one connected to the protection of the national or local environment. What is noteworthy in this context is that the presence of such an internal environmental dimension may have the consequence of changing the terms of the very difficult and delicate balancing exercise which is needed in such cases, most likely in favour of the low-carbon investments backed by climate change-related considerations, at the expenses of the ecological interests related to the protection of the environment of the land where the investment is hosted.

The 'internal environmental conflict' defined above originates from the current trend of substantially conflating all the environmental issues into climate change (and energy-related) ones. In the last decades, in fact, climate change has become at a global level the leading environmental problem, downgrading all the other environmental issues to a secondary role. However, despite the seriousness of the climate change issue, its prominence should not be used as a justification for giving a green light to every kind of investment (which may even entail a negative environmental impact), simply on the basis of real (or alleged) GHG emissions reductions or mitigations, thus dismissing or downgrading any other environmental requirement or consideration. In more detail, when a climate change-related investment is proposed, the inherent ecological features of the land where it is going to be placed should be thoroughly analyzed and assessed, and the ecological interests at stake should be carefully balanced with the expected climate change advantages, through a comprehensive and holistic evaluation process. In other words, since the projects characterized by an 'internal environmental' conflict may give rise to negative environmental externalities, possibly arising as a side-effect of climate change-related investments, a suitable framework should be in place to enable a proper assessment of the competing interest at stake. Such an assessment should be conducted, in particular, within the authorization procedure, as will be explained in greater detail below.

The quest for the (ecological) sustainability of investments

In order to properly address the 'internal environmental conflict' presented above, the traditional instruments and criteria used by international as well as national courts and tribunals to deal with the controversies between investment and environmental interests may be not fully appropriate. Therefore, I am convinced that a new and different criterion, set at an overarching level, should help to correctly steer the balance between the competing interests at stake in the cases characterized by an internal environmental conflict. This criterion should be able to fully consider and take into account the need to ensure the ecological sustainability of climate change-related investments, that is their capacity to promote global climate-related goals while, at the same time, not endangering the local environment of the land where they are located.

This new criterion should consist in the ecological sustainability paradigm, which I have identified and described in greater detail elsewhere.[19] Such a new paradigm is essentially based on the concept of ecological sustainability, which consists of 'the duty to protect and restore the integrity of the Earth's ecological systems'[20] and which may be said to refer to the need of the human civilization to live in harmony with nature and the eco-systems which enable life on the planet and support human development. The inspiration for the choice of this concept as the possible paradigm for promoting an ecologically sound application of international and national environmental law is taken mainly from the work of Klaus Bosselmann, which refers to the principle of sustainability as the fundamental concept for transforming law and governance at the global level.[21]

In my view, therefore, the concept of ecological sustainability should represent the reference concept to be relied upon in order to promote an ecologically sound application of the balancing instruments available in environmental law, so as to address and possibly solve the 'internal environmental' conflicts presented above. This is also in line with the principle of sustainable development, which is one of the overarching objectives of climate change law. The clear link between sustainable development, ecological sustainability and environmental law has been highlighted by Steffan Westerlund, who has affirmed that 'sustainable development cannot take place without ecological sustainability, which in turn is related to environmental quality and natural resources' and has strongly advocated the need to make law 'sustainable', both in its definition and its application, in so far 'unless law is made sustainable, it protects unsustainable conduct'.[22]

In this sense, drawing inspiration from Klaus Bosselmann's work, it may be argued that the ecological sustainability paradigm should constitute a *Grundnorm*, which should 'underpin and guide the interpretation of existing and the creation of new laws'. Such a *Grundnorm* should become the paradigm for creating new environmental legislation, as well as for revising and interpreting the existing environmental law provisions. To this effect, it could be aptly relied upon to promote an ecologically-based preventive assessment of all the climate change-related investment projects which are characterized by 'internal environmental' conflicts.[23]

The plea for an ecologically based preventive assessment of the projects through the Holistic Impact Assessment

Through the application of the ecological sustainability paradigm to the internal environmental conflicts that may characterize climate change-related investments, an ecologically sound assessment of those relevant investments projects may be performed, prior to their realization. To this effect, the traditional instruments devised under both international law and national laws for preventive evaluation purposes, such as the environmental impact assessment (EIA) procedure for the preventive evaluation of the possible negative effects of certain projects, as well as the strategic impact assessment (SIA) for the preventive evaluation of the possible negative effects of certain plans and programmes, could represent an appropriate starting point. However, these traditional instruments, which are nowadays widely recognized and applied both at international and national level, as specific tools to promote, both at downstream (EIA) and at upstream (SIA) level, an *ex ante* evaluation of the possible negative impacts of certain planned activities, should be aptly revised in the light of the ecological sustainability paradigm.

In my view, as I have argued in greater detail elsewhere,[24] this revision should be made by merging the existing EIA and SEA procedures into a new comprehensive instrument, so as to increase the coordination between them and to make them fully compatible with the paradigm of ecological sustainability. Such a new instrument, which ought to be based upon a holistic sustainability approach, could be named the Holistic Impact Assessment (HIA). Within such a context, the two types of existing evaluations will continue to exist and be conducted separately, still dealing respectively with the upstream (SIA) and downstream (EIA) assessment, but will be placed under a single framework, inspired by a common approach, governed by substantially the same rules and managed in a coordinated way. In such a way, the HIA, as the new comprehensive instrument for the preventive evaluation of projects as well as of plans and programmes, may become the common reference framework for the prior assessment of all the activities likely to have significant adverse effects on a certain land.[25]

Notes

1 The 1st IPCC Report was adopted in 1990 and it largely influenced the adoption of the 1992 Framework Convention on Climate Change, while the latest one, the 5th IPCC Report, was published in 2013–14.
2 M. Montini, Reshaping Climate Governance for Post-2012, *European Journal of Legal Studies* 4(1) (2011), pp. 7–24.
3 To this respect, the first remarkable step has been taken by COP 2011 in Durban which, by virtue of Decision 1/CP.17, launched a new negotiation process through the establishment of the Ad Hoc Working Group on the Durban Platform for Enhanced Action (ADP), a subsidiary body hereby created under the UNFCCC, with the mandate to develop 'a protocol, another legal instrument or an agreed outcome with legal force under the Convention applicable to all Parties' and to explore options to enhance the level of mitigation ambition (Decision 1/CP.17 paragraphs 7–8).
4 See UNEP/DTU partnership, CDM projects by type, http://www.cdmpipeline.org/cdm-projects-type.htm (accessed 1 October 2014).
5 See UNEP/DTU partnership, CDM projects by host region, http://www.cdmpipeline.org/cdm-projects-region.htm (accessed 1 October 2014).
6 See UNEP/DTU partnership, JI projects, http://www.cdmpipeline.org/ji-projects.htm (accessed 1 October 2014).
7 See 2012 Doha COP-18 Decision 1/CP.18 (§50), referring to 2011 Durban COP-17 Decision 2/CP.17 (§83), in turn referring to 2010 Cancun COP-16 Decision 1/CP.16 (§80).
8 The amount of CO_2 produced when a fuel is burned is a function of the carbon content of the fuel. The pounds of CO_2 emitted per million Btu of energy for various fuels are as follows: coal (anthracite) 228.6, coal (bituminous) 205.7, coal (lignite) 215.4, coal (subbituminous) 214.3, diesel fuel and heating oil 161.3, gasoline 157.2, propane 139.0, natural gas 117.0. See EIA, How Much Carbon Dioxide is Produced When Different Fuels are Burned?, www.eia.gov/tools/faqs/faq.cfm?id=73&t=11.
9 Greenpeace, Natural Gas and Global Warming – Methane's Contribution to Global Warming, www.greenpeace.org/usa/en/campaigns/global-warming-and-energy/science/Natural-Gas-and-Global-Warming.
10 Ecofys, *Effects of New Fossil Fuel Developments on the Possibilities of Meeting 2°C Scenarios – Final Report*, 2012, p. 9.
11 EPA, United States Environment Protection Agency, http://epa.gov/climatechange/ghgemissions/gases/ch4.html#ref1.
12 Greenpeace, Natural Gas and Global Warming – Methane's Contribution to Global Warming, www.greenpeace.org/usa/en/campaigns/global-warming-and-energy/science/Natural-Gas-and-Global-Warming.
13 Greenpeace Italia, http://m.greenpeace.org/italy/it/high/News1/blog/tutti-i-pericoli-del-fracking/blog/43951.
14 The so-called 'land consumption phenomenon' consists of the progressive loss of pristine and agriculture land for the development of industrial, commercial and residential buildings and facilities.

15 M. Montini and E. Orlando, Balancing Climate Change Mitigation and Environmental Protection Interests in the EU Directive on Carbon Capture and Storage, *Climate Law* 3 (2012), pp. 165–80.
16 For a quick, yet exhaustive, overview on the health and environmental risks, see, for instance, J. Fogarty and M. McCally, Health and Safety Risks of Carbon Capture and Storage, Commentary, *JAMA* 303(1) (2010), pp. 67–8.
17 See, for instance, M. D. Zobacka and S. M. Gorelick, Earthquake Triggering and Large-Scale Geologic Storage of Carbon Dioxide, *PNAS* 109(26) (2012), pp. 10,164–8.
18 J. Vinuales, *Foreign Investment and the Environment in International Law*, Cambridge: Cambridge University Press, 2012; V. Vadi, *Cultural Heritage in International Investment Law and Arbitration*, Cambridge: Cambridge University Press (2014); V. Vadi, *Public Health in International Investment Law and Arbitration*, Abingdon: Routledge (2013); S. Di Benedetto, *International Investment Law and the Environment*, Cheltenham: Elgar (2013); K. Tienhaara, *The Expropriation of Environmental Governance*, Cambridge: Cambridge University Press (2009).
19 M. Montini, Revising International Environmental Law Through the Paradigm of Ecological Sustainability, in F. Lenzerini and A. Vrdoljak (eds), *International Law for Common Goods: Normative Perspectives in Human Rights, Culture and Nature*, Oxford: Hart Publishing (2014), pp. 271–87.
20 K. Bosselmann, *The Principle of Sustainability*, Aldershot: Ashgate Publishing (2008), p. 53.
21 *Ibid.*, p. 53.
22 S. Westerlund, Theory for Sustainable Development, in H. C. Bugge and C. Voigt (eds), *Sustainable Development in International and National Law*, Groningen: Europa (2008), pp. 52–4.
23 K. Bosselmann, Grounding the Rule of Law, in C. Voigt (ed.), *The Rule of Law for Nature*, Cambridge: Cambridge University Press (2013), pp. 75–93.
24 M. Montini, Towards a New Instrument for Promoting Sustainability beyond the EIA and the SEA: the Holistic Impact Assessment (HIA), in C. Voigt (ed.), *The Rule of Law for Nature*, Cambridge: Cambridge University Press (2013), pp. 243–58.
25 For more details on the HIA, see Montini, *ibid.*, pp. 253–7.

Part III

Environmental legal issues in Europe and beyond

10 Evaluation and development of small island communities with special reference to uninhabited insular areas

Grigoris Tsaltas, Aristotelis Alexopoulos, Gerasimos Rodotheatos and Tilemachos Bourtzis

Overview

The Greek territory contains a large number of islands and rocks (hereafter insular formations). Approximately 9,800 of them have been registered by the Greek authorities, covering up to one-fifth of Greece's total surface, rendering Greece the country with the largest coastline in the European Union. Nevertheless, only a small portion of insular areas (around 0.01%), are inhabited and an unspecified number is used occasionally for economic purposes.

During past years, several estimations and scenarios about the development capacity of those uninhabited insular areas have been expressed, mainly as part of the wider debate on the appropriate response to the fiscal and social crisis that has been affecting the country since 2008. According to preliminary findings, the development capabilities cover a wide spectrum, mainly related to natural resources exploitation, either on the land or the surrounding marine areas.

However, all those estimations remain rough and offer an incomplete picture of the real situation. This chapter proposes the application of a methodology referred to as Greek Uninhabited Islands Development Evaluation (GUIDE) for the assessment of the current position and the development of future potentials of the Greek insular areas. GUIDE aims to examine a series of parameters that govern the development capacity of each insular area or complex of areas, highlighting the comparative advantages and proposing alternative options for their utilization, where appropriate. The main scope of this methodology is to encourage solutions that are in conformity with sustainable development principles set by the Rio Declaration on Environment and Development (1992), and to move the uninhabited insular areas one step forward to self-regulation and management.

Introduction

Due to its exceptional geomorphologic nature, Greece consists of a vast number of insular areas, the largest portion of which are uninhabited.[1] Moreover, many of these areas do not even have an official name for registration purposes. This chapter aims to:

• give a general overview of the volume and nature of the Greek Uninhabited Islands;
• explore their development prospects; and
• briefly present a methodological tool that provides for the evaluation, development, management and self-regulation of small island communities.

The exact number of Greek insular areas is very hard to calculate. Earlier evaluations estimated them at around 3,000. However, current estimations refer to 9,835 insular formations (rocks included) that cover an area of 25,019 km². Of those, few more than 100, which stand out in terms of size, are inhabited. The rest, that is 99.99 per cent, are either uninhabited or occasionally visited.

According to international law, only insular formations that have the ability to sustain human habitation or economic life of their own can be considered as Islands.[2] Where this is not so, they are deemed to be rocks. In the case of Greek insular formations, and specifically in the Aegean Sea, the enjoyment of those rights seems to be of major importance in order to highlight their significance and promote their economic, social and environmental development.

Islands and the law of the sea

According to the Geneva Convention on the Territorial Sea and the Contiguous Zone of 1958 (art. 10, para. 1) and the United Nations Convention on the Law of the Sea of 1982 (art. 121, para. 1), 'an island is a naturally formed area of land, surrounded by water, which is above water at high tide'. The 1982 Convention excludes from this definition rocks, which are those formations that 'cannot sustain human habitation or economic life of their own' and due to this 'have no exclusive economic zone or continental shelf' (art. 121, para. 3).

This definition implies that even a small ground feature can be characterized in legal terms as an 'Island', a fact that, depending on its geographical position, can expand the maritime zones of the coastal state.[3] In a geological context, a rock is 'a consolidated lithology, i.e., a solid natural mass, of limited extent, including sand, sandstone, otherwise solidified sand, or igneous matter'[4] within the marine domain.[5] Historically, attempts were made to exclude mere rocks and uninhabitable islets from the definition of the term 'island' so that the coastal state would not be entitled to claim the ocean spaces around them.[6]

The literature tends to employ a series of rather ineffective but commonly used terms such as 'rock-island', 'deserted island', 'barren island', 'insular ground' and 'islet',[7] all of which seek to interpret the 'art. 121, para. 3' rule. For the purposes of this chapter, we have adopted the term 'uninhabited island', meaning 'an insular formation that is not currently inhabited and does not host any kind of economic activity'. That understanding is aligned with the description contained in art. 121, para. 3, by including the main elements of human sustainability and economic viability, in place of the term 'rock-island'[8] which remains very popular among Greek scholars.[9]

The main source of contention in the debate of whether to employ the term 'rock' or 'island' is the unclear wording of article 121 of the LOSC. It is true that during the Third United Nations Conference on the Law of the Sea (1973–82) this article generated various disputes and caused furious contradictions between the different negotiating groups of states.[10] One group consisting of Greece, Canada, Cyprus, New Zealand, Fiji, Tonga, Trinidad and Tobago, Venezuela and Samoa claimed that the definition given in art. 10, para. 1 of the Geneva Convention on the Territorial Sea and the Contiguous Zone (1958) was still relevant.[11]

Contrary to that, another block[12] adopted the Turkish–Romanian position which was that this definition was vague, covering all types of formations, from really big islands down to islets, rocks and drying reefs, and for this reason it required further elaboration. Meanwhile, Malta proposed that the definition of an island should not include natural formations covering less than $1m^2$ of surface.[13]

The final version of the Convention defined, or at least attempted to define, the distinction between an island and a rock on the basis of the ability of the ground feature to sustain human habitation and/or economic life. However, the Convention does not go into the specifics of the actual terms used, thus encouraging a degree of uncertainty. The result is that uninhabited islands which are commonly larger than some states,[14] and small islands of which are highly populated, are both considered to be Islands, under the Convention.

The issue of the appropriateness of the relevant terms is also met in Greek waters,[15] where many islands that used to be inhabited in the past were later abandoned due to temporary or longstanding changes in living conditions. The island of Ro is such an example. It hosted a permanent population until the early 1960s, but since then only three persons have been living there.[16] The island of Spinaloga was a leprosarium until 1954 but has been abandoned since then, and without being guarded or otherwise protected, it receives, as a tourism site, a substantive amount of visitors on a daily basis.

Like the physical occupancy, the economic life or autonomy criterion also entails a certain degree of interpretative difficulties. It is often disputed whether the exploitation of marine living resources in proximity to a rock or an island should be considered as a genuine proof of 'economic life'. The island of Kira Panagia, which is owned by the Megisti Lavra Monastery (at Mount Athos), has systematically been cultivated by monks, while the Island of Oxia was leased to cattlemen, who used it as grazing fields.

All of the above lead us to the suggestion that for the sake of effectiveness and clarification, the definition of 'rock' should be connected to the geographical rather than the legal criterion, in order to be distinguishable from an island. Thus, a formation surrounded by water at all times should not exceed $1m^2$ of surface in order to be classified as a rock.

The Greek waters

In geographical terms, an island is a land mass smaller than a continent that is surrounded by water, irrespectively of its volume. Hence both Greenland, the biggest island on the planet (covering 2,175,600 km^2) and the uninhabited island of Stroggyli at the eastern-most tip of Greece (covering 0.978 km^2) are considered to be islands. Their joint classification as islands demonstrates the dissonance between law and geography. Further, international law grants a different status to artificial islands and structures as well as to other natural formations such as reefs. Moreover, islands situated on rivers and lakes enjoy a different status still.

Islands are capable of generating all types of maritime zones in the same sense that continental land mass does so. This rule also applies to island states.[17] Today this rule is considered to be customary, as it is repeated in the outcomes of the Hague Conference for the Codification of International Law (1930), the Geneva Convention (1958) and the LOSC (1982).

Greece has the distinctive feature of belonging to a limited group of countries that possess thousands of insular formations. Several attempts to list them have resulted in variable results. The systematic work of the Hellenic Navy Hydrographic Service came to the conclusion that Greece hosts 9,835 insular formations, including those that exist in freshwaters.[18] This list is subject to change because of, among other things, phenomena such as submergence, emergence, segregation, corrosion, unification with continental mass, manmade modification of insular formations, or even omissions or double registrations during past attempts.[19]

Table 10.1 Greek Islands according to geographical region.

Geographical region	No. of islands, islets and rocks	No. of inhabited islands (census of 1991)	Area (in sq. miles)
Ionian and Adriatic Sea	1102	17	6337 islets < 0.000386
Argo-Saronic Gulf	17	8	
Northern Sporades	711	8	
Cyclades	2242	26	
North-Eastern Aegean Sea	492	13	
Dodecanese Islands	1139	21	
Euboea Archipelagos	593	3	
Cretan Archipelagos	717	3	
Other Islands*	2822	14	
Total	9835	113	99 islands > 1.93

Source: Data compiled by the authors based on Hellenic Navy Hydrographic Service, *Ploigos*, Athens, 1996; G. Giagakis, *Isolario of the Inhabited Greek Islands 1940–1991*, Agistri, 1995, p. 33–45; and A. B. Alexopoulos, The Legal Regime of Uninhabited Islets and Rocks in International Law: The Case of the Greek Seas, *Revue Hellenique de Droit International* 56 (2003): 131–51.

* Including those islands and islets that are situated in sweet or brackish waters. It may be said that some insular formations previously regarded as 'islands' are no longer maintaining insular status, i.e. the islet of Makronisi has been linked with a bridge to the nearby coast of Kithira island, forming the new port of Diakofti.

Property regime

Greek uninhabited islands have historically been tied to special circumstances. The characteristics of remoteness and isolation have at times served as attraction for population groups and special uses (e.g. monastic life, national security activities, confinement institutions, sanitariums etc.) while today they stand as ideal places for recreation.

The modern trend for exploiting the Greek uninhabited islands began in the late 1980s and continued throughout the next decade, when national authorities and a considerable amount of private investors were keen on engaging in this market. The phenomenon reached its climax in the mid-1990s, when it was largely presented by the mass media as an issue of 'national interest'. The matter has re-surfaced again, as local and international real estate firms have spotted a chance for investment. The rise of interest largely derives from two types of investors: (1) wealthy persons coming from various parts of the world; and (2) off-shore companies, often of arguable incentives and activities. Both actors are not usually interested in permanent habitation, but rather in economic exploitation. One potential direct result of this trend is the growth in 'private islands'. Suffice it to say that the attractive landscapes are commonly owned by economic elites and host recreational facilities, away from indiscreet eyes and mass tourism resorts. Apart from that, private islands usually cover rather small surfaces and are heavily guarded. Some typical examples of private islands in the Greek Seas are the following:

- Scorpios: located in the Ionian Sea, east of Lefkada island. It has a surface of 80 hectares and used to belong to the Onassis family (sold to D. Rybolovlev), who bought it in 1963, along with the Island of Castri, from the Filippa Family, for the price of 1,800,000 Drachmas. A few months later, in the same area, the Onassis family also bought the Islets of Tsokari (0.9 hectares) and Sparta (70 hectares).
- Spetsopoula: located in the Saronic Gulf, close to the island of Spetses. It was bought in 1958 by the Niarchos family, at the price of US$150,000. Much work has been done on the island in order to make it more attractive to its inhabitants, for example, the construction of a man-made beach and the installation of technical incubators of rare species for hunting, etc.
- Petali: a complex of 10 islets located in the South Evian Gulf, once owned by the Empricos family, and with the exception of one, sold to the Karnessis family, which further on sold the Xero Islet to the Picasso family, at the price of 380,000,000 Drachmas.

In most of these cases, preservation and protection infrastructures and works have been installed,[20] while at the same time the owners are mostly keen on conserving the environment in its pristine form.[21]

Where private citizens do not own islands, islands are reserved as assets by (a) the central government and local administration entities, such as the Greek Tourism Organization and other municipalities respectively, and (b) not-for-profit entities, such as the Church of Greece and other religious institutions. Those privately owned assets have been obtained through one of the following ways:

- islands which were formerly owned by a person or a group of persons (i.e. Farmakonisi, Seskli, Alimnia etc.) and were eventually returned to the Greek State;
- islands that have been sold during the past two centuries by their previous private owner(s);
- islands that historically belong to local administration entities (i.e. Makra, Stroggyli, Astakida, Iounia, Litsa, Pentikonisia, Plati, and Kiourka all belong to the Municipality of Kassos, and Agios Georgios and Ipsili belong to the island of Megisti) or central administration entities (i.e. Spinaloga and Pezonisi that belong to the Greek Tourism Organization);
- islands that belong or are claimed by religious institutions (i.e. the Strofades complex belongs to the Diocese of Zakynthos[22] and Trikeri belongs to the Diocese of Hydra); and
- state-owned islets and rocky-islands that are usually low-lying and not exceeding 20 hectares.[23]

Despite the above classifications, the property status is not always so clear, especially when multiple owners are engaged or when property titles date back to more than three centuries ago. Hence, several times public property has been claimed or illegally occupied due to those uncertainties. Obviously the absence of a fully updated National Cadastre does not seem to stand as an obstacle to the sale of insular formations between persons, and at the same time the Greek Public Properties Company cannot intervene, since this type of public property is not fully recorded.[24]

Sustainable development and preservation of Greek uninhabited islands

High risk actions and uncontrolled exploitation of resources by owners of Greek uninhabited islands may lead to uncertain and potentially catastrophic consequences that will degrade the land as well as the marine environment. Mainly due to their size and/or location, those formations constitute fragile ecosystems that often serve as habitats or refuges for various species, and have limited restoration capabilities if they are harmed.

Eventually, more and more small islands and Greek uninhabited islands will become accessible to a large number of visitors, mainly through the use of recreation vessels. If this phenomenon expands, environmental protection and preservation should be of key importance for the central government. For the time being, some of those formations are protected under various schemes such as:

- Special Protection Areas[25] or Areas of Special Conservation Interest;[26]
- Ramsar Sites; and

- Natural Parks, Wild Life Refuges and Areas of Outstanding Natural Beauty.[27]

Apart from having many things in common, diversity is another essential characteristic of the Greek uninhabited islands and a critical factor for their development prospects. Their small size usually highlights the scarcity of resources that ought to be sensibly managed. The accessibility factor raises the cost of communication and transportation, which demand special infrastructure. And last but not least, every area has a unique natural and cultural background that would benefit from being safeguarded. To sum up, there is no 'silver bullet' that will manage to solve all the problems and provide for their sustainability. Thus each island has to be separately assessed and proposals need to be drafted on the basis of their uniqueness.

Tourism and insular areas

The tourism sector is commonly referred to as one of Greece's heavy industries,[28] contributing 16 per cent of the national GDP. It is considered to be one of the driving forces that will help the country overcome the fiscal crisis it has been facing since 2008. Among the various scenarios and possible solutions that have been proposed, many are directly or indirectly connected to insular areas. Plans for investment in tourism, real estate and the exploitation of land and marine resources have been proposed and examined by the administration and (possible) investors.

The influences of tourism on the environment (positive and negative) were first studied in the 1970s. During the next decade the quality of the environment as a major component of touristic attractiveness was recognized.[29] The two pillars of the environmental component are:

- protection and preservation of natural and cultural characteristics, which are under threat by some careless and rapidly expanding elements of the tourism industry; and
- the gradual improvement of the environment through the execution of infrastructure works, serving tourists as well as the permanent population.[30]

The environmental problems of insular areas can be split into four categories, among which there is a great volume of interaction:[31]

- environmental pressures with either internal or external character (which are intensified by both the limited surface and the coastline);
- resources scarcity (mainly freshwater and food. Energy demands can be covered, up to a point, by renewable sources);
- waste and sewage management; and
- coastal erosion and sea level rise effects.[32]

Those multifaceted threats ought to be tackled through a series of well-managed attempts by the relevant authorities; attempts which will have to elaborate and implement cohesive policies on the following issues:

- integrated coastal zone management and maritime spatial planning;
- biodiversity conservation measures in compatibility with human presence;
- assessment, monitoring and evaluation of environmental and development data that can be easily accessed and used by policy makers and society; and finally,
- promotion of low-impact economic activities, such as eco-tourism.

Implementation of development policies

The optimum starting point for policy development would be the systematic collection of data and their analysis.[33] Obviously, the most important features are those that respond to the interaction of the natural environment with human activities. So, apart from natural environment features,[34] man-made environment elements should also be registered.[35] Moreover, distinct criteria, processes and guidelines[36] should be elaborated for those cases of islands that should be put under a special protection regime.

Since those small insular areas form a peculiar web of natural and cultural units, where each one has its own traditions and social characteristics but where all of them face the same threats and pressures, their protection and preservation should spearhead every attempt.[37] This effort should focus on three aspects:

- data compilation for each island, which is tactically updated;
- integration of information into geographical information systems (GIS) that could be of benefit to decision makers, local societies, visitors and investors; and
- promotion of eco-tourism, which combines high-level recreation, protection, profit and participation.

Undoubtedly, an effort to apply all these actions to the thousands of Greek islands would not only be highly costly, but in some cases even useless. However, there is a substantial number of areas that could benefit from them, and some additional work can be done in order to improve their evaluation and development prospects.[38] For those cases, extra studies could be done aiming at (a) the evaluation of exploiting natural resources, and (b) the motivation and characteristics behind historic and contemporary human presence.

GUIDE: a new methodology is introduced

For the last three years, researchers at the European Centre for Environmental Research and Training of Panteion University have been trying to draw attention to those issues by designing one methodological tool, the so-called GUIDE, for the evaluation and development of small island communities. The tool can function in three stages.

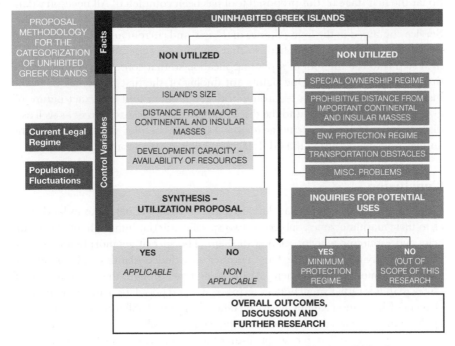

Figure 10.1 Tool for the assessment and evaluation of uninhabited small islands.

1st stage: data compilation and elaboration of methodology

- Global state of the art overview
- Literature review
- Collection of data on Greek uninhabited islands
- Building up the GUIDE methodology

2nd stage: running the tool

- Selection of specific insular complexes or groups, either utilizable or not
- Running of the GUIDE methodology
- Analysis of output
- Selection of insular complexes or groups for stage 3

3rd stage: application/exploitation

- Evaluation of stage 1 and 2 results
- Analysis of development prospects for each insular complex or group
- Proposals to decision makers and local societies

So far the first stage of this proposed tool has been completed. All necessary data have been obtained (using available databases of the Hellenic Navy Hydrographic Service contained in the four volumes of *Ploigos*) and current and future needs and applications have been identified, so that the tool can operate accordingly. The next step is to use the tool as a pilot application so that some first results can be discussed and evaluated. One significant finding of the first phase has been the inconsistent nature of data regarding both the number and the exact nature of Greek uninhabited islands and the disparities between various sources as well as a distinct lack of theme-specific academic literature, besides that relating to the authors' own work during the GUIDE methodology development.

Conclusion

Contemporary international law has attributed to islands equal status as land areas as to their maritime zones, subject to two specific criteria: human habitation capacity and economic activity relating to them. The special geomorphology of the Greek territory means that the large number of included islands can be a decisive factor for economic development. At the same time, though, the vast majority of Greek islands is rather small both in terms of size and population, meaning that the element of habitation cannot always be taken for granted. All the above call for the application of sustainable management policies, especially in cases of small uninhabited islands as well as in cases of insular formations with special regimes.

The economic potential of small insular formations can be important, but so are the resulting challenges. The environmental pressures caused by the main economic activities in small islands such as tourism, infrastructure projects and resource exploitation should not be underestimated, particularly during the planning period of any undertaking. An additional factor to be taken into account includes the special environmental regime of many insular formations under national and EU law. For the above mentioned reasons, analytical research tools such as the designed GUIDE methodology need to be developed and applied. In that sense, the second stage of GUIDE application, which includes the actual testing of specific insular complexes, should provide some early results on the actual applicability of sustainable policies focusing on specific needs.

Notes

1 According to the Population Census 2011. Data available by the Hellenic Statistical Authority at www.statistics.gr/portal/page/portal/ESYE/PAGE-census2011 (accessed 11 August 2014).
2 A fact that entitles them the right to claim all maritime zones, in the same sense as continental land mass can do, see art. 121, para 2, United Nations Convention on the Law of the Sea of 1982 (LOSC, hereafter).
3 J. Symonides, The Legal Status of the Islands in the New Law of the Sea, *3rd Scientific Symposium: The Aegean through the Centuries, 4–5 June 1987*, Athens: Aegean Foundation, pp. 118–19.
4 G. Walker (ed.), *Definitions for the Law of the Sea: Terms Not Defined by the 1982 Convention*, Martinus Nijhoff Publications, Leiden, 2012, p. 286.

5 It has been suggested by some writers that rocks can have a measured area that is less than 1 km². See D. W. Bowett, *The Legal Regime of Islands in International Law*, Oceana Publications, 1979, pp. 251–5.

6 E. D. Brown, *The International Law of the Sea, Volume I: Introductory Manual*, Dartmouth Publications, 1994, pp. 148–9.

7 Those terms are in conflict, since they signify different substance. 'Rock-island' is a small, rocky elevation, surrounded by water. 'Deserted island' is a term that mostly highlights the infertility of the soil. 'Barren island' focuses on the incapacity of sustaining human habitation. 'Insular ground' has a wider meaning, including rocks and lighthouses (but not drying reefs, fringing reefs, roadsteads and harbour works). 'Islet' is considered to be a very small island, with a total surface not exceeding 1 km², but not identical to a rock.

8 The phrase 'rock-island' originated from *Rockall*, a barren rock situated northwest of the British Isles that was a matter of dispute among UK, Denmark and Iceland due to the rights of the surrounding marine space.

9 Professor Antonis Bredimas comments on the use of the terms 'island', 'islet' and 'rock-island' in the book S. Perrakis (ed.), *Aegean Sea: Evolutions and Perspectives of Greek–Turkish Disputes' Settlement*, Ant. Sakkoulas, Athens, 2003, pp. 41–72. Krateros Ioannou refers to 'uninhabited rocks' in T. Kariotis, (ed.), *Greece and the Law of the Sea*, Martinus Nijhoff, Leiden – Boston, 1997, p. 123. Haralambos Athanasopulos mentions 'small islets' and 'coral islands' in his book *Greece, Turkey and the Aegean Sea*, McFarland, Jefferson, NC, 2001, p. 65.

10 J. M. Van Dyke and R. A. Brooks, Uninhabited Islands: Their Impact on the Ownership of the Oceans' Resources, *Ocean Development and International Law* 49 (1983), pp. 267–268.

11 C. R. Symmons, *The Maritime Zones of Islands in International Law*, Martinus Nijhoff, The Hague, 1979, p. 17.

12 The Delimitation Group supporting Equitable Principles, M. Nordquist, (ed.), *The United Nations Convention on the Law of the Sea: A Commentary*, vol.1, Martinus Nijhoff, Dordrecht, 1985, p. 78.

13 C. R. Symmons, Some Problems relating to the Definition of Insular Formations in International Law – Islands and Low-Tide Elevations, *Maritime Briefing* 1(5) (1995), pp. 1–30

14 Such as the Canadian Archipelago islands.

15 Aegean Sea, Ionian Sea as well as other Greek Islands in the region of Eastern Mediterranean.

16 Data compiled by the 10-year National Population Census. Provided by the Hellenic Statistical Authority at www.statistics.gr (accessed 11 August 2014).

17 States that do not possess continental lands (i.e. Barbados, Cyprus and Ireland). Additionally, some island states that fulfil specific conditions can be characterized as Archipelagic States, thus gaining some extra rights (i.e. Indonesia, Philippines and the Bahamas). See also D. M. Sodik, The Indonesian Legal Framework on Baselines, Archipelagic Passage, and Innocent Passage, *Ocean Development and International Law* 43(4) (2012), pp. 330–41.

18 Hellenic Navy Hydrographic Service, *Ploigos*, Athens, 1996.

19 A. B. Alexopoulos and S. Politis, The Strategic Importance and the International Status of the Greek Islands (Outermost Points) Serving as the Basic Criteria for Delimiting Maritime Zones According to International Law, paper presented at 2nd International Conference: Maritime History-Geopolitics at Sea, 2002, University of the Aegean.

20 Such as roads, electricity, telephone and water networks, etc.

21 It is often assumed that due to the remoteness or size of the islands, those kinds of measures would not have ever been taken by the local authorities or the central administration.

22 A. B. Alexopoulos, G. Karris and A. Kokkali, The Environmental Use, the Socio-economic Development and the Integrated Management of Remote Islets under the Concept of Sustainable Development: The Case of Strofades, in *Proceedings of 10th National Symposium of Oceanography*, National Center of Marine Research, 2012, pp. 1–7.

23 H. Theoharatos, *Ethnos* newspaper, 11 November 1991.

24 In the late 1980s, an effort to register all Greek uninhabited islands was initiated by the Greek Cadastral Mapping Service (under Law 972/1979) and was supported by the Hellenic Navy Hydrographic Service, which had already listed all the Greek uninhabited islands in a collection called *Ploigos* that was drafted in order to provide navigational information for the Greek Seas. Unfortunately, this project has yet to come to a conclusion.

25 Under the EU Birds Directive 147/2009.

26 Under the EU Directive on the Conservation of Natural Habitats 43/1992. Ibid.

27 Under L. 1650/1986 as amended by L. 3937/2011.

28 The other one being the shipping sector.

29 H. Kokkosis, Tourism–Environment: An Interactive Relation, in N. S. Margaris (ed.), *Ecology and Environment in Greece*, Fillipotis Publications, 1995, pp. 87–93.

30 A. B. Alexopoulos, N. Konstantopoulos and D. Sakkas, Evaluating the Prospects for Alternative Forms of Tourism: Applying a Strategic Plan for the Small Cyclades, in *Proceedings of International Conference on Tourism Development and Management*, Scientific Events, 2009, pp. 565–69.

31 A. B. Alexopoulos and I. Theotokas, Quality Services in the Coastal Passenger Shipping Sector and its Contribution to the Development of Tourism in Small Islands: The Case of Psara island, in *Proceedings of International Conference Tourism in Island Areas and Special Destinations*, University of the Aegean, 2000, pp. 1–11.

32 Some low-lying formations are facing the event of partial or total submergence. Apart from losing a piece of land, this event would also affect the ability of generating maritime zones. For a discussion see G. Tsaltas, G. Rodotheatos and T. Bourtzis, Artificial Islands and Structures as a Means of Safeguarding State Sovereignty against Sea Level Rise: A Law of the Sea Perspective, paper presented at the 6th ABLOS Conference Contentious Issues in UNCLOS – Surely Not?, International Hydrographic Bureau, Monaco, 25–27 October 2010.

33 For the time being, the only comprehensive data that are available have been compiled by the Hellenic Navy Hydrographic Service (concerning position and accessibility) and the National Statistical Service of Greece (concerning population and administrative affiliation).

34 Eligible either for protection or resource utilization.

35 Those elements are present either due to human habitation in the past, or current temporary/seasonal uses and can have the form of road and footpath networks, bridges, water and electricity networks, production facilities, freshwater resources, churches and monasteries, historical monuments, lighthouses, cultivations, etc.

36 For example, within SPAs only specific activities can take place: (1) traditional farming (fishing, herding, cultivation, etc.), (2) repairing or restoration of existing infrastructure, and (3) construction of installations compatible with the natural environment and only for national security, navigation, archaeology and scientific research purposes.

37 A. Alexopoulos, G. Karris and A. Kokkali, Sustainable Research and Development of the Small Uninhabited Islands: The Case of Ionian Oinousses, *Journal of Global Nest*, 15(1) (2013), pp. 111–20.

38 S. Mazi, G. Karris, A. Martinis, A. B. Alexopoulos and A. Kokkali, Determination of Priorities for Education for Sustainable Development in the Islands of Paxi-Antipaxi, in *Proceedings of International Conference for Organic Agriculture and Agro-Eco Tourism in the Mediterranean*, 2011, pp. 1–4.

11 Access to justice in environmental matters in the European Union legal order

Is there a need for a more coherent and harmonized approach?

Vasiliki Karageorgou

Introduction

Environment has no voice of its own[1] and environmental interests are, to a large extent, diffuse, fragmented and collective.[2] In particular, due to the collective nature of many environmental interests and the clashes of the diverging interests that emerge within the framework of the implementation of the environmental legislation, it is more urgent in environmental law than in any other field of law to introduce mechanisms that ensure the effective implementation of the relevant substantive rules.[3] In this context, access to justice in environmental matters is viewed by both citizens and environmental interest groups as the '*ultimum refugium*' from acts and omissions of public entities and private persons that violate the relevant provisions and pose significant threats to the environment.

Due to the particularities of environmental law, one central question that arises is *who* is entitled to take legal action in order to ensure the enforcement of the relevant substantive or procedural rules, as traditional approaches concerning legal standing fall short. Furthermore, the second central question relates to the *means* for taking legal action and the *remedies* provided against decisions, acts and omissions violating environmental provisions.[4]

Although access to justice in environmental matters was, thus, a subject of significant consideration in the majority of the national legal systems, it was not until the 1992 Rio Conference that increased attention was given to this issue due to the emphasis placed on public participation as a feature of environmental governance (Principle 10 of the Rio Declaration). Furthermore, another significant cornerstone was the adoption of the Aarhus Convention, which constitutes the most developed legal instrument at a global level[5] in terms of establishing three closely inter-related procedural rights relating to the protection of the environment (access to environmental information, public participation in environmental-related decision-making procedures and access to judicial review procedures in cases of violations of the environmental legislation).

Taking into account that the Aarhus Convention is a 'mixed agreement'[6] and its provisions are binding[7] for both the EU and Member States (hereinafter MS) as an 'integral part of the legal order of the European Union' (art. 216 TFEU), the main aim of this chapter is to explore whether the current EU framework concerning access to judicial review procedures at MS level in cases of violation of national environmental law originating from EU Law is sufficient in terms of implementing the access to justice provisions of the Aarhus Convention (art. 9 para. 3) or whether there is a need for an EU legislative intervention, in order to ensure compliance with international obligations and effective enforcement of EU environmental law. To this end, the first part of the analysis will examine the regulative context of the third pillar of the Aarhus Convention, the relevant EU regulatory framework in place and the jurisprudence of the EU Courts (b). In the second part, the analysis will focus on the question of whether there are sufficient reasons to support the adoption of a principal instrument setting minimum standards for access to justice in environmental matters at the national level and whether the EU has the competence to introduce such rules (c). Finally, the main message of the chapter as regards the need for an EU legislative intervention in the respective field is illustrated in the epilogue.

The third pillar of the Aarhus Convention and the relevant EU regulatory framework

The third pillar of the Aarhus Convention

The third pillar of the Aarhus Convention concerns access to justice in environmental matters with the aim to entitle individuals and environmental NGOs to challenge decisions through effective judicial mechanisms. Art. 9, in particular, sets the framework for reasonable conditions for the legal review of decisions, acts and omissions in three types of situations:

- the refusal or the inadequate handling of the relevant requests for access to environmental information by the public authorities (art. 9 para. 1);
- the decisions, acts and omissions on specific issues subject to public participation and covered by art. 6, namely, those acts or omissions regarding the authorization of installations and infrastructure projects (art. 9 para. 2) that can be challenged not only for reasons relating to the impairment of 'public participation rights', but also for reasons relating to their procedural and substantive legality;[8] and
- the violation of the national environmental law provisions that are not covered by the previous two categories (art. 9 para. 3).

Contrary to the very relaxed *locus standing* conditions provided in cases with regard to environmental information ('any person'), art. 9 para. 2 limits the right to challenge the relevant decisions only to a group of persons, namely to the 'public concerned.'[9] Furthermore, while art. 9 para. 2 is formulated in a quite precise way and leaves relatively narrow margin to the Contracting Parties to define the 'public concerned' mainly by obliging them to recognize, under certain conditions, the standing rights of the environmental NGOs, art. 9 para. 3 is drafted in more general terms, thereby leaving significant room as regards the implementation of its requirements.[10] The margin of discretion provided in art. 9 para. 3 cannot, however, be exercised in such a way as to contradict the object of the Convention to contribute to the enforcement of the environmental law by holding public authorities accountable through judicial proceedings.[11] Finally, art. 9 para. 4 sets minimum standards applicable to the above types of access to justice procedures, namely 'the provision of adequate and effective remedies, the fairness, the equity and the timeliness of the relevant procedures and the not prohibitively expensive costs'.

The relevant EU regulatory framework

Since the EU signed the Aarhus Convention in 1998, it has adopted certain pieces of legislation, in order to implement its provisions into the EU legal system. In particular, the first two pillars of the Convention were implemented by the adoption of two Directives, namely Directive 2003/4[12] and Directive 2003/35.[13] These Directives also inserted provisions, which incorporated the access to justice provisions of the first two pillars of the Aarhus Convention (art. 9 paras 1 and 2) in the EU legal system and have similar wording with the relevant provisions of the Convention.[14] Moreover, the EU has adopted Regulation 1367/2006, which aims at subjecting the European bodies to the provisions of the Aarhus Convention, including, subsequently, the requirements of art. 9.[15]

The only legislative proposal relating to the incorporation of the Aarhus Convention requirements in the EU legal order that was not adopted was the Proposal for a Directive on Access to Justice in Environmental Matters,[16] because certain MS claimed that it was incompatible with the subsidiarity principle. The objective of the Draft Directive, which was recently withdrawn under the exercise of simplification,[17] was to establish a framework of minimum standards for access to national courts in cases of alleged violation of national environmental law, the origins of wihch lay in the EU legislation. Its scope of application was therefore wide, as it covered all areas of EU environmental law. Moreover, the Draft Directive contained provisions that granted rights to individuals and environmental NGOs to have access to judicial review procedures, including interim relief, to challenge the substantive and procedural legality of administrative acts and omissions for alleged breaches of environmental law (art. 4).

From a general point of view, it is worth noting that the 'fragmented' implementation of the access to justice requirements of the Aarhus Convention by the revision of certain Directives[18] and the lack of a principal instrument setting the framework for access to national courts in cases of alleged violations of national environmental law founded on EU law raises issues not only as regards the compliance of the EU with its international obligations, but also as regards the effective enforcement of the various EU environmental rules at the MS level.

This can be intensified in cases where the national procedural systems are underpinned by a restrictive approach as regards access to justice. Moreover, the difficulties in implementing the existing access to justice provisions of the relevant EU Directives is demonstrated by the increasing number of requests for preliminary rulings submitted by the national courts to the Court of Justice of EU (hereafter CJEU).

The evolution of the jurisprudence

The significant body of jurisprudence developed by CJEU that deals mainly with the interpretation of the provisions of the EIA and the IPPC/IED Directive,[19] including those that relate to access to judicial review procedures, covers a wide range of issues that relate to the entitlement of third parties to take legal action to review the decision of a competent authority to subject a project to an EIA or not,[20] the standing of NGOs before national courts,[21] the costs of the relevant judicial procedures,[22] the remedies[23] and the scope of the judicial review.[24]

Moreover, the Court was asked to rule on whether art. 9 para. 3 of the Aarhus Convention has a direct effect within an EU Member State's legal order (*Slovac Brown Bear Case*). The ruling[25] is not only very relevant for the topic of the paper, but also significant in many aspects. First of all, the Court departed from the relevant case law and allocated itself the competence to rule on the direct effect of this provision, although the EU has not legislated on the issue.[26] Furthermore, although the Court did not confer direct effect on art. 9 para. 3 on the grounds that it is not clear and precise enough to regulate directly the legal position of individuals (para. 40), it gave concrete guidelines to the national court as regards how to interpret national procedural requirements consistent with the Aarhus Convention and the principle of the effectiveness of EU law (paras 45, 51), gripping thereby indirectly in the national procedural autonomy of the MS.[27]

In conclusion, the CJEU jurisprudence has played a significant role in healing, at least to some extent, the shortcomings of the national procedural systems and the vacuum left by the EU legislator, as it has clarified the content and the scope of the access to justice provisions of the relevant EU Directives and thereby indirectly of the Aarhus provisions, also by adopting a teleological approach for their interpretation and by invoking general principles of EU law, such as the principle of effectiveness for the sake of the environment.[28] In this way, it has set limits on the procedural autonomy of the MS[29] in adopting or having in place rules that are disproportionally restrictive from several aspects (*locus standi*, costs, remedies) for the affected individuals and the environmental NGOs, in order to exercise their rights arising by EU law before national courts.[30] Finally, the relevant jurisprudence opened a window for a fruitful dialogue between the CJEU and the Compliance Committee of the Convention, mainly through the references of the Advocate Generals to its findings for the purposes of the interpretation of the relevant provisions of the Convention.[31]

Examining the need for an EU legislative intervention under the prism of the recent developments

The recent legal developments and the reasons justifying the adoption of a more harmonized approach

Besides the growing body of the relevant case law, certain legal and political developments at the EU level revived the discussion as regards the need to adopt a legal instrument setting minimum standards for access to justice in environmental matters at MS level. The first legal development relates to the elevation of certain general principles concerning access to justice, already developed by case law, to fundamental rights and norms of the highest rank by virtue of the Lisbon Treaty. In particular, the recognition of a right of access to justice as a fundamental right in the EU legal order (art. 47 of the Charter of Fundamental Rights of EU) and the obligation of the MS to provide sufficient remedies in the fields covered by EU law (art. 19 para. 1 lit. b of TEU) have created a new legal ground as regards the way that national judicial systems should be organized, in order to ensure effective legal protection in cases of alleged violations of norms arising from EU law. The second relevant development relates to the recognition of access to justice in environmental matters as an issue of renewed political priority within the framework of the seventh Environment Action Programme (EAP).[32]

These developments created an increased momentum for the adoption of a legislative instrument in this field and there are several reasons speaking thereof. One significant reason lies in the particularities of the environmental interests that either relate to the protection of the 'commons' or are diffuse and fragmented by nature, so that they cannot be defended before courts by individual claimants in cases of diffuse pollution or when the damage is restricted only to the elements of nature itself.[33] Subsequently, the adoption of a principal legislative instrument at the EU level is more necessary than in other areas of law, in order to ensure that national procedural systems do not put unreasonable hurdles for the representation of the collective interests in the judicial review procedures[34] and also provide sufficient remedies (e.g. injunctive relief) responding to the particularities of the environmental litigation. The claimed argument, according to which the expansion of the possibilities of the NGOs to initiate court proceedings will result in the overloading of the courts, is not persuasive, first because it is not supported by empirical findings and second because it does not sufficiently take into account the urgent need for representation of the environmental interests before the courts.[35]

Another reason that speaks for the necessity of such a legislative intervention relates to its contribution as a means for ensuring sufficient compliance both with the access to justice requirements of the Aarhus Convention and the relevant provisions of EU primary law in terms of providing effective legal protection in cases of alleged violations of EU environmental law at national level. Such a thesis is supported by the fact that the great disparities among the procedural systems of the MS and the leeway left to them concerning the transposition of the relevant requirements[36] cannot even ensure the effective and harmonized implementation of the existing access to justice provisions of the EU Directives, as is demonstrated by the increasing number of requests for preliminary rulings. In addition, the Aarhus Convention Compliance Committee found that certain MS are in breach of their obligations arising from the various provisions of art. 9 of the Convention.[37] The insufficient implementation of the access to justice requirements of the Aarhus Convention and the shortcomings of the legal protection provided in cases of violations of EU environmental law is also confirmed by the relevant empirical findings, according to which there are several barriers for access to judicial review procedures in environmental matters at the national level that relate to the standing criteria, the costs of the procedures, the intensity of review and the effectiveness of the provided remedies.[38]

Furthermore, a legislative intervention for the further harmonization of the access to justice requirements seems to be necessary for addressing the implementation shortcomings of EU environmental law at MS level[39] that are demonstrated by the increasing number of infringement procedures and can also be attributed to weak enforcement mechanisms.

Finally, a legislative intervention in the form of a Directive is regarded as the most appropriate choice among the options that were considered for addressing the existing deficiencies and shortcomings. The alternative options range from the adoption of a soft law approach and the reliance on the development of the CJEU jurisprudence and the subsequent national adaptation to the initiation of infringement procedures for bringing the national procedural rules in line with CJEU case law. The reasons that support such a choice relate both to the comprehensiveness and effectiveness of this kind of approach in terms of ensuring effective legal protection, as required by EU primary law, and to the predictability and legal certainty that it can create for all relevant actors.[40]

EU competence on access to justice in environmental matters

Having the above remarks as a starting point, the central question that arises is whether the EU has the competence to adopt a principal instrument setting minimum standards for access to judicial review procedures in environmental matters at national level. To answer this central question, certain clarifications as regards the competence of the European Union should be made.

The first clarification relates to the *nature* of EU competence in the field of environmental policy, which is classified as one of the shared competences (art. 4 para. 2. lit. e TFEU). This means that both the EU and MS have the power to legislate in this field, while MS intervention can be envisioned only to the extent that the EU has not exercised its competence.[41] Subsequently, EU environmental legislation leads mainly to a minimum harmonization, thereby leaving the MS with the opportunity to adopt more stringent environmental regulations in line with EU law. Furthermore, due to its nature as a shared competence, its exercise is subject to the application of the principle of subsidiarity (art. 5 para. 3 of TEU), which is a fundamental principle of EU law focusing on the regulation of the use of powers between EU and MS.[42]

The second clarification relates to the *extent* of EU environmental competence. Art. 192 of TFEU constitutes the relevant legal basis in the field of environmental policy by conferring to the EU the competence to adopt relevant measures. Any effort to set the contours of EU competence should have as a starting point the fact that its content is determined in relation to a wide area, such as the environment, while there is no relevant definition in EU primary law.[43] Moreover, the only decisive criterion for determining the extent of the relevant competence concerns the capacity of the designed measures to contribute to the realization of the objectives set in art. 191 of TFEU.[44]

The above remark about the *extent* of EU competence leads to the conclusion that the respective competence, based on art. 192 of TFEU, is not limited to the adoption of legislative instruments setting substantive rules for environmental protection, but also includes the power to rule on procedural matters, in the case that they are necessary to ensure the effective implementation of EU (substantive) environmental law.[45] Subsequently, it is undoubtedly recognized that the EU has the competence to introduce rules extending access to justice in relation to a specific legislative act, not only when this is necessary for ensuring compliance with International Obligations (e.g. Public Participation Directive), but also when it aims to safeguard the effective implementation of a certain piece of legislation (e.g. Directive on Environmental Liability). In the same line of argumentation, it should be recognized that art. 192 of TFEU confers on the EU the competence to adopt a principal instrument for access to judicial review procedures at a national level to the extent that it constitutes a necessary implementation instrument for the effective application of EU environmental law and subsequently to the achievement of the objectives set in art. 191 of TFEU.[46] Such a competence cannot, though, go beyond what is absolutely necessary for achieving the objectives of EU environmental policy, so that it would encompass the adoption of minimum requirements concerning judicial review procedures only for violations of EU environmental law or of national environmental law founded on EU law.[47]

Furthermore, the exercise of the relevant competence would also satisfy both criteria (negative and positive) of the principle of subsidiarity. In particular, the first one that requires that the proposed action cannot be sufficiently achieved by MS is satisfied because the implementation deficit of EU environmental law demonstrates that national implementation measures are not sufficient to ensure its effective implementation.[48] The second criterion that requires that the proposed action can be better achieved at union level by reason of its scale or its effect is also satisfied, as, due to the significant discrepancies of the procedural systems of the MS, a minimum harmonization of the relevant standards has an added value in terms of contributing to the achievement of the objectives of EU environmental policy and of ensuring effective judicial protection in the respective field of EU law.[49]

Finally, the adoption of a horizontal instrument on access to justice in environmental matters would not contradict the principle of the procedural autonomy of the MS, since this principle, already limited by the principles of the non-discrimination and the effectiveness of EU law, would not even apply in this case. This is due to the fact that MS enjoy a certain level of autonomy as regards the choice of the appropriate measures for the implementation and enforcement of EU law, only to the extent that they are not bound by specific EU legislation in this field.[50]

Epilogue

The entitlement of the individuals and NGOs to challenge acts or omissions concerning violations of environmental legislation through effective judicial mechanisms constitutes a very important instrument for ensuring its effective implementation and addressing the imbalances of the current legal systems concerning the protection of the 'commons'. The Aarhus Convention has made a significant contribution in terms of enshrining this procedural right and establishing concrete obligations to the Contracting Parties for its effective exercise.

Several reasons ranging from the implementation deficit of EU environmental law to the effective compliance with the relevant international obligations support the view that the existing 'fragmented' approach of the EU legal order as regards access to justice in environmental matters is not sufficient. Consequently, it can be persuasively argued that there is a need for an EU legislative intervention in the form of a horizontal instrument that would introduce only the minimum standards that are essential for the achievement of the EU environmental objectives and would not override national procedural rules, where it is not sufficiently justified. In such a context, it would not be in conflict with the principle of subsidiarity either.

It remains to be seen whether the relevant standards for access to justice in environmental matters will still be determined by the existing 'patchwork' of norms at the EU and national level and the relevant developments of the case law or by a horizontal instrument aiming at their minimum harmonization. The reaction of the MS to a future Commission proposal[51] will be critical for defining the direction of the EU in terms of ensuring effective environmental protection.

Notes

1 See L. Krämer, The Environmental Complaint in the EU, *JEEPL* 6 (2009), pp. 13, 25.
2 See B. Ebbesson, Comparative Introduction, in J. Ebbesson (ed.), *Access to Justice in Environmental Matters in the EU*, Kluwer Law International, 2009, p. 4.
3 See C. Ponchlelet, Access to Justice in Environmental Matters: Does the European Union Comply with its Obligations?, *JEL* 24 (2012), pp. 287, 289.
4 These issues are discussed by J. Ebbesson (*supra* note 2, pp. 25ff.) in a comparative perspective as regards EU Member States.
5 Convention on Access to Information, Participation in Decision-Making and Access to Justice in Environmental Matters (adopted 25 July 1998 and entered into force 30 October 2001).
6 This means that it relates to a policy field in which both the EU and the MS share competence to accede to international agreements. See A. Rosas, The Future of

Mixity, in C. Hillion and P. Koutrakos (eds), *Mixed Agreements revisited: The EU and its Member States in the World*, Hart Publishing, 2010, pp. 367ff.

7 EU became party to the Convention through the Council Decision 2005/370, OJ 2005 L 124/1.

8 See J. Ebbesson, Access to Justice at the National Level: Impact of the Aarhus Convention and the European Union Law, in M. Pallemaerts (ed.), *The Aarhus Convention at Ten: Interactions and Tensions between Conventional International Law and EU Environmental Law*, Europa Law Publishing, 2011, pp. 247, 259.

9 As 'public concerned' are defined those natural persons that have either a sufficient interest or claim an impairment of a right, where national law requires this as a precondition, while NGOs that meet the requirements of national law are deemed to have either a sufficient interest or a right capable of being impaired (art. 2 para. 5).

10 See M. Hedemann-Robinson, EU Implementation of the Aarhus Convention's Third Pillar: Back to the Future over Access to Justice? Part 1, *European Energy and Environmental Law Review* 23 (2014), pp. 102,105.

11 See B. Dette, Access to Justice in Environmental Matters: A Fundamental Democratic Right, in M. Onida (ed.), *Europe and the Environment*, Europa Law Publishing, 2004, pp. 1, 7; Findings and Recommendations of the Aarhus Convention Compliance Committee (hereinafter ACCC), with regard to Communication ACCC/C/2005/11 concerning Compliance by Belgium, paras 34–6.

12 Directive 2003/4 on public access to environmental information and repealing Directive 90/313, OJ 2003, L41/26.

13 Directive 2003/35 on public participation in respect of the drawing of certain plans or programmes relating to the environment, OJ 2003, L/156/1.

14 Art. 6 para. 2 of Directive 2003/04 establishes clear minimum standards that must underpin the national review procedures in cases of refusal or inadequate handling of requests for access to environmental information (Ebbesson, *supra* note 8, pp. 253–6). Furthermore, the Public Participation Directive (2003/35) inserted new provisions as regards access to justice for the 'public concerned' both into the Environmental Impact Assessment Directive (art. 11 of Directive 2011/92, as amended by Directive 2014/52) and in the previous Integrated Pollution and Prevention Directive (IPPC), which is incorporated in the Industrial Emissions Directive (IED) (art. 25 of the 2010/75 Directive). According to these provisions, MS must offer to the 'public concerned' access mainly to judicial review procedures to challenge the procedural and substantive legality of decisions, acts and omissions subject to public participation provisions (Ebbesson, *supra* note 8, pp. 256–62). Finally, an access to justice provision with similar content to those of the relevant provisions of the EIA and the IED Directive respectively was inserted in the SEVESO III Directive on industrial hazards (art. 23 of the 2012/18 Directive).

15 Regulation 1367/2006, OJ 2006, L264/13. See also N. de Sadeleerand and C. Ponchelet, Protection Against Acts Harmful to Human Health and the Environment Adopted by the EU Institutions, *The Cambridge Yearbook of European Legal Studies 2011–2012*, pp. 177, 196ff., putting forward well-documented arguments as regards the incompatibility of the Regulation with the access to justice standards of the Aarhus Convention

16 EU Commission, Proposal for a Directive of the European Parliament and of the Council on Access to Justice in Environmental Matters, COM (2003), 624 final. For the basic contours of the Draft Directive see Hedemann-Robinson, *supra* note 10, p. 107.

17 Withdrawal of Obsolete Commission Proposals, OJ 2014, C-153/03.

18 See Ponchelet, *supra* note 3, p. 291. It is worth noting that the 2004/35 Directive on Environmental Liability includes certain provisions that ensure access to judicial review procedures for the public and environmental NGOs, defined in a similar way as the 'public concerned', in order to challenge the competent authority's conduct

concerning remediation of possible environmental damage (art. 12, 13), covering a field that falls within the scope of art. 9 para. 3 of the Aarhus Convention.

19 In the *Janecek* Case the Court did not base itself on the Aarhus Convention, but on the objective of the Air Quality Directive to protect public health, in order to recognize standing for persons that might be affected by the wrong application of EU law (CJEU C-237/07, 2008, I.6221, paras 38–9).

20 CJEU C-77/08, *Christopher Mellor*, 2009, I. 03709, paras 57, 58 and 59.

21 In the *Djürgården* Case the Court ruled that the minimum number of members that an environmental NGO must have in order to be classified as 'public concerned' in accordance with the national law, cannot be fixed at such a level that it runs counter to the objective of the EIA Directive to give the 'public concerned' wide access to justice (CJEU, C-263/08, *Djürgården*, 2009, ECR I-09967, paras 42–52). Moreover, in the *Trianel* Case the Court ruled that access to courts has to be granted even where national procedural law does not permit to rely on rules aiming to protect only the interests of the general public and not the interests of the individuals, so that NGOs have standing rights of their own to challenge decisions according to national rules implementing EU legislation (CJEU, C-115/09, *Trianel*, 2011, I-3673, paras 45, 46–48, 50). Subsequently, the Court held that the German restrictive system of access to courts based on the so-called 'Schutznormtheorie', is incompatible with the relevant access to justice provisions of the Aarhus Convention and the EIA Directive respectively. See A. Schwerdfeger, Schutznormentheorie and Aarhus Convention-Consequences for the German Law, *JEEPL* 4 (2007), pp. 270ff.

22 CJEU C-260/11, *Edwards*, Judgement of 11 April 2013, where the Court ruled that the requirement that litigation in environmental matters should not be 'prohibitively expensive' means that persons affected by a decision falling within the scope of the EIA or the IPPC Directive should not be prevented from seeking a review by the courts by reason of the financial burden that might arise as a result (para. 35).

23 In the *Krizan* Case the Court ruled that despite the silence of the relevant provision of the IPPC Directive (art. 15a) on the issue of *interim relief*, the effective exercise of the rights conferred under this provision presupposes that the 'public concerned' also has the right to request interim measures by the national courts (CJEU C-416/10, *Krizan*, Judgement of 15 January 2013, para 109).

24 In the *Alptrip* Case, the Court ruled that national law cannot limit access to courts only in cases where no EIA was undertaken, but instead members of the 'public concerned' must be able to invoke any procedural defect in support of an action challenging a decision under the EIA Directive. (CJEU C-72/12, *Alptrip*, Judgement of 7 November 2013, paras 48, 53.)

25 CJEU C-240/09, *Slovac Brown Bears*, 2011, I-1255.

26 In contrast to the *Merck Genricos* Ruling (CJEU, C-431/05, 2007, I-7001), the Court justified the competence to rule on the direct effect of art. 9 para. 3 on the grounds that it falls within the scope of EU law, as it relates to a field (protection of Brown Bear) covered in large measure by it (para. 40). See the relevant critique for making obsolete the Declaration of Competence signed by EU (J. Jans and H. Vedders, *European Environmental Law*, 4th edn, Europa Law Publishing, 2012, p. 74) and for 'stepping into legislator shoes' (Opinion of the AG Sharpston, Case 240/09, para. 70).

27 See C. Eckes, Environmental Policy 'Outside In': How the EU's Engagement with International Environmental Law Curtails National Autonomy, *GLJ* 13(11) (2012), pp. 1151, 1167–8.

28 See Ponchelet, *supra* note 3, p. 294.

29 The principle of procedural autonomy relates to the freedom of MS to introduce the appropriate institutional and procedural arrangements, including the access to justice provisions, which are necessary for the implementation of EU law (CJEU Judgment C-33/76, *Rewe*, 1976, I-1523). The application of the principle is subject to two significant limitations, the non-discrimination principle and the principle of

effectiveness. See U. Galetta, *Procedural Autonomy of EU Member States: Paradise Lost?*, Springer Verlag, 2010, pp. 7ff.

30 For an overview of the relevant case law and its full reception by the national courts see J.-F. Brakeland, Access to Justice in Environmental Matters – Developments at EU Level, paper presented at the Towards an Effective Guarantee of Green Access conference held at Osaka University, March 2013, available at http://greenaccess.law. osaka-u.ac.jp/wp-content/uploads/2013/04/10pjp_brakeland.pdf.

31 See Opinion of the AG Kokott in *Edwards* Case (*supra* note 22), paras 36, 43. The Findings of the ACCC are non-binding, but they can be considered valid sources of interpretation of the Convention from the perspective of international law after their endorsement by the subsequent Meeting of the Contracting Parties (Hedemann-Robinson, *supra* note 10, p. 109).

32 Decision 1386/2013 on a General Union Environment Action Programme to 2020, OJ 2013 L354/190–191.

33 See German Advisory Council on the Environment, *Access to Justice in Environmental Matters: The Crucial Role of Legal Standing for Non-Governmental Organizations*, February 2005, pp. 7–8; Hedemann-Robinson, *supra* note 10, p. 106.

34 Opinion of the AG Sharpston in the *Trianel* Case (*supra* note 21), para. 51 and CJEEU Judgment in Case *Edwards* (*supra* note 22, para. 32) emphasizing the role of the NGOs and the public in environmental protection.

35 See N. de Sadeleer, G. Roller and M. Dross (eds), *Access to Justice in Environmental Matters and the Role of NGOs: Empirical Findings and Legal Appraisal*, Europa Law Publishing, 2005, p. 168; Hedemann-Robinson, *supra* note 10, p. 106.

36 The need for an EU legislative intervention is intensified due to the emerging regulatory trend for the authorization of large-scale projects either at the higher level of administrative hierarchy or by specific legislative act at the national level, so that the public participation and access to justice rights of those affected can be significantly restricted. See J. Därpö, *Effective Justice? Synthesis of the Report of the Study on the Implementation of Articles 9.3 and 9.4 of the Aarhus Convention in the Member States of the European Union*, 11 October 2013 final, p. 10.

37 See Findings and Recommendations of the ACCC with regard to Communication ACCC/C/2005/11 concerning compliance by Belgium and Findings and Recommendations of the ACCC regarding Communication ACCC/C/2006/18 concerning compliance by Denmark relating both to cases of non-compliance with art. 9 para. 3.

38 These barriers were identified in a series of national reports that relate to the implementation of art. 9 paras 3 and 4 of the Aarhus Convention (http://ec.europa. eu/environment/aarhus/access_studies.htm). The main findings of the national reports were included in the synthesis report entrusted to Professor Daprö, President of the Access to Justice Task Force of the Aarhus Convention (Därpö, *supra* note 36).

39 See N. de Sadeleer, Enforcing EUCHR Principles and Fundamental Rights in Environmental Cases, *Nordic Journal of International Law* 81 (2012), pp. 39, 58. The harmonization of access to justice requirements can also have a preventive effect, as it will force both public authorities into more consistent enforcement of environmental law and operators into increasing compliance with the relevant regulations (German Advisory Council on the Environment, *supra* note 33, p. 7).

40 Professor Därpö argued in the synthesis report that the development of the CJEU jurisprudence through preliminary rulings and national adaptation is too uncertain and slow to be considered as an alternative to legislative intervention, while also the reliance on the initiation of infringement procedures for bringing national provisions for access to justice in line with the CJEU case law is too ineffective, time-consuming and piecemeal. See Därpö, *supra* note 36, p. 25. Furthermore, one of the significant results of the study that related to the socio-economic implications of widening access to justice in environmental matters and was elaborated under the Commission's

assignment, was that the adoption of a Directive is the only viable choice for providing legal certainty. See Faure *et al.*, Final Report, *Possible Initiatives on Access to Justice in Environmental Matters and Their Socio-economic Implications*, 9 January 2013 final, pp. 68ff.

41　See Jans and Vedder, *supra* note 26, p. 67.
42　See N. de Sadeleer, Principle of Subsidiarity and the EU Environmental Policy, *JEEPL* 9 (2012), p. 63ff.
43　See A. Epiney, *Umweltrecht der Europäischen Union*, 3rd edn, Nomos Verlag, 2013, pp. 34ff., underlining that the notion of environment in EU law is very wide.
44　*Ibid.*, p. 99 (Rdnr. 4); Calliess in: Calliess/Ruffert, EUV/AEUV, 4. Auflage, 2011, Art. 192, Rdnr. 2.
45　See I. Pernice and V.Rodenhoff, Die Gemeinschaftskompetenz für eine Richtlinie über den Zugang zu Gerichten im Umweltangelegenheiten, *ZUR*, 2004, pp. 149, 150; German Advisory Council on the Environment, *supra* note 33, p. 14ff; Epiney, *supra* note 43, p. 99 (Rdnr. 4).
46　See A. Schwerdtfeger, *Der deutsche Verwaltungsrechtschutz unter dem Einfluss der Aarhus Konvention*, Tübingen 2010, pp. 298ff.; Epiney, *supra* note 43, p. 99 (Rdnr. 4). For the opposite opinion see T. von Danwitz, Aarhus-Konvention: Umweltinformation, Öffentlichkeitsbeteiligung, Zugang zu den Gerichten, *NVwZ*, 2004, pp. 272, 277f.
47　See Epiney, *supra* note 43, p. 99 (Rdnr. 4). The opposite view is defended by Pernice and Rodenhoff, *supra* note 45, p. 150.
48　See A. Epiney, Verbandsklage und Gemeinschaftsrecht, *NVwZ*, 1999, pp. 485, 491ff.; German Advisory Council on the Environment, *supra* note 33, p. 15.
49　The position expressed by the Commission in the Roadmap Document on an Initiative on Access to Justice in Environmental Matters is that such an initiative is consistent with the principle of subsidiarity, because it satisfies both the negative criterion for a variety of reasons and the positive criterion, as it has an added value in terms of achieving harmonization (see http://ec.europa.eu/smart-regulation/impact/planned_ia/docs/2013_env_013_access_to_justice_en.pdf).
50　See Pernice and Rodenhoff, *supra* note 45, p. 151; German Advisory Council on the Environment, *supra* note 33, pp. 15–16.
51　The relevant studies (Darpo's Synthesis Report, *supra* note 36, and Faure *et al.*, *supra* note 40), constituted the basis for a stakeholder consultation that was completed in September 2013. The Roadmap Document that was produced for the purposes of an impact assessment review (*supra* note 49) has not yet been approved by the Impact Assessment Board of the European Commission. Such an approval constitutes a prerequisite for the Commission, in order to come forward with a new legislative proposal.

12 Unconventional gas mining – what a fracking story!

Policy, regulation, law and trade agreements

Janice Gray

Introduction

There is a widespread global uptake of unconventional gas mining activities and associated technologies. Simultaneously, unconventional gas mining has also attracted much media coverage; government, legal and regulatory agency attention; social commentary; and academic scholarship in recent years. Reporting, discussion and analysis of unconventional gas mining has tended to focus on four main issues – those of the environment, human health, property rights and the economy – with the role of science underpinning much of the debate. This chapter will provide an update on selected issues, in particular, the effectiveness of legal and regulatory tools in the unconventional gas space such as those relating to approvals and monitoring and the nexus between trade law and unconventional gas governance as exemplified by the Trans-Pacific Partnership Agreement (TPP). This chapter begins with an explanation of what constitutes unconventional natural resources before moving to a discussion of the Australian experience of one unconventional natural resource, that of coal seam gas (CSG), and the associated technique of hydraulic fracturing (fracking).

The chapter argues that:

*   CSG mining is not the energy panacea it is mooted to be;
*   even if the Australian government pursues the exploitation of CSG as a transitional energy on the path towards renewable energy, there are regulatory gaps that need plugging; and
*   it may be difficult to plug those gaps and adhere to ESD principles if Australia becomes a signatory to the TPP.

What are unconventional natural gas resources?

Unconventional natural resources are natural resources which require unconventional means to extract them. 'Unconventional' in this context simply means greater than industry-level standards of technology are needed to exploit the resource. Unconventional gas includes coal seam gas (CSG),[1] shale gas and tight gas[2] and like conventional gas it is largely composed of methane. When unconventional and conventional gases are frozen and converted into liquid form, they are known as liquefied natural gas (LNG). In some jurisdictions, such as Alberta, Canada, the unconventional resource activity has been in the exploitation of another hydrocarbon, that of oil, and in particular oil that is recovered from tar sands. In the United States, the focus has mainly been on the exploitation of shale gas. In Australia, however, CSG has, to date, been the main hydrocarbon that has been exploited unconventionally (GeoScience Australia undated b).

Hydraulic fracturing (fracking) is a technique that helps release unconventional gas from where it is held underground. It involves drilling down and injecting water, sand and chemicals into the earth's surface at high pressure (Gray 2014).

The Australian experience

While Australia has a long history of mining, the scale of gas (and mineral) exploitation has seen unprecedented growth in the last decade or so. The scale is changing the landscape visually and affecting food production, water security and communities throughout the country (ANEDO 2013: 1). In Queensland, CSG gas mining approvals for both exploration and production are numerous, and extensive infrastructure will support a range of processing plants including ones being constructed on Curtis Island in the environmentally sensitive Great Barrier Reef World Heritage Area (*ibid.*). A key driver for such expansion seems to date back to a Queensland government decision in 2000 which required 13 per cent of all power supplied to the state electricity grid to be generated by gas by 2005. That requirement was later increased to 15 per cent by 2010 and 18 per cent by 2020 (GeoScience Australia undated a). However, in some other Australian states the attitude, although broadly favourable, has been somewhat tempered. In Victoria, for example, a moratorium on hydraulic fracturing was introduced in August 2012 and will remain in place until at least July 2015.

An even more robustly oppositional approach appears to be underway in Nova Scotia, Canada, where there are plans to introduce legislation prohibiting high volume fracking for onshore shale gas (Service Nova Scotia 2014). Meanwhile, in New South Wales (NSW), the key focus of this chapter, there is also some evidence of the government beginning to 'hasten slowly', having:

- imposed a temporary moratorium on CSG mining in drinking water catchments in 2013;
- extended a freeze on new CSG Petroleum Exploration Licence (PEL) applications until September 2015; and
- in 2014, cancelled three CSG exploration licences originally granted in 2009 and renewed in 2013 (ABC News 2014; Higginson 2014).

Some of these decisions are arguably in response to widespread, vocal community concern and resistance (Gray 2014).

Yet, despite there being some recent examples of government reticence in relation to the uptake of CSG mining in NSW, numerous PELs have already been issued and exploration has begun in the Sydney, Gunnedah, Clarence-Moreton and Gloucester basins; basins which include the Pilliga State Forest, Barrington Tops and the highly populated city of Sydney (GeoScience Australia undated a). Production, although less well developed in NSW than exploration, has, neverthe-less, been in progress at Camden (near Sydney) for some years (NSWOCSG undated a; McDonald-Smith 2014; NSW Department of Trade and Investment undated a).

The gas (both conventional and unconventional) supply chain in Australia is divided into a number of separate stages but broadly speaking there is an upstream supply chain and a downstream supply chain. 'Upstream' refers to exploration, development and production while 'downstream' refers to processing, distribution, storage, wholesale and retail. The upstream sector has been dominated by well-known companies: in 2011, six major producers (Santos, BHP Billiton, ExxonMobil, Origin Energy, Woodside and Apache Energy) met 65 per cent of domestic gas demand; in 2012, six companies (Chevron, Shell, ExxonMobil, BG, Inpex and Woodside) had the largest shares of 2P gas reserves in Australia and together they held 61 per cent of the total (85,120 petajoules). In 2012, the six companies with the largest shares in 2P gas reserves in the eastern market (BG, Origin, ConocoPhillips, Santos, PetroChina and Shell) together held 66 per cent of the total (48,858 petajoules), and in 2012, in relation to NSW, just four companies held 99.4 per cent of total 2P gas reserves in that state (2,824 petajoules). They were Santos, AGL, Metgasco and EnergyAustralia (Haylen and Montoya 2013: paras 2.3, 2.4 and 2.4.3).[3]

Some of the companies above (including Metgasco and Santos) are involved in CSG mining and they, together with the government, have attracted criticism from resistors and others, who demand that the potential impacts of CSG mining and fracking on the environment; water quality and security; agriculture; land use and food production, as well as on Australian communities, receive greater attention. Thus calls for more effective laws and regulation have been made (Owens 2012; NWC 2012; Climate Commission undated: 2).

In response to public concerns and as a matter of administrative pragmatism, the NSW Office of Coal Seam Gas (NSWOCSG) was set up in 2013 to administer CSG licensing – a sign that CSG mining is very much on the agenda. That office has portrayed CSG mining as a clean and safe industry, actively differentia-ting it from shale gas and shale gas fracking (NSWOCSG). But given that large shale gas deposits have also been discovered in Australia; gas prices are rising; improved extraction techniques for shale gas have been developed; and the likelihood of industry wishing to exploit shale gas deposits is high, it may not be possible for the NSW government to distance itself from the negative press associated with shale gas mining indefinitely. One company, Santos, has already begun producing shale gas in the Cooper Basin, having started in 2012 (AER 2013: 86).

The pattern in NSW in relation to CSG and fracking is one of mixed messages. On one hand, the NSW government has taken some significant steps to slow the uptake of CSG mining through important policy decisions (bans and moratoria) but on the other, it has highlighted in Departmental information a range of positive features associated with CSG mining, suggesting it is striving to balance competing views.

The NSWOCSG's website information serves to highlight some of the underlying complexities in CSG discourse in NSW. There are few clear-cut party political positions in this space either, and, as discussed elsewhere, the forces of opposition are disparate, with unlikely alliances having been forged. Farmers and environmentalists, for example, have joined forces in calling for legislative and regulatory change. In response, government and other political parties have offered splintered, inconsistent and arguably half-baked responses. Indeed, at times, the very same government or political party has both embraced and rejected CSG mining, exposing the difficulties that politicians face in being able to understand and tap into the diverse views of their constituents, making coherent and robust legislation and regulation difficult to develop (Gray 2014).

Constituents have, for example, diverse views about the imposition of bans and restrictions on the uptake of CSG mining. Some critics focus on their temporary nature, arguing that, in reality, they have changed little (Gorrey 2013; Stop CSG Sydney undated) and were introduced in an attempt to quell vocal resistance, and to create a more conducive atmosphere for the later approval of CSG mining activities. Others argue that resistance to CSG activities manifests itself in legal and regulatory changes that have serious, positive and ongoing impacts on the industry and society more generally.

The dominant and competing narratives and the legal process

In this context a dual, dominant narrative has emerged. It is that unconventional gas exploitation (particularly CSG) and fracking are needed for economic stability, and there is a looming domestic energy crisis which can only be satisfactorily addressed by the uptake of unconventional gas mining and fracking.

The two prongs of the dominant narrative have helped facilitate the uptake of unconventional gas mining and fracking and seen CSG companies enter the market and consequently engage with relevant law and regulation (Grudnoff 2014). They have, for example, applied for Petroleum Exploration Licences (PELs) under legislation (Petroleum (Onshore) Act 1991 (P(O)A)). In this context law and regulation have evolved, responding to and accommodating the dominant narrative. However, some claim that reform is still needed in relation to the CSG-related fields of biodiversity conservation, GHG emissions, air quality, water resources, chemical use and access to land still arguably needs improvement (ANEDO 2013: 3). The degradation of groundwater systems; the lack of a safe system to dispose of water and salt produced from CSG activities; cumulative gas emissions from wells and pipelines; health impacts from air, water and soil pollution; the diminution of agricultural land; and seismic activity associated with fracking and aquifer re-injection all present potential risks requiring effective governance.

New policies and codes of practice, such as those relating to hydraulic fracturing and aquifer interference, have been developed to support the legal regime in

NSW (NSW Department of Primary Industries undated). But still weaknesses in the legal and regulatory framework have emerged. They include: the standard of environmental inquiries undertaken before CSG drilling and other works are approved; the effectiveness of environmental monitoring, particularly regarding cumulate effects, seismic activity and aquifer interferences; and the inclusion and effectiveness of public participation. Others relate to land use, property right and health issues (Williams *et al.* 2012; Grudnoff 2014).

The exposure of weaknesses has given rise to a competing narrative in which the over-arching, purported economic benefits of CSG mining and fracking have been challenged and concerns have been raised as to the manner in which localized (as opposed to meta) economies and industries including tourism, hospitality and housing may suffer as unconventional gas exploitation takes off. The emerging counter narrative also emphasizes health concerns associated with unconventional gas mining. It is a narrative that sounds in resistance and has found voice in grass roots level oppositional groups (Stop CSG Sydney undated; see also www.lockthegate.org.au).

The following section considers and analyses aspects of two selected weaknesses or areas of concern: environmental protection, approvals and monitoring; and the economy and trade in relation to the Trans Pacific Partnership and its connection with the environment, particularly CSG mining and fracking.

Weaknesses in the legal and regulatory regimes

Environmental protection – mechanisms and responses

One perceived weakness of the present legal and regulatory regime is its ability to offer an effective system for the granting of CSG approvals. Approvals granted for exploration subject to a Review of Environmental Factors (REF) rather than an Environmental Impact Statement (EIS) under the Environmental Planning and Assessment Act 1979 (EPAA) have been criticized as being too easy to acquire. (AGL's four pilot/exploration wells were subject to REFs).

A REF involves an in-camera review and addresses issues including the potential impacts of the proposal on the environment, water resources and the community but it is less onerous and less comprehensive than a fully-fledged EIS, which is a merit-based review that considers air quality, noise, transport, flora and fauna, surface and ground-water management, methods of mining or petroleum production, landscape management and rehabilitation (EPAA §§24, 26).

The repeal of Pt 4A and incorporation of Pt 5 (EPAA) have brought about some improvements but the changes do not apply retrospectively and exploration activities approved under the earlier, less demanding provisions remain valid. Yet drilling involved in pilot/exploration projects may be just as harmful as that involved in production, which is subject to more a rigorous (EIS) assessment. Hence the approval of pilot projects remains problematic (*Fullerton Cove*).

However, even the use of an EIS presents problems. In theory, an EIS involves extensive public consultation and consideration of community submissions and expert reports before a decision is made. Whether this occurs in practice and whether stringent regulatory goals are met is contestable, as whistle-blower information reveals (Eastley 2013).

AGL–Waukivory

Further, reports claiming that AGL's Waukivory project in NSW (which includes fracking) was approved two months before the NSWOCSG received the relevant laboratory tests confirming the nature of fracking chemicals has undermined confidence in the CSG approvals process (Hasham 2014). Even if the NSW government did not technically breach its obligations in granting an early approval (because only identification of, not the testing of, chemicals was required) the allegation potentially undermines public confidence in the approvals process and deters public participation because it relates to a sensitive issue. The identification of proppants[4] and fracking fluids has been controversial because mining companies have commonly closely guarded the nature of chemicals and other materials used as a trade secret, making it difficult for litigants relying on medical torts to establish the requisite levels of causation.

Santos case

If a project is approved, it will be conditioned to minimize potential negative environmental impacts, including in the case of all PELs issued under the P(O)A, rehabilitation and environmental performance conditions (P(O)A s23). Whether the conditions imposed achieve the desired effect is debatable, as the Santos experience at Pilliga indicates.

In 2014 Santos was fined $1500 by the Environmental Protection Agency because a CSG project it had taken over contaminated an aquifer with uranium, lead, aluminium, arsenic and other chemicals (EPA 2014). The chemicals reached the aquifer because of a leaking storage pond (Nicholls 2014). The Santos case serves to demonstrate that conditions do not stop accidents. It also shows that potentially serious harm may result in relatively weak sanctions, raising the effectiveness of those sanctions as regulatory tools.

Leichhardt Resources

In 2014, a breach of compliance conditions caused the cancellation of three of Leichhardt Resources, PELS. The breaches were:

- a failure to consult (which was part of the approval conditions);
- lack of technical and financial qualifications to hold a petroleum exploration licence;
- failure to conduct the required minimum work program, including groundwater studies;

- failure to conduct community consultation and report on this to the NSW government;
- failure to comply with reporting obligations; including lodging an annual exploration report; and
- failure to inform the Moree Local Aboriginal Land Council of the licence renewal (Higginson 2014).

That these breaches occurred undermines the integrity of the regulatory system.

It is notable that the cancellations come shortly after the NSW Chief Scientist identified major weaknesses in the regulatory system for CSG mining, including the failure by some regulators to check that mandatory reports and data had been delivered. She recommended reforms including the introduction of one Act for onshore subsurface resources except water (NSW Chief Scientist 2014).

As these examples demonstrate, while there has been a strong uptake of CSG activities in NSW, many regulatory gaps exist. Whether they can be easily plugged either in the manner recommended by the Chief Scientist or in other ways is problematic in light of the Trans-Pacific Partnership, which may act as a blocker on the road to legal and regulatory reform in the unconventional gas space.

The Trans-Pacific Partnership

The Trans-Pacific Partnership (TPP) is a free trade agreement concerned with developing strong economies by smoothing the path for investment and encouraging conditions conducive to trading and production. It is also arguably an obstacle on the path towards a more sustainable future.

The agreement is being negotiated between 12 parties: Australia, Brunei, Canada, Chile, Japan, Malaysia, Mexico, New Zealand, Peru, Singapore, the United States and Vietnam (DFAT undated: 3). The first round of negotiations was held in 2010 and by 2014 there had been 19 more rounds, but no final agreement had been reached.

The proposed agreement covers issues such as trade in goods, trade remedies, market access, investment, intellectual property and, very importantly, for the purposes of this chapter, the environment. However, it is extremely difficult to analyse the TPP because the agreement, like most other trade agreements, is being negotiated in secret. The rationale for this clandestine approach is that it facilitates more effective negotiation by safeguarding negotiating positions and strategies, some of which are likely to deal with sensitive issues of national interest, particularly in relation to markets and trade. But such secrecy has elicited heavy criticism from those (particularly consumer groups) who claim it is an undemocratic process which excludes public consultation in relation to the development of commitments that will bind Australian constituents for long into the future (Wade 2014). In that sense the negotiating process surrounding the TPP reflects a cognitive dissonance with the notion of parliamentary democracy, a notion which encourages participation. Elected representatives in

Australian state and Commonwealth parliaments have been denied access to the draft TPP documents. In the United States the situation is similar, but in that jurisdiction, 600 (unelected) corporate advisors have had access to the draft text. The secret nature of negotiations has also led to civil society protests in the United States with a grand coalition of civil society organizations, the Sierra Club, speaking out against the secrecy and the leaked content of the TPP (Sierra Club undated).

In Australia, although the Federal government has claimed that there has been a lot of consultation over a range of industry sectors, 'consumer groups [have maintained] that they have been excluded from any meaningful dialogue' (Wade 2014). Given that one key principle of ecologically sustainable development (ESD) is public participation, it would seem that exclusion from, or at least the inadequacy of, public consultation and participation in negotiations on the Environment Chapter reflects at best, government inconsistency and at worst, hypocrisy. How can ESD principles be embraced in both policy and legislation (such as in §3 of the Water Management Act 2000 (NSW)) but be so blatantly flouted in negotiating a key agreement dealing with the environment? Further, given that cyberspace developments such as Twitter, Instagram and Facebook, which rely on and cultivate a culture of sharing, are so embedded in modern life, the high dependence on secrecy and confidentiality in the context of the TPP negotiations is arguably an anachronism; an anachronism which serves to benefit corporations more readily than the wider public and the environment. Indeed, Kelsey argues that the environment is not well supported by the TPP. The other (non-Environment) chapters of the TPP 'subordinate the environment, natural resources and indigenous rights to commercial objectives and business interests. [Accordingly] the corporate agenda wins both ways' (Kelsey 2014). The environment loses out because it is not privileged in other chapters and weak obligations are created for its protection in its own chapter (discussed below).

Despite the intended secrecy, the full draft of the Environment Chapter became available, when leaked by Wikileaks in January 2014 (TPP Environment undated). What was revealed, in terms of environmental compliance, was a 'toothless tiger'. Although the Chapter deals with conservation, environment, biodiversity, indigenous knowledge and resources, over-fishing, and climate change – all areas which are potentially linked to CSG activities and fracking – the TPP reveals weak obligations that can be avoided because of inadequate compliance mechanisms (Kelsey 2014). Avoiding environmental obligations, especially weak obligations, in relation to unconventional gas mining may potentially have serious negative impacts on the environment, particularly in relation to accidents, cumulative effects, seismic activities and aquifer interferences – all CSG-related areas of interest identified by the Chief Scientist (NSW Chief Scientist 2014).

The draft Environment Chapter appears wanting in several respects, including the fact that its specific scope remains ill-defined. It mentions environmental laws, policies, practices and proceedings, for example, but only defines 'environmental laws' narrowly (a) by way of Article SS.33: General Commitments where it refers to protection of the environment and (b) by way of clauses which prevent danger to human life and health where the purpose of the law concerns pollutants or environmental contaminants, for example.

Further, while Article SS.2, covering the objectives of the TPP, refers to 'enhanced co-operation to protect and conserve the environment and sustainably manage [signatory countries'] natural resources bring[ing] benefits which can contribute to sustainable development, strengthen their environmental governance and complement the objectives of the TPP', there is some concern that this objective may not be readily fulfilled. For example, 'cumulation' encourages global supply chains based on member states contributing different elements to the production of goods from their different geographic locations (DFAT undated: 3). This may result in effectiveness of supply and production but the transportation and re-transportation of component elements may also increase travel miles, GHG emissions and the carbon footprint.

It may also be difficult to strengthen CSG and fracking governance, for example, if investor-state dispute settlement (ISDS) provisions are ultimately (as some supranational corporations would like) included in the agreement (Faunce 2012).

ISDS provisions are common in international trade agreements (Bates 2009), but have not been proven to enhance economic benefit (Australian Productivity Commission 2010). They were originally introduced to help guard against corporations having their investments unfairly expropriated by governments but because ISDS provisions commonly permit corporations to sue national governments on the basis that national (and/or state) laws or policies have deleterious effects on investments, the ISDS provisions also undermine national sovereignty. They effectively allow foreign companies to override the application of Australian law and simultaneously permit corporations to seek damages from the Australian government directly in the International Centre for the Settlement of Investment Disputes for restrictions on trade (ICSID undated).

Under ISDS provisions, a government may arguably be curtailed in its regulation of CSG activities and fracking for fear that it could be sued by a supranational corporation, many of which operate in the CSG space (see above). Legislation and policies such as the Aquifer Interference Policy (NSW Department of Primary Industries undated), the Code of Practice for Coal Seam Gas-Fracture Stimulation Activities (NSW Department of Trade and Investment undated b), the Environmental Planning and Assessment Act 1979 (NSW) and the Petroleum (Onshore) Act 1991 (NSW) as well as future Acts or policies, could potentially be the subject of disputes resolved under ISDS provisions. Resolution of such disputes may involve long and expensive proceedings resulting in significant liability for damages for national governments, such as the Australian government. The availability of ISDS mechanisms may act as a deterrent to more effective regulation of CSG and fracking (Richard Denniss, quoted in Australia Institute 2013) and governments may be marginalized, with power for significant environmental decisions effectively shifting to corporations. In other cases, governments which have close affiliations with corporations may join together with those corporations in an attempt to use the ISDS provisions to 'defeat' the legislation of another, different national government.

Discussion of the ISDS provisions also serves to demonstrate how it is not merely the Environment Chapter of the TPP which needs to be assessed and reviewed carefully. The ISDS provisions (above) are not housed in the draft Environment Chapter. They are, instead, part of the draft Investment Chapter (TPP Investment 2012). Hence the non-Environment Chapters may also potentially impact on unconventional gas and fracking governance, indicating that a holistic and thorough assessment of the TPP is needed to reveal its potential impact on environmental governance. Piecemeal analyses of separately leaked, disparate Chapters may not necessarily reveal how the TPP could impact on ecological sustainability. Other member state governments who become signatories to the TPP, particularly those seeking to govern unconventional gas, unconventional oil (such as tar sands) and fracking should perhaps also be mindful of the inter-connection between the Investment Chapter and environmental governance. They, too, will be subject to it.

Concerns about the effectiveness and suitability of the ISDS provisions have been expressed by the public, member states and by the European Union Trade Commissioner, Karel De Gucht, who acknowledged 'deep dissatisfaction' with the Commission about the secrecy of ISDS adjudication and the 'vulnerability of the system to corporate abuse' (Seccombe 2014). However, in the face of growing international opposition to ISDS provisions (particularly from Germany and previously from the Rudd and Gillard governments in Australia) the Abbot government (in Australia) has not ruled out the inclusion of ISDS provisions in the TPP, stating that it would consider them on a 'case by case' basis. To many this is a puzzling position given that the Australian government has had some experience with ISDS provisions already, albeit in a different context, that of tobacco.

In *JT International SA v Australian Commonwealth* [2012] HCA 43 (5 October 2012), the Philip Morris tobacco company initially challenged the constitutionality of the Tobacco Plain Packaging Act 2011 (Cth), a statute which requires the introduction of plain-packaged cigarettes. The Australian statute which was enacted in compliance with obligations under the World Health Organization's Framework Convention on Tobacco Control was designed to deter smokers from purchasing cigarettes and consequently help address health issues affecting individuals and society more generally. The Philip Morris constitutional challenge was unsuccessful. The Court referred to Philip Morris's proposition that its commercial and economic position gave rise to a distinct proprietary interest as being based on 'delusive exactness' (*JT International SA v Australia*, paras 47 and 48).

Meanwhile, Philip Morris rearranged its assets to become a Hong Kong investor, allowing it to rely on the ISDS provisions in the Australia–Hong Kong Bilateral Investment Treaty (BIT). Under this treaty, the dispute is to be resolved by a panel of three arbitrators ruling on whether Australia is liable to pay Philip Morris damages for passing legislation that has already been judged to be constitutional by the Australian High Court. Whether the arbitrators' decision will be in Australia's best interest is debatable. The international investment lawyers comprising the panel are arguably more likely to focus their decision on how the law affects investment rather than on whether it is in the public interest.

This raises the question of whether private or public interests should be privileged; a question commonly considered by courts. But in the case of courts the issue is resolved in a transparent manner using the rules of evidence. Further, unlike legally constituted courts, member states have no recourse to appeals from arbitrator decisions. Additionally, because arbitrators may also be advocates in other proceedings, bias, real or perceived, may possibly exist.

The Phillip Morris case represents an attempt by a supra/multinational corporation to override Federal legislation and assert the primacy of trade and investment interests. That a corporation can do this is cause for concern (Patricia Ranald, convenor of the Australian Fair Trade and Investment Network, quoted in Australia Institute 2013). It highlights how hollowed out the state has become and reinforces the view that real power to govern increasingly rests in the hands of (unelected) corporations rather than elected governments, which serves to undermine the rule of law. Such a realization has prompted calls for the reassertion of the state within the body politic (Sierra Club undated; Westra 2013). Given that the environment has historically been difficult to protect under the law (Bonyhady 2012), the fact that law's role in regulating fracking and unconventional gas could be further eroded by ISDS provisions if included in the TPP is a major cause for concern.

In Canada, but under a different trade treaty, the North Atlantic Free Trade Agreement (NAFTA), this issue has come to the fore in the unconventional resources and fracking space. Provincial law effectively stands to be overridden, exposing the familiar tension between economic and environmental objectives. That trade treaties may deter governments from passing and maintaining laws which are designed to strengthen environmental protections may be seen in the *Lone Pine Resources* case, which involves arbitration of a dispute concerning the Quebec government's Bill 18, for the suspension of all oil and gas exploration including fracking and the revocation (without compensation) of Lone Pine's permits to mine for oil and gas under the St Lawrence River. The matter was initiated by the investor, Lone Pine, in accordance with Article 3 of the arbitration rules of the United Nations Commission on International Trade law of 1976 (UNICTRAL Arbitration Rules) and Articles 1117 and 1120 of NAFTA as an alleged breach of NAFTA's Articles 1110 and 1105. It involves a $250 million lawsuit (NAFTA Notice of Arbitration). At the time of writing, the matter remained unresolved.

Bearing in mind the *Lone Pine* case, it is interesting to observe that in NSW (Australia), three CSG PELS (exploration licences) were revoked by the relevant Minister in 2014 (mentioned above). Should trade treaties to which Australia is already a signatory be breached by such a revocation, the Australian government may find itself vulnerable to expensive lawsuits regarding the impostions of restrictions on fracking and gas exploration as has the Canadian government. Legislative responses to environmental concerns such as those related to fracking may be met with a roadblock in the form of the ISDS provisions. That supra/multinational corporations are not shy about using provisions such as the ISDS provisions may also be observed in claims by some sources that 'Exxon Mobil,

Dow Chemical, Chevron, and others have filed more than 500 cases against more than 90 governments' (Sierra Club undated).

It is perhaps not surprising that economists have drawn conclusions such as the following:

> When agreements like the TPP govern international trade – when every country has agreed to similarly minimal regulations – multinational corporations can return to the practices that were common before the Clean Air and Clean Water Acts became law (in 1970 and 1972, respectively).
>
> (Stiglitz 2014)

Such a position is unlikely to strengthen ecological integrity.

Conclusion

Although CSG mining is commonly presented as a cleaner, greener energy source and a viable alternative to coal-fired fuels, it remains a non-renewable energy source that emits GHGs and contributes to global warming (Grudnoff 2014: 52). It is also unlikely to prove to be a viable transitional fuel on the path to a renewable energy future because its increased use is likely to impede Australia's capacity to reach its GHG reduction targets on time (UNEP 2013). Envisioning CSG as an energy panacea is, therefore, neither realistic nor wise, although talk of a looming energy crisis causing consumers to pay higher energy prices helps makes CSG seem, at least superficially, attractive.

In this context, CSG companies have entered the market and engaged with law and regulation, exposing certain weaknesses. Regulatory and institutional arrangements have potentially allowed CSG mining companies to breach a range of licence conditions because regulators have not checked for compliance. This not only raises issues about the integrity of the regulatory model itself but has left the environment (and communities) vulnerable to harm. Approvals have been granted under processes which may technically be legally correct but which serve to undermine public confidence by, for example, leaving the impression that data collection has been inadequate and decisions have been made prematurely. This has not encouraged a climate for meaningful public participation in decision-making, and law and policy implementation.

Meanwhile, the legislative provisions governing subsurface natural resources span several Acts including, but not limited to, P(O)A and the EPAA. Compliance is complex and tricky and according to the Chief Scientist, one Act is needed to cover all subsurface resources (except water). She also recommended a number of changes including that: the government use its planning powers to designate those areas of the state in which CSG activity is permitted to occur; the government separate the process for the allocation of rights to exploit subsurface resources (excluding water) from the regulation of exploration and production activities; the government establish a single independent regulator with high levels of scientific and engineering expertise; the government move towards a target- and outcome-focused regulatory system with regularly reviewed environmental impact and safety targets; and appropriate and proportionate penalties for non-compliance and automatic monitoring processes be implemented (NSW Chief Scientist 2014; Higginson 2014).

Yet if the Australian government agreed to the inclusion of ISDS provisions in the TPP and also wished to implement the reforms above, it may find it very difficult to do so. Foreign corporations could sue the national government if those reforms had a negative impact on the corporation's trade and investment. Governments could be tied up in lengthy litigation and ultimately be liable for large sums in damages.

Introducing law and policy to reform unconventional gas governance may become too difficult in the face of hostile corporations. National sovereignty, which usually provides an over-arching power to make laws for a country, could be eroded by corporations seeking to protect profits. Private commercial goals could trump public goals that seek to protect the common heritage of mankind, in the form of the environment.

Australia, and indeed all potential signatories to the TPP who wish to safeguard the environment, should think twice (or even thrice) before signing the agreement.

Notes

1 Coal seam gas is known in the United States as coal bed methane.
2 Tight gas is gas that is held in low permeability reservoirs such as sandstone. See GeoScience Australia (undated b).
3 '2P' is a measure of proven energy reserves. It stands for 'proven' and 'probable'. 2P gas may be conventional or unconventional.
4 A proppant is a propping agent (commonly sand) that, along with fracking fluids (including water and often chemicals), is injected into a well to keep open fissures (in the coal or shale seam, for example), allowing the gas more readily to flow to the surface.

References

ABC News (2014) Community Welcomes Petroleum Exploration Licence Cancellation. Available at www.abc.net.au/news/2014-10-15/community-welcomes-petroleum-exploration-licence-cancellation/5815110 (accessed 16 October 2014).

AER (2013) *State of the Energy Market 2013*. Melbourne: Australian Energy Regulator. Available at www.aer.gov.au/sites/default/files/Complete%20report%20A4.pdf (accessed 11 February 2015).

ANEDO (2013) *Coal and Gas Mining in Australia: Opportunities for National Law Reform*. Report prepared for the Australia Institute. Technical Brief No. 24. August 2013. Available at www.tai.org.au/system/files_force/TB%252024%2520Coal%2520and%2520gas%2520mining%2520in%25 (accessed 29 October 2014).

Australia Institute (2013) Aussies in the Dark about Risky TPP Trade Deal. Media release, 13 December. Available at http://tai.org.au/content/mr-aussies-dark-about-risky-tpp-trade-deal (accessed 21 October 2014).

Australian Productivity Commission (2010) *Report on Regional and Bilateral Trade Agreements*. Available at www.pc.gov.au/__data/assets/pdf_file/0010/104203/trade-agreements-report.pdf (accessed 27 October 2014).

Bates, R. (2009) The Trade in Water Services: How Does GATS Apply to the Water and Sanitation Services Sector? *Sydney Law Review* 31(1): 121–42.

Bonyhady, T. (2012) Putting the Environment First? *Environmental and Planning Law Journal* 29(4): 316.

Climate Commission (undated) *The Critical Decade: Generating a Renewable Australia.* Available at www.climatecouncil.org.au/uploads/9cd594c54774fc0536b4550ccb68c 324.pdf (accessed 19 October 2014).

DFAT (undated) *Trans-Pacific Partnership Agreement.* Department of Foreign Affairs, Australian Government. Available at www.dfat.gov.au/fta/tpp/tpp-overview.pdf (accessed 20 October 2014).

Eastley, T. (2013) *Questions Raised Over Environmental Assessments for Billion Dollar CSG Developments.* 1 April. Available at www.abc.net.au/am/content/2013/s3727070.htm (accessed 27 October 2014).

EPA (2014) Santos Fined $1500 for Water Pollution. Available at www.epa.nsw.gov.au/ epamedia/EPAMedia14021802.htm (accessed 29 October 2014).

Faunce, T. (2012) An Affront to the Rule of Law: International Tribunals to Decide on Plain Packaging. *The Conversation* (29 August). Available at https://theconversation. com/an-affront-to-the-rule-of-law-international-tribunals-to-decide-on-plain-packaging-8968 (accessed 23 October 2014).

GeoScience Australia (undated a) *Coal Seam Gas.* Available at www.ga.gov.au/scientific-topics/ energy/resources/petroleum-resources/coal-seam-gas (accessed 19 October 2014).

GeoScience Australia (undated b) *Unconventional Petroleum Resources.* Available at www.ga.gov. au/scientific-topics/energy/resources/petroleum-resources/unconventional-resources (accessed 19 October 2014).

Gorrey, M. (2013) Temporary Ban on CSG Mining. *Wollondilly Advertiser* (19 November). Available at www.wollondillyadvertiser.com.au/story/1917974/temporary-ban-on-csg-mining (accessed 17 October 2014).

Gray, J. (2014) Frack off! Law, Policy, Social Resistance, Coal Seam Gas Mining and the Earth Charter. In L. Westra and L. Villela (eds), *The Earth Charter, Ecological Integrity and Social Movements,* Abingdon: Routledge.

Grudnoff, M. (2014) *Fracking the Future: Busting the Myths about Coal Seam Gas.* Australia Institute Paper no. 16. Available at www.tai.org.au/content/fracking-future (accessed 29 October 2014).

Hasham, N. (2014) CSG Project Approved before Chemicals Tested for Safety. *Sydney Morning Herald* (29 October): 9.

Haylen, A. and Montoya, D. (2013) *Gas: Resources, Industry Structure and Domestic Reservations Policies.* Briefing Paper no. 12/2013. NSW Parliamentary Research Service. Available at www.parliament.nsw.gov.au/prod/parlment/publications.nsf/0/BEA3EE2904867594 CA257C3F00136087/$File/Gas%20-%20resources,%20industry%20structure%20 and%20domestic%20reservation%20policies.pdf (accessed 17 October 2014).

Higginson, S. (2014) CSG Licence Cancellation Shows Need for Regulatory Overhaul. Available at www.edonsw.org.au/csg_licence_cancellations_show_need_for_regulatory_ overhaul (accessed 29 October 2014).

ICSID (undated) The International Centre for Settlement of Investment Disputes [7.4.1]. Available at www.ictregulationtoolkit.org/en/toolkit/notes/practicenote/1880 (accessed 25 October 2015).

Kelsey, J. (2014) TPPA Environment Chapter and Chair's Commentary Posted on WikiLeaks – Issues for NZ. Available at https://wikileaks.org/tppa-environment-chapter.html (accessed 21 October 2014).

McDonald-Smith, A. (2014) New South Wales Government Extends Freeze on New South Wales Petroleum Exploration Licence Applications. *Sydney Morning Herald* (25 September). Available at www.smh.com.au/business/the-economy/nsw-government-extends-freeze-on-petroleum-exploration-licence-applications-20140925-10lpzt.html (accessed 16 October 2014).

Nicholls, S. (2014) Coal Seam Gas Project Contaminates Aquifer. *Sydney Morning Herald* (8 March). Available at www.smh.com.au/environment/santos-coal-seam-gas-project-contaminates-aquifer-20140307-34csb.html (accessed 29 October 2014).

NSW Chief Scientist (2014) *Final Report of the Independent Review of Coal Seam Gas Activities in NSW.* September. Available at www.chiefscientist.nsw.gov.au/reports (accessed 23 October 2014).

NSW Department of Primary Industries (undated) *Aquifer Interference Policy.* Available at www.water.nsw.gov.au/Water-management/Law-and-policy/Key-policies/Aquifer-interference/Aquifer-interference (accessed 26 October 2014).

NSW Department of Trade and Investment (undated a) *NSW Titles.* Available at www.resourcesandenergy.nsw.gov.au/miners-and-explorers/applications-and-approvals/titles-services/tasmap (accessed 16 October 2014).

—— (undated b) *Code of Practice for Coal Seam Gas-Fracture Stimulation Activities.* Available at www.nsw.gov.au/sites/default/files/csg-fracturestimulation_sd_v01.pdf (accessed 26 October 2014).

NSWOCSG (undated a) FAQ on Coal Seam Gas. Available at www.resourcesandenergy.nsw.gov.au/landholders-and-community/coal-seam-gas/the-facts/faqs#_what-exploration-and-production-is-underway-in-_n_s_w__003f (accessed 11 October 2014).

—— (undated b) *About the Office of Coal Seam Gas.* Available at www.resourcesandenergy.nsw.gov.au/landholders-and-community/coal-seam-gas/office-of-coal-seam-gas (accessed 16 October 2014).

NWC (2012) *Position Statement – Coal Seam Gas.* June. Available at www.nwc.gov.au/nwi/position-statements/coal-seam-gas (accessed 19 October 2014).

Owens, K. (2012) Strategic Regional Land Use Plans: Presenting the Future for Coal Seam Gas Projects in New South Wales. *Environmental and Planning Law Journal* 39: 113–28.

Seccombe, M. (2014) Big Tobacco's Plan to Stub Out Plain Packaging. *The Saturday Paper* (8 March). Available at www.thesaturdaypaper.com.au/opinion/topic/2014/03/08/big-tobaccos-plan-stub-out-plain-packaging/1394197200#.VErYUBaTA2w (accessed 25 October 2014).

Service Nova Scotia (2014) Government to Prohibit Hydraulic Fracturing. News release, 3 September 2014. Available at http://novascotia.ca/news/release/?id=20140903005 (accessed 19 October 2014).

Sierra Club (undated) *Responsible Trade Program: Trans-Pacific Partnership.* Available at http://vault.sierraclub.org/trade/trans-pacific-partnership-agreement.aspx (accessed 21 October 2014).

Stiglitz, J. (2014) On the Wrong Side of Globalisation. *New York Times* (15 March). Available at http://opinionator.blogs.nytimes.com/2014/03/15/on-the-wrong-side-of-globalization/?_php=true&_type=blogs&_r=0 (accessed 25 October 2014).

Stop CSG Sydney (undated) How CSG Threatens Drinking Water Catchments. Available at http://stop-csg-illawarra.org/csg-in-water-catchments (accessed 17 October 2014).

TPP Environment (undated) *Trans-Pacific Partnership Treaty (Secret) Environment Chapter, Consolidated Text.* Available at https://wikileaks.org/tpp2/static/pdf/tpp-treaty-environment-chapter.pdf (accessed 20 October 2014).

TPP Investment (undated) *Trans-Pacific Partnership Draft Investment Chapter*. Available at www. huffingtonpost.com/2012/06/13/obama-trade-document-leak_n_1592593.html (accessed 23 October 2013).

UNEP (2013) *Emissions Gap Report* 2013. Available at www.unep.org/pdf/UNEPEmissions GapReport2013.pdf (accessed 27 October 2014).

Wade, W. (2014) Trans-Pacific Partnership is a Big Deal, but Hardly Anyone Knows. *Sydney Morning Herald* (17 February). Available at www.smh.com.au/action/printArticle? id=5168994 (accessed 20 October 2014).

Westra, L. (2013) *The Supranational Corporation: Beyond the Multinationals*. Abingdon: Routledge.

Williams, J., T. Stubbs and A. Milligan (2012) *An Analysis of Coal Seam Gas Production and Natural Resource Management in Australia*. Report prepared for the Australian Council of Environmental Deans and Directors. Available at http://acedd.org.au/wp-content/ uploads/2013/05/CSG-Analysis-Report.pdf (accessed 29 October 2014).

Cases and notices

Fullerton Cove Residents Action Group Incorporated v Dart Energy Ltd (No 2) [2013] NSWLEC 38 (28 March 2013).

JT International SA v Australian Commonwealth [2012] HCA 43 (5 October 2012). Available at www.austlii.edu.au/au/cases/cth/HCA/2012/43.html (accessed 25 October 2014).

Lone Pine Resources v Government of Canada, Notice of Intent to Submit a Claim of Arbitration Under Chapter Eleven of the North American Free Trade Agreement (8 November 2012). Available at www.international.gc.ca/trade-agreements-accords-commerciaux/assets/pdfs/disp-diff/lone-01.pdf (accessed 25 October 2014).

Notice of Arbitration under the Arbitration Rules of the United Nations Commission on International Trade Law and Chapter Eleven of the North Atlantic Free Trade Agreement, cl 58 (25) p18. Available at www.italaw.com/sites/default/files/case-documents/italaw1596.pdf (accessed 26 October 2014).

13 Hidden and indirect effects of war and political violence

Yuliya Lyamzina

Introduction

Past and present conflicts continue to affect the lives of millions of people around the world every day. It has been recognized that all types of violence and different kinds of security threat are not only external, but also internal. They cover all aspects of our lives (political, social and economic concerns such as poverty, deprivation of resources, unemployment, environmental degradation, different social tensions, displacement, immigration, etc.). They have not only obvious consequences on people's lives, but also hidden and indirect effects on the mental wellbeing of millions of people around the world.

Violence

The phenomenon of violence occurs globally and in all parts of society. In many countries, it is the norm; even worse, it is a part of widely tolerated and supported cultural practices. We see violence everywhere, not only in armed conflicts, where it is used as a weapon, but also during peacetime. Daily, there are news reports about emergencies, natural disasters, terrorist attacks and mass violence. The rise in mental and psychosocial problems is therefore not surprising.

Despite the very well-known fact that stress does influence human health (Erikson and Torssander 2008) and that different types of stressors (i.e. socio-economic stress, domestic stress and prenatal stress) contributed independently to emotional and social problems (Helm *et al.* 2010), not much has been done in this matter.

Also it is assumed that 'each stressor type provides its own contribution to the emergence of emotional or behavioral problems to which their effects do not just add up but interact over-additively' (Helm *et al.* 2010). Unfortunately, scientific evidence regarding mental health and psychosocial support during such events remains very poor (Inter-Agency Standing Committee 2007). Already in 1996 the World Health Assembly in its 1996 resolution WHA49.25 declared violence a leading global public health problem. Nevertheless, violence continuingly occurs in every country and in different parts of society. In many countries, it is the norm, or a part of historical and cultural practice that is widely tolerated and supported.

A global study on homicide conducted by the United Nations Office on Drugs and Crime in 2011 confirms that 'women are the most frequent victims of intimate partner violence and they are often killed by family members in all countries and across all cultures. Home is where women are most at risk of been killed, while men are more at risk in the street' (UNODC 2011).

Data related to gender-based violence remain mostly hidden due to the problems related to measurements and the comparability of data on reported crime and/or type of crime that is measured. In terms of gender-based and sexual-based violence we usually see only the tip of iceberg, while most of the data remain hidden due to all the aforementioned problems.

Terrorism as a form of collective violence

In the 1970s Brian Jenkins (in Schmid 2000) said, 'Terrorists want a lot of people watching, not a lot of people dead.' This is true. In comparison to civil war, genocide or ethnic cleansing, most non-state terrorism as we have experienced it since 1960s has generally not been a major killer.

Terrorism is one of the key security threats today. However, despite debates since 1972 the United Nations has not managed to define 'terrorism' conclusively to the satisfaction of all member states. There are over 100 definitions of terrorism, but Schmid and Jongman in 1988 created the most used one:

> Terrorism is an anxiety-inspiring method of repeated violent action, employed by a (semi-) clandestine individual, group or state actors for idiosyncratic, criminal or political reasons, whereby – in contrast to assassination – the direct targets of violence are not the main targets. The immediate human victims of violence are generally chosen randomly (targets of opportunity) or selectively (representative or symbolic targets) from a target population, and serve as message generators. Threat- and violence-based communication processes between a terrorist (organization), (imperiled) victims, and main targets are used to manipulate the main target (audience(s)), turning it into a target of terror, a target of demands, or a target of attention, depending on whether intimidation, coercion, or propaganda is primarily sought.
>
> (Schmid and Jongman 1988)

Currently, 13 international conventions and three protocols serve as a major framework in the fight against terrorism. In addition, a comprehensive convention on international terrorism is being elaborated under the auspices of the United Nations. UN Security Council resolves to call upon all Member States to ratify and fully implement them into domestic legislations, in order to fulfil obligations imposed on them by international conventions (UNODC 2011).

The Terrorism Knowledge Base provided data in recent trends in international terrorism through the RAND-MIPT database. Right now, the database is one of two comprehensive compilations of international terrorism events; another one is ITERATE data (see www.systemicpeace.org).

Current trends in armed conflicts

> Between 1985 and 1995 there were about 8 million deaths caused by ongoing wars (this figure includes deaths that took place before 1985 in wars that continued after this year), and almost 25 million internally and 10 million externally displaced populations. If we add to these figures those from wars that ended before 1985, we have a total of 19 million refugees, and possibly as many as 38 million internally displaced people.
>
> (Ugalde *et al.* 1999)

The latest evidence evidence clearly shows that quantity of armed conflicts displays a downward global trend. However, currently there are 24 states directly affected by ongoing wars (32 wars total, up from 27 at the end of 2006). Of these 24 states, more half (13) are affected by protracted wars, that is, armed conflicts persisting for more than ten years: Afghanistan (33 years), Colombia (36), Democratic Republic of Congo (19), India (59), Iraq (31), Israel (46), Myanmar (63), Nigeria (14), Pakistan (14), Philippines (39), Somalia (23), Sudan (28) and Turkey (27). Sri Lanka ended its protracted war with ethnic Tamil separatists in 2009. The remaining protracted wars continue to defy concerted efforts to gain settlement or resolution, although several of these conflicts have diminished substantially in magnitude in recent years. On average, during the contemporary period, interstate wars lasted about three years; civil wars lasted just over five years; and ethnic wars lasted nearly 10 years.

War and public health

Among social scientists and public health experts, there is a growing recognition of the health impact of armed conflicts and political violence. Political violence and war have become identified as a major public health problem, not only because of the deaths and disability they cause, but also because of their longer term and more indirect effects on the health, well-being, and livelihoods of individuals, families, and communities (Lyamzina 2014). In addition, wars and political violence shift large amounts of scarce resources from social and health services to the military. Third world countries spent more than US$400 billion on arms between 1960 and 1987. Wars and political violence have become an important health issue because they:

- cause much morbidity and mortality among civilians;
- have important and long-lasting negative health effects on the population;
- disrupt the provision and organization of health services; and
- do not seem to decrease with the passing of time; as soon as some conflicts are resolved, new violence explodes elsewhere (Ugalde *et al.* 1999).

Hidden and indirect effects of war and political violence

Very often the impact of war on health takes shape indirectly; it cannot be observed immediately and most likely, it will surface only some time later, in some cases even in subsequent generations. These are so called hidden and indirect costs and effects of war and political violence on health, such as:

* additional human suffering caused by deterioration of quality and availability of health services and health services organization (scarcity of medicines, delays of surgical interventions and complications resulting from the delays, increased problems of referrals, reduction of physical and human resources, etc.);
* mental health disorders from rapes and from exposure to horrors and brutalities;
* negative impact on nutrition produced by reductions of arable land because of the destruction of physical infrastructure;
* conflicts over resources/deprivation of resources, antipersonnel land mines, and degradation of the environment;
* delayed effects from exposure to biological and chemical agents, and from radiation;
* indirect effects from war-induced violent behaviour, alcoholism, and drug abuse; and
* indirect health effects, which cannot be measured with precision, from the contamination of air, water, and soil (Ugalde *et al.* 1999).

Physical health consequences

Victims (direct and proximate)

For the victims who are taken hostage, kidnapped, injured or killed, conventions and distinctions have secondary importance. Unfortunately, there is also no classification of terrorist victims (Schmid 2000). The UN's Declaration of Basic Principles of Justice for Victims of Crime and Abuse of Power from 1985 defines victims as:

> persons who, individually or collectively, have suffered harm, including physical or mental injury, emotional suffering, economic loss or substantial impairment of their fundamental rights, through acts or omissions that are in violation of criminal laws operative within Member States, including those laws proscribing criminal abuse of power . . . The term "victim" also includes, where appropriate, the immediate family or dependants of the direct victim and persons who have suffered harm in intervening to assist victims in distress or to prevent victimization.
>
> (UN General Assembly 1985)

Communicable and non-communicable diseases

Conflicts do lead to the displacement of large populations into temporary settlements or refugee camps with overcrowding, rudimentary shelters, inadequate safe water, sanitation, and increased exposure to all types of diseases during the acute phase of the emergency. In protracted and post-conflict situations, populations may have high rates of illness and mortality due to breakdown of health systems, flight of trained staff, failure of existing disease control programmes, and destroyed infrastructure. These populations are usually more vulnerable to infection and disease because of high levels of under-nutrition or malnutrition, low vaccine coverage and long-term stress. Long-term consequences of civil war do affect entire countries, for example Angola, the Democratic Republic of the Congo or Afghanistan, because of chronic lack of investment in health, education, and public works. These conditions, which are encountered during or after war and conflict, favour emergence of infectious diseases. Examples of emerging infectious diseases in conflict situations, where several overlapping risk factors are often involved, are numerous (Gayer *et al.* 2007).

Despite our knowledge, very little attention has been paid to the ways the post-conflict environment increases risks of other non-communicable diseases. First, high levels of psychological distress contribute to harmful health behaviour, such as hazardous drinking and increased smoking, which in turn increase the future burden of non-communicable diseases. Second, post-conflict countries commonly experience rapid urbanization, also associated with increased alcohol and tobacco use, as well as higher levels of obesity and reduced physical activity. Third, tobacco, alcohol and food companies often take advantage of weakened post-conflict trading systems (Roberts *et al.* 2012).

Detection, containment, and control of emerging infectious diseases in conflict situations are major challenges because of multiple risk factors that promote disease transmission and hinder control even more than those in many resource-poor settings. Beyond the global public health imperative to prevent the emergence and international spread of infectious diseases, there is also a moral imperative to alleviate the effects of these diseases on already vulnerable conflict-affected populations (Gayer *et al.* 2007).

Refugees and internally displaced persons

According to the latest figures released by the Geneva-based Internal Displacement Monitoring Centre (IDMC), there were 28.8 million internally displaced persons (IDPs) around the world in 2012, up from 2011. The global number of IDPs has steadily increased from around 17 million in 1997. The 28.8 million internally displaced civilians recorded in the IDMC report included more than 6.5 million newly displaced, almost twice as many as the 3.5 million recorded the year before. The conflicts in Syria and the Democratic Republic of the Congo (DRC) were responsible for around half of the new displacements, with 2.4 million and one million respectively, while an estimated 500,000 people fled their homes in both

Sudan and India. The largest regional increase in the number of internally displaced people in 2012 was in the Middle East and North Africa, where 2.5 million people were forced to flee their homes. There were almost 6 million IDPs in the region at the end of 2012, a rise of 40 per cent on the 2011. The figure continued to rise in 2013. Asia showed the second highest increase in new displacement after the Middle East and North Africa, with 1.4 million people forced to flee their homes during 2012. The region with the largest total number of IDPs was sub-Saharan Africa, which was hosting 10.4 million, an increase of 7.5 per cent compared with the year before, thus reversing the downward trend recorded since 2004. The Americas region hosted the second largest number of IDPs in 2012 with 5.8 million, an increase of 3 per cent. Colombia remains the country with the highest number of IDPs in the world, with a total of between 4.9 and 5.5 million, according to the IDMC. The country's internal armed conflict forced an estimated 230,000 people to flee their homes during the year (UNHCR 2014).

Civilians

According to the World Health Organization director general, 'armed conflict always amplifies factors that lead to increased incidence of infectious diseases among civilians':

> Mass movement of populations, overcrowding, lack of access to clean water, poor sanitation, lack of shelter, and poor nutritional status all increase the population's vulnerability to disease. In addition, the collapse of public health infrastructure and the lack of health services hamper control programs such as vaccination or vector control . . . Despite the advances in medicine and weapons technology throughout the 20th century, disease has continued to be the most formidable enemy, at least for the civilian population.
>
> (World Health Organization 2003a)

In comparison with the wars of the past centuries, in recent wars, civilians become the main target of violence. More than 107 million people have died in wars the last century. Whereas earlier in the century 90 per cent of all casualties were military, in recent wars 90 per cent have been civilians (Ugalde *et al.* 1999). Unfortunately, despite ongoing scholarly efforts focused on collecting data on war, only limited data have been collected regarding violence against civilians. The existing datasets are limited to genocide or mass killings (Harff 2003; Valentino *et al.* 2004; Rummel 1994), interstate wars (Downes 2004), or rely only on a proxy for violence (Azam and Hoeffler 2002). Although there is a growing academic emphasis on non-state groups, no global study has focused on violence against civilians by rebel groups (Mkandawire 2002; Humphreys and Weinstein 2006; Kalyvas 2006; Weinstein 2006). Comparable data for rebel and government violence does not exist, therefore most of violence in conflict remains unstudied and hundreds of civilians are continuing being injured, every day.

Children

The consequences of war are not limited to impaired physical health. War may also impair the mental health of affected individuals. The negative consequences for mental health can be significant, especially when the victims are children. Living in an unstable environment during childhood has detrimental effects on children (Ghazi *et al.* 2012). War is inevitably having a huge impact on civilians, and particularly on children. Civilian injuries have a double impact on children – whether or not they are the direct victims, they may find themselves without a mother, a father, brothers or sisters. Many of the 'military casualties' we hear about are men whose children will now have to grow up without them. Those who are injured and survive usually end up in hospital needing treatment for third degree burns, having limbs amputated and never again being able to run properly or play (World Health Organization 2003b). While the physical impact of the conflict is possible to measure, psychological impact of conflict cannot be measured (fear, the loss of family members, displacement, etc.). Usually physical and psychological damages of conflicts take years to heal, in most cases though they are very likely to leave many permanent scars. For example, United Nations humanitarian agencies in 2003 have appealed for US$325 million to cover the immediate health, nutrition, water and sanitation needs of the most vulnerable populations, particularly women and children. We should not forget that children are our future and it should become a priority for everyone to ensure the rapid reinstatement of a safe environment for children to grow up in. The health of children is their fundamental right, and they simply cannot grow properly in areas affected by high levels of violence.

Nutrition

Wasting and bilateral oedema are severe forms of malnutrition – resulting from acute food shortages and compounded by illness. About 1.5 million children die annually due to wasting. Rising food prices, food scarcity in areas of conflict and natural disasters diminish household access to appropriate and adequate food, all of which can lead to wasting. Wasting demands emergency nutritional interventions to save lives. Additionally, the use of herbicides, antipersonnel land mines, the targeting of markets and the destruction of the rural infrastructure greatly reduce the availability of arable land, food production and food distribution. In low-income nations, the reduction of food availability is frequently accompanied by large protein, calorie, and micronutrient deficiencies, and these deficiencies can produce severe malnutrition. In the presence of disease, nutritional intake is lowered and malnutrition diminishes the body's ability to fight infection. The symbiotic interaction between disease and malnutrition often lead to high levels of mortality, especially in the early phases of complex emergencies. Wars cause malnutrition by increasing food prices, which reduces food accessibility. Also it is recognized that the delivery of humanitarian aid, including food supplies, can be controlled by competing factions in their effort to manage the conflict and undermine their opponents and/or using food resources for sale in exchange for arms (Ugalde *et al.* 1999).

Environmental degradation

The experimentation, testing, and production of weapons have a heavy impact on human health. Only recently have scientists started paying attention to the environmental impacts of the production, testing and stockpiling of arms, and the environmental damage caused by maintaining armies in readiness for combat. Unfortunately, only few studies have assessed the health consequences of these exposures. Governments have made few efforts to increase our knowledge of the health risks posed by the presence in the environment of the many chemicals and radioactive materials released by the military, but it is suspected that:

> human exposure (to fuels, paints, solvents, heavy metals, pesticides, polychlorinated by-phenyls, cyanides, phenols, acids, alkalis, propellants, and explosives) through drinking, skin absorption, or inhalation may cause cancer, birth defects, and chromosome damage, or may seriously impair the function of the liver, kidneys, blood, and central nervous system.
>
> (Renner, cited in Ugalde *et al.* 1999)

Measuring long-term health effects is further complicated due to the difficulties of determining levels of exposure, poor hospital records, and scientific limitations on separating the contribution that other risk factors have on a disease (*ibid.*).

With the passing of time, some types of antipersonnel land mines break down before exploding; when this is the case rains and floods spread the toxic chemicals over large tracts of land, contaminating rivers, aquifers and soil. The environmental problem posed by the mines is so grave that are considered by some authors a 'major ecological disaster'. According to McGrath (1994: 121) the mines left in the fields after conflicts are denying not only 'vital land to farmers, pastoralists and returning refugees, but have covered large tracts of the earth's surface with non-biodegradable and toxic garbage'. An understudied impact of war is its effect on animal and plant life: the 1997 Zairian civil war destroyed national parks and reserves, placing some endangered species in an extremely critical situation (Ugalde *et al.* 1999).

More studies that examine the health consequences produced by the environmental degradation caused by wars are needed.

Mental health consequences

Scott pointed out that despite all insights, ten years after the first World Health Organization report on the global burden of disease; mental health remains a low priority especially in most low and middle income countries. For example, in the USA more than $70 billion is invested annually toward global public health research, less than 10 percent is devoted to research into the health problems that account for 90 percent of the global disease burden. Despite the findings and recommendations of both the World Health Assembly and the World Health Organization on violence and its effects on people of all ages and both sexes around the world, prevention, especially in developing countries, is just beginning to take hold as a global issue (Scott 2008). Prince *et al.* warn that mental disorders cannot be underestimated because they overlap with all health conditions, increase risk for communicable and non-communicable diseases and contribute to unintentional injury; however, developing countries still prefer control and eradication of infectious diseases that cause early death above those that cause years of disability, such as mental disorders, dementia and stroke (Prince *et al.* 2007).

Mental disorders are an important cause of long-term disability and dependency; while depression predicts the onset and progression of both physical and social disability (Bruce *et al.* 1994; Pennix *et al.* 1998). According to Mollica *et al.* (2004), depression is the fourth leading disease burden in 1990 and is predicted to move to second place in 2020; of the 10 leading disabilities worldwide, five are psychiatric conditions. Other main contributors to disability and mortality from non-communicable disease are cancer and cardiovascular disease. Evidence from population-based research in the USA reported moderate to strong prospective associations between depression, anxiety and coronary heart disease (Hemingway and Marmot 1999; Kuper *et al.* 2002).

Mental disorders are associated with risk factors for chronic disease such as smoking, reduced activity, poor diet, obesity, and hypertension; however, these lifestyle factors have not yet been shown to mediate associations with morbidity and mortality (Lyamzina 2011). In addition, some health conditions can affect the risk of mental disorders (Prince *et al.* 2007):

• Some disease processes directly affect the brain, for example infections like HIV, malaria or tuberculosis, cerebrovascular diseases (cortical stroke and progressive subcortical damage), diabetes, alcohol and substance use, neurodevelopmental disorders. The consequences of such effects depend on extent and location of brain damage and can include cognitive impairment, behaviour disturbance, mood disorders, delusions and hallucinations.
• Many chronic diseases create a psychological burden, which arises from factors such as the acute trauma of diagnosis; the difficulty of living with the illness; stigma, which can lead to guilt, loss of social support or breakdown of key relationship.
• Disability associated with chronic health conditions might mediate risk for depression and other common mental disorders.

Prince and colleagues continue by stating that for many health conditions mental illness makes an independent contribution to disability and quality of life. Despite of all this knowledge, mental health is missing from the policy framework for health improvement and poverty reduction. It is also missing from social and health research, its targets and interventions. Furthermore, awareness about mental health has to be integrated into social and health policies, health system planning, and health care delivery. It needs to be recognized as an integral component of practice in primary and secondary health care. Health-care workers needs to be trained in recognition and evidence based treatment of mental disorders and given sustainable supervision and support. Primary and secondary care providers should overcome their reluctance to treat patients with severe mental illnesses and learn effective ways to interact and communicate with these patients (Prince *et al.* 2007).

Moreover, mental health has not been acknowledged as an obstacle in the achievement of the Millennium Development Goals, notably, promotion of gender equality and empowerment of women, reduction of child mortality, improvement of maternal health and reversal and spread of HIV/AIDS, malaria and other diseases (World Health Organization 2002). There is a lack of mental health specialists; due to the range of the problem, those who are available might not be sufficient to meet the need, especially in low-income countries. This is our moral and ethical obligation redressing the imbalance in provision for people with mental disorders, which can brook no delay (Patel *et al.* 2006). The Universal Declaration of Human Rights, which enshrines the equal value of human life, and the Geneva Convention, which protects civilians and medical personal during the conflicts, can be good starting points to improve the medical and humanitarian response during the conflict (Lancet 2009).

Hidden economic costs of violence

Violence has serious negative implications on the economic and social development of countries and society as a whole. It has negative effects such as macroeconomic instability, losses in production capital (destruction in infrastructure, houses), depleted financial capital (inflation and decreasing investments) and eroded human capital (reduced nutrition, different mental diseases, displacement, diminished education), among others (Geneva Declaration 2008). There are several studies that have made models of the economic cost of violence including value of life methodology to estimate the social value of violence reduction (Soares 2006; Hess 2003; Londoño and Guerrero 1999; Restrepo *et al.* 2008). The true costs of different forms of violence remain unknown because the human costs in grief and pain cannot be calculated.

The absence of a system of punishment around the world, inadequate data, unsuccessful implementation of laws, widespread lack of political will and immunity for perpetrators make situation difficult to solve (Geneva Declaration 2008). Linda Bilmes and Joseph Stiglitz suggested that between US$4 trillion and US$6 trillion are currently needed to treat veterans with different health disorders (Bilmes and Stiglitz 2008). They stated that 'the cost of medical care and disability benefits for veterans has a price tag that the US government is not prepared for' (*ibid.*), and that no matter which steps the federal government will take, it has to be done now in order to prepare for the next several decades of providing care to OIF/OEF veterans. The cost of veterans' care is approaching the price USA are currently spending on combat operations. Two years after their book was published, Bilmes and Stiglitz considered their estimates to have been too conservative: 'The tale of this war – the tale of any war – is very, very long' (US Medicine 2010).

Furthermore, it is obvious that women do not have access to medical services in almost all low-income countries for various reasons. People in those countries have a massive need for mental health structures. However, there is a huge lack of financial and human resources. Medical personnel would need to work under heavy pressure and to use clinical psychological services, which are poorly developed or are not developed at all. The issue of mental health and psychosocial problems are vital in all countries. It seems that the only way to deal with the problem is for the international community to recognize the problem and start paying more attention to the mental and psychosocial problems in the communities affected by decades of wars, conflicts and other types of violence. A broad range of efforts is needed in order to rebuild these communities and to include women in all areas of the reconstruction can be a good solution and a good start.

Deeper research in order to generate baseline knowledge on the negative impact of health because of armed conflicts, organized crime and political violence is needed. Only then will we be able to prepare and implement public health and legal intervention in the most affected areas. Therefore a large-scale study among the affected population should be carried out, not only among refugees, displaced persons, veterans or victims of sexual and gender based violence, because obviously not only these people are affected in those countries, but not everybody is getting necessary attention, help and health care services.

Unfortunately usually these types of projects and researches are not be cost-effective from the viewpoint of politicians and government and they would hardly fund research to set up or to support a local rehabilitation centre if only 100 traumatized women were found in an area with 2 million people. That is why the international community needs to recognize and start paying attention to all communities affected by conflicts. In the beginning, it will be costly and difficult. However the author strongly believes that this investment is worth the risk and that the real result is to be found in the appearance of healthy and prosperous communities and countries, a growing gross domestic product, a decline in poverty, economic growth for the country and healthy future generations, as violence affects the whole community. In addition, several more conflicts would be over, so it is worth the risk.

References

Azam, J.-P., and A. Hoeffler (2002) Violence Against Civilians in Civil Wars: Looting or Terror? *Journal of Peace Research* 39(4): 461–85.

Bilmes, L. J., and J. E. Stiglitz (2008) *The Three Trillion Dollar War: The True Cost of the Iraq Conflict*. New York: W. W. Norton & Co.

Bruce, M. L., T. E. Seeman, S. S. Merrill and D. G. Blazer (1994) The Impact of Depressive Symptomatology on Physical Disability: MacArthur Studies of Successful Aging. *American Journal of Public Health* 84: 1796–9.

Downes, A. B. (2004) Targeting Civilians in War. PhD dissertation, Department of Political Science, University of Chicago.

Erikson, R. and J. Torssander (2008) Social Class and Causes of Death. *European Journal of Public Health* 18(5): 473–8.

Gayer, M., D. Legros, P. Formenty and M. A. Connolly (2007) Conflict and Emerging Infectious Diseases. *Emerging Infectious Diseases* 13(11): 1625–31.

Geneva Declaration (2008) *Global Burden of Armed Violence (GBAV)*. Geneva: Geneva Declaration.

Ghazi, H. F., Z. M. Isa, S. Aljunid, S. A. Shah, A. M. Tamil and M. A. Abdalqader (2012) The Negative Impact of Living Environment on Intelligence Quotient of Primary School Children in Baghdad City, Iraq. *BMC Public Health* 12: 562.

Harff, B. (2003) No Lessons Learned from the Holocaust? Assessing Risks of Genocide and Political Mass Murder Since 1955. *American Political Science Review* 97(1): 57–73.

Helm, D., D. Laussmann and D. Eis (2010) Assessment of Environmental and Socio-economic Stress. *Central European Journal of Public Health* 18(1): 3–7.

Hemingway, H., and M. Marmot (1999) Evidence Based Cardiology: Psychosocial Factors in the Etiology and Prognosis of Coronary Heart Disease. Systematic Review of Prospective Cohort Studies. *British Medical Journal* 318: 1460–67.

Hess, G. (2003) *The Economic Welfare Cost of Conflict: An Empirical Assessment*. Working Papers in Economics. Claremont: Claremont Colleges.

Humphreys, M. and J. M. Weinstein (2006) Handling and Manhandling Civilians in Civil Wars: Determinants of the Strategies of Warring Factions. *American Political Science Review* 100(3): 429–47.

Inter-Agency Standing Committee (2007) *Guidelines on Mental Health and Psychosocial Support in Emergency Settings*. Geneva: Inter-Agency Standing Committee.

Kalyvas, S. N. (2006) *The Logic of Violence in Civil Wars*. Cambridge: Cambridge University Press.

Kuper, H., M. Marmot and H. Hemingway (2002) Systematic Review of Prospective Cohort Studies of Psychosocial Factors in the Etiology and Prognosis of Coronary Heart Disease. *Seminars in Vascular Medicine* 2: 267–314.

Lancet (2009) Violent Conflict: Protecting the Health of Civilians. Editorial. *The Lancet* 370: 859–77.

Londoño, J. L., and R. Guerrero (1999) *Violencia en America Latina: Epidemiología y Costos*. Washington, DC: IADB.

Lyamzina, Y. (2011) *Political Violence and its Impact on Female Civil Population: Prevention of Health Damage as a Result of Terrorism, Political Violence and its Impact on Civil Population, with Special Focus on Women*. Saarbrücken: Lambert Academic Publishing.

Lyamzina, Y. (2014) *Hidden and Indirect Effects of War and Political Violence*. Saarbrücken: Lambert Academic Publishing.

McGrath, R. (1994) Landmine: The Global Problem. In P. Davies and N. Dunlop (eds), *War of the Mines: Cambodia, Landmine and the Impoverishment of a Nation*, 121–6. London: Pluto Press.

Mkandawire, T. (2002) The Terrible Toll of Post-Colonial 'Rebel Movements' in Africa: Towards an Explanation of the Violence Against the Peasantry. *Journal of Modern African Studies* 40(2): 181–215.

Mollica, R. F., B Lopes Cardozo, H. J. Osofsky, B. Raphael, A. Ager and P. Salama (2004) Mental Health in Complex Emergencies. *The Lancet* 364: 2058–67.

Patel, V., B. Saraceno and A. Kelinman (2006) Beyond Evidence: The Moral Case for International Mental Health. *American Journal of Psychiatry* 163: 1312–15.

Pennix, B. W. *et al.* (1998) Depressive Symptoms and Physical Decline in Community Dwelling Older Persons. *JAMA* 279: 1720–26.

Prince, M., V. Patel, S. Saxena, M. Maj, J. Maselko, M. R. Phillips and A. Rahman (2007) No Health without Mental Health. *The Lancet* 370(9590): 859 77.

Restrepo, J., B. Ferguson, J. M. Zúñiga and A. Villamarin (2008) Estimating Lost Product Due to Violent Deaths in 2004. Unpublished background paper for the Small Arms Survey. Geneva: Small Arms Survey.

Roberts, B., P. Patel and M. McKee (2012) Noncommunicable Diseases and Post-Conflict Countries. *Bulletin of the World Health Organization* 90: 2–2A. Available at www.who.int/bulletin/volumes/90/1/11-098863/en.

Rummel, R. J. (1994) Power, Genocide and Mass Murder. *Journal of Peace Research* 31(1): 1–10.

Schmid, A. P. (2000) Magnitudes of Terrorist Victimization: Past, Present and Future. Paper presented at the Ancillary Meeting on Terrorist Victimization: Prevention, Control and Recovery held in conjunction with the Tenth UN Congress on the Prevention of Crime and the Treatment of Offenders, Vienna, Austria.

Schmid, A. P., and A. J. Jongman (1988) *Political Terrorism: A New Guide to Actors, Authors, Concepts, Data Bases, Theories, and Literature.* New Brunswick, NJ: Transaction Publishers.

Scott, K. A. (2008) *Violence Prevention in Low- and Middle-Income Countries: Finding a Place on the Global Agenda.* Washington, DC: National Academies Press.

Soares, R. (2006) The Welfare Cost of Violence across Countries. *Journal of Health Economics* 25: 821–46.

Ugalde, A., P. L. Richards and A. Zwi (1999) Health Consequences of War and Political Violence. *Encyclopedia of Violence, Peace, and Conflict* 2: 103–21.

UN General Assembly (1985) Declaration of Basic Principles of Justice for Victims of Crime and Abuse of Power. UN Doc. A/RES/40/34, 29 November, 96th plenary meeting. Available at www.un.org/documents/ga/res/40/a40r034.htm.

UNHCR (2014) *Global Trends Report 2014.* Available at http://unhcr.org/globaltrends june2013/UNHCR%20GLOBAL%20TRENDS%202012_V08_web.pdf.

UNODC (2011) *Global Study on Homicide.* Vienna: United Nations Office on Drugs and Crime.

US Medicine (2010) Lifetime Cost of Treating Latest Generation of Veterans Higher than Predicted. Available at www.usmedicine.com/agencies/department-of-veterans-affairs/lifetime-cost-of-treating-latest-generation-of-veterans-higher-than-predicted.

Valentino, B., P. Huth and D. Balch-Lindsay (2004) Draining the Sea: Mass Killing and Guerrilla Warfare. *International Organization* 58(2): 375–407.

Weinstein, J. M. (2006) *Inside Rebellion: The Politics of Insurgent Violence.* Cambridge: Cambridge University Press.

World Health Organization (WHO) (2002) *The World Report on Violence and Health.* Geneva: World Health Organization.

World Health Organization (2003a) Bioterrorism and Military Health Risks. G20 Health Ministers Forum, World Economic Forum, Davos, Switzerland, 25 January. Available at www.who.int/dg/brundtland/speeches/2003/DAVOS/en.

World Health Organization (2003b) WHO Health Briefing on Iraq. Available at www.who.int/features/2003/iraq/briefings/monday7/en.

14 Commonly unrecognized benefits of a human rights approach to climate change

Benjamin A. Brown and Donald A. Brown

Introduction

There is a large and growing literature that examines links between human rights law and climate change.[1] This chapter summarizes the main conclusions of this literature while identifying additional practical benefits of examining climate change as a human rights problem that are not widely discussed in the relevant literature. These insights about human rights law are well known by human rights law practitioners but less well known by environmental lawyers and policy professionals.

This chapter will conclude that there are several practical lessons to be learned from human rights law and its philosophical foundations that should help achieve a greater global response to climate change These lessons will include:

- substantive conclusions about national obligations that follow when human rights are violated;
- procedural lessons about how to achieve greater compliance by nations with their climate change related obligations; and
- several other specific ideas about how to get nations to take their ethical and equity obligations seriously in formulating climate change policies.

Although the chapter acknowledges challenges and limitations of a human rights approach to climate change, the paper will explain that despite these limitations, greater use of a human rights approach could be very helpful in moving toward a climate change solution.

Highlights of the literature on human rights with respect to climate change

As mentioned, there is a growing literature on the significance of human rights law on climate change. This literature usually focuses on the following issues:

- What are human rights and where do they come from?
- What human rights are violated by climate change?

- The advantages of a human rights approach to climate change?
- The disadvantages of a human rights approach to climate change?

What are human rights and where do they come from?

Individual human rights are widely acknowledged to be derived from the idea that all human beings should be treated with dignity and respect. Furthermore, specific core human rights that have been widely acknowledged by almost all nations are simply deductions from the obligation to treat every human being with respect.

The Universal Declaration of Human Rights (UDHR) adopted by the international community in 1948 is widely understood to be the seminal document in modern human rights law. It begins with the following in the Preamble:

- Whereas recognition of the inherent dignity and of the equal and inalienable rights of all members of the human family is the foundation of freedom, justice and peace in the world.
- Whereas it is essential, if man is not to be compelled to have recourse, as a last resort, to rebellion against tyranny and oppression, that human rights should be protected by the rule of law.

(UDHR 1948: Preamble)

And so human rights are understood to apply to all people. Furthermore, governments have a duty to protect all citizens' rights by law. This duty is understood to impose a responsibility for nations to adopt laws to protect human rights. If climate change is violating human rights, therefore, nations have duties to pass laws to prevent climate change. The duty to do this does not depend upon prior national law if basic human rights are violated. The duty precedes legislative action.

Article 1 of the UDHR reinforces the idea that human rights should apply to all human beings because each human being is equal in dignity and rights. It says: 'All human beings are born free and equal in dignity and rights. They are endowed with reason and conscience and should act towards one another in a spirit of brotherhood' (UDHR 1948: Article 1). The duty to protect the human rights of citizens does not depend upon the nationality of citizens, nor other distinguishing features.

Article 2 of the UDHR provides:

Everyone is entitled to all the rights and freedoms set forth in this Declaration, without distinction of any kind, such as race, colour, sex, language, religion, political or other opinion, national or social origin, property, birth or other status. Furthermore, no distinction shall be made on the basis of the political, jurisdictional or international status of the country or territory to which a person belongs, whether it be independent, trust, non-self-governing or under any other limitation of sovereignty.

(UDHR 1948: Article 2)

Article 3 of the UDHR makes it clear that the basic rights to life and security, rights which are clearly violated by climate change, apply to all human beings no matter where they are: 'Everyone has the right to life, liberty and security of person' (UDHR 1948: Article 3).

That human rights exist prior to any legislative action that creates them has been grounded in several philosophical ethical traditions including, for instance, natural law theories, the deontological theories of Kant and others, some utilitarian or consequentialist theories such as the theory of John Stuart Mill, and theories of justice derived from theories of fair process including John Rawls's theory of justice as fair process (Moeckli *et al.* 2010).

And so human rights are often claimed to be self-evident truths discoverable by reason. Therefore, there is no need for government legislative action to claim that governments have a duty to take action to prevent threats to basic human rights including life, food security, and human health.

Many of the human rights that have now been widely acknowledged as binding obligations of states to their citizens have been recognized for hundreds of years, including in documents such as the Magna Carta in Great Britain in 1215, the English Bill of Rights in 1689, the French Declaration on the Rights of Man in 1789, the US Bill of Rights in 1791 and the United Nations Charter in 1945.

Since the Universal Declaration of Human Rights was adopted in 1948, numerous international and regional treaties on human rights have been adopted around the world.

Although the Universal Declaration of Human Rights was a non-binding legal document, in 1966 the UN adopted the International Covenant on Civil and Political Rights (ICCPR) and the International Covenant on Economic, Social, and Cultural Rights (ICESCR), which created legal duties for most of the rights in the Universal Declaration to ratifying nations, making the rights in the UDHR binding on nations which have adopted these treaties and thereby creating new binding human rights law for most of the world. Since then, there have been five other international human rights treaties that have been widely adopted by most nations.[2] These treaties include:

- Convention on the Rights of Persons with Disabilities (CRPD, 2006)
- Convention on the Rights of the Child (CRC, 1989)
- Convention Against Torture and Other Cruel, Inhuman or Degrading Treatment or Punishment (CAT, 1984)
- Convention on the Elimination of All Forms of Discrimination Against Women (CEDAW, 1979)
- International Convention on the Elimination of All Forms of Racial Discrimination (ICERD, 1966).

Concurrent with the development of international human rights law, significant regional human rights mechanisms developed at the regional level. These major instruments included but are not limited to:

- European Convention on Human Rights (ECHR, 1950)
- American Declaration of the Rights and Duties of Man (ADRDM, 1948)
- American Convention on Human Rights (ACHR, 1969)
- African Charter on Human and People's Rights (ACHPR, 1981).

Most nations have domestic laws, usually set out in constitutions, which protect basic human rights. Consequently, assessing human rights protection potential involves examining international, regional, and domestic laws. The rights protected under these instruments vary but are usually organized into civil and procedural rights and economic, social, and cultural rights.

Civil and political rights include rights to life, freedom, and security of person; freedom from slavery, servitude or torture; equality before the law, protection from arbitrary detainment, incarceration or exile; right to a fair trial; right of property; right to political participation; freedom of thought, conscience, religion, opinion and expression; right of peaceful assembly and association; right to participate in national government, either directly or by means of elected representatives (UN 2014).

Economic, social and cultural rights include right to work, right to equal remuneration for work of equal value, right to form and to join trade unions, right to adequate standard of living, right to education, right to take part in cultural life (*ibid.*).

According to human rights law, all rights have:

- a right-holder, usually a citizen, that is the party who has the right;
- an object, or what the right is; and
- an addressee, the party who must make the right available (Moeckli *et al.* 2010).

For human rights violated by climate change, the right-holder is all citizens whose rights are being violated, and the object is the rights to life, health, and sustainable food among other human rights infringed by climate change. And the addressee is the nation or other government entity, courts, or legislatures.

Human rights are understood to be the solemn promises of nations to citizens to which they are already bound. And so nations already have obligations to individuals to prevent climate change from violating their human rights.

Human rights obligations are understood to include the duty to respect human rights, to protect citizens from human rights violations, and the duty to fulfil human rights enjoyment (OHCHR 2014).

The duty to protect requires governments to protect them not only from the acts of the government that would deprive them of human rights but also to protect citizens from the climate change-causing activities of entities within the nation's jurisdiction. And so, under a human rights approach to climate change, nations have a duty to take action to prevent high-emitting entities within their jurisdiction to reduce their greenhouse gas (GHG) emissions so that total national GHG emissions levels do not exceed the nation's fair share of global emissions that will cause human rights violations.

The duty to fulfil means that nations must enact laws that are necessary to assure that citizens will enjoy human rights. Therefore governments have duties to pass laws on climate change that will ensure that all citizens enjoy the free and full exercise of their rights.

What human rights are violated by climate change?

Climate change violates many human rights, including three of the most fundamental and least controversial rights, namely: the right to life, the right to health, and the right to subsistence (Caney 2010).

Climate change violates the right to life because climate change will and is killing people through more intense storms, floods, droughts and killer heat waves.

Climate change violates the right to health by increasing the number of people suffering from disease, death and injury from heat waves, floods, storms and droughts, increases in the range of malaria and the burden of diarrhoeal diseases, cardio-respiratory morbidity associated with ground-level ozone, and the number of people at risk from dengue fever.

Climate change violates the right to subsistence by increasing: droughts, which will undermine food security; water shortages and sea level rise, which will put some agricultural areas under water; and flooding, which will lead to crop failure.

Caney explains that other human rights are affected by climate change but an understanding that climate change violates these three rights puts the claim that climate change violates human rights on the most uncontroversial grounds (Caney 2010). Caney also explains that climate change is also morally objectionable on other grounds than human rights, including non-anthropogenic and utilitarian moral grounds (*ibid.*).

In addition to these three human rights, climate change has been claimed to violate the right to take part in cultural life, the right to use and enjoy property, the right to an adequate standard of living, the right to food, and the right to the highest attainable standard of physical and mental health (Male Declaration 2007).

What are the advantages of a human rights approach to climate change?

Because climate change violates some core human rights – most of the human rights which have been acknowledged to be binding on governments for a very long time – a strong argument can be made that once climate change science had concluded that human activities causing climate change were threatening human lives, health, and food security, nations and others causing climate change have a duty to take action. That is, they must reduce the threat of climate change despite the absence of a treaty or other consensually arrived at international agreement that would require them expressly to reduce GHG emissions. The mere violation of core human rights creates legal obligations for states to prevent human rights violations.

The extant literature on human rights and climate change concludes that once climate change is understood to be a human rights problem, certain additional implications for policy follow, including:

- Because human rights are violated, costs to those causing climate change entailed by policies to reduce the threat of climate change are not relevant for policy (Caney 2010). That is, if a person is violating human rights, he or she should desist even if it is costly. The abolition of slavery was immensely costly for slave owners yet because basic human rights were violated, costs to the slave owner of abolishing slavery were not relevant.
- If climate change is a human rights problem, compensation is due to those whose rights have been violated. The human rights approach generates both duties for mitigation and adaptation. It also generates duties of compensation for harm (Caney 2010).
- Human rights apply to each and every human being as they are based on the idea that all human beings are born free and entitled to certain rights.
- If one has a right not to suffer a particular harm, then it is wrong to violate that right just because one can pay compensation. It is, for instance, wrong to assault someone even if the person assaulted can be paid compensation for the harm.
- If the human rights of the most vulnerable are being violated they need not bear the burdens of mitigating the threats. Human rights usually take priority over other human values such as efficiency and promoting happiness (Caney 2010).
- Governments' duty to protect human rights requires governments to protect citizens from the acts of individuals and high-emitting entities within their jurisdiction that would deprive citizens of their human rights.

What are the disadvantages of a human rights approach to climate change?

There are at least four limitations of a human rights approach to climate change, addressed in the following sections.

Not all nations have ratified the relevant human rights treaty

The United States, for instance, has not ratified the International Covenant on Economic, Social and Cultural Rights nor accepted the jurisdiction of the Inter-American Court on Human Rights, the regional tribunal in which individuals in the Americas may seek to redress human rights violations.

Problem of extraterritoriality

Because states are usually duty holders only to their citizens within their jurisdiction, an African may not usually, under existing law, file a human rights complaint against the United States. His or her only remedy is against his or her specific government under current human rights law. This is known as the problem of extraterritoriality of human rights law. The law dealing with problems of extraterritoriality in protecting human rights contains some complex exceptions which under certain circumstances may allow citizens of a foreign jurisdiction to sue a nation for violation of their human rights; however, the problem of extraterritoriality may limit a person who is a citizen of one country from bringing an action against another nation for climate change caused violations of human rights. For a discussion of the problem of extraterritoriality in human rights law in regard to climate change, see Humphries (2010).

Problems of enforcement

The enforceability of human rights law is a complex subject which requires analyses of the specific human rights instrument on which the human rights violation is based. Human rights treaty enforcement capability has been accomplished in the UN system initially with the creation of a body, usually a committee or tribunal to monitor national human rights performance. These committees and tribunals usually have the authority to hear individual complaints and require nations to report on their human rights compliance on a regular basis. For instance, the ICCPR is implemented through the Human Rights Committee (HRC), which was created to promote compliance with its provisions. The HRC frequently expresses its views as to whether a particular practice is a human rights violation, but it is not authorized to issue legally binding decisions. Other treaties and bodies exist within the UN system with varying enforcement and implementation powers and duties to implement human rights goals. For the most part, these enforcement powers are weak and improvements in human rights violations are best achieved through holding offending nations to the court of international opinion by creating a clear record of national compliance with their human rights obligations.

Even though citizens are usually able at minimum to get a declaration that their nation is responsible for a violation of his or her rights under the relevant human rights treaty by filing a complaint against the nation with the relevant treaty body, if the nation ultimately refuses to correct the violation often not much can be done other than shaming the nation. In addition, after a citizen files a complaint with the relevant human rights tribunal or committee, a process of regular review of the national compliance with its human rights obligations can maintain pressure on the nation to correct human rights compliance. For a description of the complexities of enforcing human rights, see Koh (1998) and Moscrop (2014).

Problem of a remedy

All litigation to find remedies for climate change have thus far been unsuccessful because courts have had difficulty in allocating responsibility for climate change damages as almost all nations and most of the people in the world are contributing to raising GHGs in the atmosphere. Although each nation has a duty to reduce GHG emissions to its fair share, determination of the amount of any nation's fair share is very challenging (for a discussion of the litigation history in attempts to find a remedy for climate change, see Brown 2013). Human rights theory does not provide guidance on how to allocate responsibility among all entities and governments emitting GHGs.

Thus far, no one has successfully brought a human rights claim in a court for climate change-caused damages, although the Inuit Peoples filed such a claim in the Inter-American Commission on Human Rights. This claim was eventually dismissed for lack of a remedy (for a discussion of this case see Ansari and Campagna 2014).

Before a successful human rights claim can be brought into an existing legal forum in regard to climate change, several potential legal hurdles need to be overcome that have little to do with whether a nation or an individual has committed a human rights violation. These hurdles include jurisdictional issues, questions of proof, and authority of the relevant forum. For this reason, the failure to successfully bring legally recognized human rights claims may have little to do with whether the offending behaviour has created a violation of a protected right and more to do with the limitation of the existing legal regime. As stated above, improving human rights compliance by nations has been most successful in developing a record of a nation's non-compliance with its legal rights obligations and publicly shaming the nation.

Unrecognized benefits of a human rights approach

The following are features of human rights that are relevant to an understanding of why a human rights approach to climate change could be an important tool to reduce the threat of climate change despite the limitations of human rights law discussed above.

Substantive policy consequences of a human rights approach to climate change

LIMITATIONS ON COST JUSTIFICATION FOR NON-ACTION

Most nations emitting high levels of GHGs have been justifying inadequate national climate change policies on the basis of economic self-interest.[3] As we discussed above, human rights obligations must usually be fulfilled without a consideration of costs to those who are violating human rights. This is not to say that costs of preventing climate change will never be relevant to climate change and human rights controversies because human rights sometimes conflict and, particularly in such cases, human rights tribunals have sometimes balanced the costs to prevent the human rights violations to the environment with the costs to those who may be contributing to an environmental problem while exercising another human right (for a discussion of conflicts between human rights, see ICHRP 2008). Yet the sheer fact that preventing human rights violations may impose costs on nations causing human rights violations is not an acceptable justification under human rights law for a nation emitting GHGs above its fair share of safe global emissions.

Many arguments made by opponents of climate change policies have been based upon claims that adoption of climate change policies would reduce total global welfare often measured by gross domestic product (GPD). Although most economists agree that costs to the economy of doing nothing exceed the costs of preventing climate change (Nordhaus 2012), if nations are violating human rights, welfare maximization arguments are not relevant to whether a nation should take action to prevent enjoyment of human rights. The most recent report of the IPCC dealt with this issue directly:

> Economics is not well suited to taking into account many other aspects of justice, including compensatory justice. For example, a CBA might not show the drowning of a Pacific island as a big loss, since the island has few inhabitants and relatively little economic activity. It might conclude that more good would be done in total by allowing the island to drown: the cost of the radical action that would be required to save the island by mitigating climate change globally would be much greater than the benefit of saving the island. This might be the correct conclusion in terms of overall aggregation of costs and benefits. But the island's inhabitants might have a right not to have their homes and livelihoods destroyed as a result of the GHG emissions of richer nations far away. If that is so, their right may override the conclusions of CBA. It may give those nations who emit GHG a duty to protect the people who suffer from it, or at least to make restitution to them for any harms they suffer.
>
> (IPCC 2014: 12–13)

And so a human rights approach to climate change can help citizens understand ethical problems with national justifications for climate policies. Even if there are practical limitations in legally forcing nations to reduce their emissions to their fair share of safe global emission through a human rights approach to climate change, a human rights approach can help citizens around the world to change unacceptable national responses to climate change by helping citizens make ethical arguments about unacceptable national climate change policies.

A growing number of nations have included rights to a healthy environment as a human right. In fact, 130 constitutions in the world, including the overwhelming proportion of those amended or written since 1970, include a state obligation to protect the environment or the right to a safe, healthy, and ecologically sustainable environment (Anton and Shelton 2011). In addition, many of the international and regional human rights treaties have been construed to include a right to the environment that can be deduced from other expressly enumerated rights (Shelton 2014). Some national level courts have also found human rights to environmental protection on the basis of constitutional provisions which do not expressly guarantee a human right to the environment. For instance, the Pennsylvania Supreme Court recently said in *Robinson Township v. Commonwealth* that there is a human right to clean air and water and the state's natural resources.

The significance of a human rights approach to protect the environment is potentially profound because of several legal doctrines entailed by a rights approach including the duties of governments to respect, fulfil and protect rights, along with the procedural rights that are discussed below. A human rights approach to environmental protection, without doubt, strengthens environmental law in many ways.

New procedural tools for reducing the threat of climate change

GOVERNMENTS HAVE AN AFFIRMATIVE RESPONSIBILITY TO ACT TO PREVENT CLIMATE CHANGE-CAUSED HARMS

Under human rights law, nations have an affirmative duty to prevent human rights violations, which means they have a duty to take action against those within their jurisdiction who are responsible for climate change. This duty, as mentioned above, includes the affirmative responsibility to enact laws preventing climate change.

NEW REMEDIES FOR CITIZENS TO FIGHT CLIMATE CHANGE

Because citizens of nations that have adopted human treaties have a right to file complaints with the relevant treaty committee or tribunal against governments that are responsible for climate change, usually only after they have exhausted national procedures, human rights treaties create new remedies for many citizens around the world. As we have seen, although the enforcement powers of international human rights treaties are still weak in many respects, the committees formed by most international human rights treaties usually have the power to make quasi-judicial determinations of whether a nation is responsible for violation of human rights, and to continue to review national compliance with such determinations. In addition, all treaty bodies have developed procedures for following up on recommendations that are triggered when human rights problems have been adjudicated by them. In these proceedings, NGOs can often contribute to the national reporting process by submitting 'alternative', 'parallel' or 'shadow' reports which often challenge national reports.

Given that the treaty bodies that have been established under international law are standardizing reporting methods, and allowing states to prepare 'common core documents' that are applicable to more than one treaty, some standardized approaches to climate change-caused human rights violations are likely to arise.

Some of the regional human rights tribunals have judicial enforcement authority to actually order nations which have accepted their jurisdiction to comply with their human rights obligations, including the European Court of Human Rights, the Inter-American Court of Human Rights and the African Court of People's and Human Rights (for a discussion of the enforcement power of these regional human rights tribunals, see Cavallaro and Stephanie 2008). Although there are still enforcement limitations if the nation still refuses to abide by the tribunal's order, the adjudication of the tribunal is legally binding on the nation.

And so international human rights law provides potential remedies for those who seek to protect themselves from climate change harms that did not exist otherwise. Given that almost all nations have agreed to protect the core human rights identified by the main international human rights treaties that are violated by climate change, tens of millions of people around the world now have some legal redress to protect themselves from climate change that did not exist before their nations agreed to be bound by human rights treaties.

Those nations that provide legal protection for human rights in constitutions and national law often have an even more direct legal remedy to protect human rights threatened by climate change through legal action in national courts.

STRONG PROCEDURAL RIGHTS TO INFORMATION AND PARTICIPATION IN
DECISION MAKING

Developing human rights jurisprudence around the world in cases considering human rights to environmental protection has acknowledged rights to access to information on environmental matters held by public authorities, the opportunity to participate in decision-making processes, and the right to justice (Shelton 2014). Tribunals around the world, although especially in Europe, have found that where rights to environmental protection exist, citizens have strong rights to participate directly in environmental decision-making, obtain information, and to have their rights adjudicated by law. (*ibid.*). These rights have sometimes expanded the rule of standing that has given citizens more access to courts to litigate their grievances about harms to the environment (Anton and Shelton 2011).

Other benefits of a human rights approach

Many of the nations that have the highest per capita and total levels of GHGs emissions have historically proclaimed that they are champions of human rights. The United States, for instance, proudly proclaims its seminal role in global recognition of human rights. Yet, it has yet to officially recognize the human rights dimensions of climate change policy formulation. At the same time, the most frequent arguments made by opponents of climate change policies have included various claims about adverse costs to the United States (Brown 2012).

As we have seen, using cost to the United States alone as the basis for the United States violating other people's human rights is inadmissible under human rights theory and law. Yet there has been little discussion of this conflict in the United States (*ibid.*). And so a human rights approach to climate change has the rhetorical benefit of exposing the hypocrisy of nations that claim to be champions of human rights while providing a legal and ethical basis for undermining very common arguments made by opponents of climate change policies around the world.

Conclusion

We have identified several limitations of a human rights approach to climate change. These limitations raise important issues about the efficacy of using human rights as a legal tool to force nations to reduce their GHG emissions, at least under existing international law. Yet we have also seen numerous potential uses of a human rights approach to climate change that don't turn on judicial enforcement of human rights. They include new substantive and procedural obligations of nations and rhetorical uses of a human rights approach.

Perhaps the most important of these advantages is the power that a human rights approach gives the international community to require that nations regularly justify their climate change policies without reliance on reasons that are inadmissible under a human rights approach. Nations have thus far been able to formulate their climate change policies on climate change without public scrutiny informed by the ethical and justice problems with their policies. A human rights approach to climate change could force nations to justify their climate change policies congruent with what ethics, justice, and human rights require of them. For this reason, even if human rights law enforcement capability remains weak, a human rights approach to climate change has the potential to force nations to justify their climate change policies on ethically acceptable grounds.

Notes

1 Some of the most comprehensive literature on the subject include Bodansky (2010), CIEL (2010), ICHRP (2008), Humphries (2010), Knox (2009), OHCHR (2009) and World Bank (2011).
2 For a list of the major human rights instruments see: Cornell University School of Law, Legal Information Institute, Human Rights Overview, www.law.cornell.edu/wex/human_rights
3 This conclusion is based upon the results of a joint project of Widener University School of Law and the University of Auckland that is examining how ethics and justice considerations have been taken into account or ignored in setting national climate policy. See Nationalclimatejustice.org for a description of the project.

References

ACHPR (1981) African Charter on Human and People's Rights. CAB/LEG/67/3 1rev. 5, 21.L.M.58.
ACHR (1969) American Convention on Human Rights. 1144 UNTS 143.

ADRDM (1948) American Declaration of the Rights and Duties of Man. OAS Res. XXX, reprinted in Basic Documents Pertaining to Human Rights in the Inter-American System, OAS/Ser. L/V/I.4 Rev. 9, 43 AJIL Supp. 133.

Ansari, A. and R. Campagna (2014) Climate Change and Human Rights. Available at http://guides.brooklaw.edu/content.php?pid=82208&sid=631740.

Anton, D. and D. Shelton (2011) Problems in Environmental Protection and Human Rights: A Human Right to the Environment. Available at http://scholarship.law.gwu.edu/cgi/viewcontent.cgi?article=2050&context=faculty_publications.

Bodansky, D. (2010) Climate Change and Human Rights: Unpacking the Issues. Available at http://papers.ssrn.com/sol3/papers.cfm?abstract_id=1581555.

Brown, D. (2012) *Climate Change Ethics, Navigating the Perfect Moral Storm.* London: Routledge.

Brown, D. (2013) The Climate Change Disinformation Campaign: What Kind of Crime against Humanity, Tort, Human Rights Violation, Malfeasance, Transgression, Villainy, or Wrongdoing Is It? Available at http://blogs.law.widener.edu/climate/?s=Climate+Change+Disinformation+Campaign.

Caney, S. (2010) Climate Change, Human Rights and Moral Thresholds. In S. Gardiner, S. Caney, D. Jamieson and H. Shue (eds), *Climate Ethics, Essential Readings.* New York: Oxford University Press.

CAT (1984) Convention Against Torture and Other Cruel, Inhuman or Degrading Treatment or Punishment. 1465 UNTS 85.

Cavallaro, J., and E. Stephanie (2008) Reevaluating Regional Human Rights Litigation in the Twenty-First Century: The Case of the Inter-American Court. Available at http://papers.ssrn.com/sol3/papers.cfm?abstract_id=1404608.

CEDAW (1979) Convention on the Elimination of All Forms of Discrimination Against Women. 1249 UNTS 13.

CIEL (2010) *Climate Change and Human Rights: A Primer.* Center for International Environmental Law. Available at www.ciel.org/Publications/CC_HRE_23May11.pdf.

Cornell University School of Law (2014) Human Rights Overview. Available at www.law.cornell.edu/wex/human_rights.

CRC (1989) Convention on the Rights of the Child. 1577 UNTS 3.

CRPD (2006) Convention on the Rights of Persons with Disabilities. 2515 UNTS3.

ECHR (1950) European Convention on Human Rights. Available at http://human-rights-convention.org/the-texts/the-convention-in-1950/?lang=en.

Humphries, S. (2010) *Human Rights and Climate Change.* Cambridge: Cambridge University Press.

ICCPR (1966) International Covenant on Civil and Political Rights. S. Treaty Doc. No. 95–20, 6 ILM 368, 999 UNTS 171.

ICERD (1966) International Convention on the Elimination of All Forms of Racial Discrimination. 660 UNTS 195.

ICESCR (1966) International Covenant on Economic, Social and Cultural Rights. S. Treaty Doc. No. 95–19, 6 ILM 360, 993 UNTS.

ICHRP (2008) *Human Rights and Climate Change, A Rough Guide.* International Council on Human Rights Policy. Available at www.ichrp.org/files/reports/45/136_report.pdf.

IPCC (2014) *Mitigation of Climate Change, Social and Economic Dimensions.* Working Group III, Intergovernmental Panel on Climate Change. Available at www.ipcc.ch/report/ar5/wg3.

Knox, J. (2009) Climate Change and Human Rights. Available at http://works.bepress.com/cgi/viewcontent.cgi?article=1002&context=john_knox.

Koh, H. (1998) How is International Human Rights Law Enforced? *Indiana Law Journal* 74(3): 1397–1417.

Male Declaration (2007) Male Declaration on the Human Dimensions on Climate Change. Available at www.ciel.org/Publications/Male_Declaration_Nov07.pdf.

Moeckli, D., S. Shah and S. Sivakumaran (2010) *International Human Rights Law*. New York: Oxford University Press.

Moscrop, H. (2014) Enforcing International Human Rights Law: Problems and Prospects. E-International Relations Students. Available at www. e-ir.info/2014/04/29/enforcing-international-human-rights-law-problems-and-prospects.

Nordhaus, W. (2012) Why the Global Warming Skeptics are Wrong. *New York Review of Books* (22 March). Available at www.nybooks.com/articles/archives/2012/mar/22/why-global-warming-skeptics-are-wrong.

OHCHR (2009) OHCHR Study on the Relationship between Climate Change and Human Rights. Office of the High Commissioner on Human Rights. Available at www.ohchr.org/EN/Issues/HRAndClimateChange/Pages/Study.aspx.

OHCHR (2014) Human Rights Law. Office of the High Commissioner on Human Rights. Available at www.ohchr.org/en/professionalinterest/pages/internationallaw.aspx.

Robinson Township v. Commonwealth of Pennsylvania (2012) 52 A.3d 463 (Pa. Cmwlth. 2012). Available at http://blogs.law.widener.edu/envirolawcenter/files/2013/12/RobinsonTp-v.-Commonwealth-CommCt2012.pdf.

Shelton, D. (2014) Human Rights and Environment Issues in Multilateral Treaties Adopted Between 1991 and 2001. Office of the UN High Commissioner for Human Rights. Available at www2.ohchr.org/english/issues/environment/environ/bp1.htm.

UDHR (1948) Universal Declaration of Human Rights. Available at www.un.org/en/documents/udhr.

United Nations (2014) Frequently Asked Questions about Human Rights. Available at www.unfpa.org/derechos/preguntas_eng.htm.

UNFCCC (1992) United Nations Framework Convention on Climate Change. S. Treaty Doc No. 102–38, 1771 UNTS 107.

World Bank (2011) Human Rights and Climate Change. Available at http://siteresources.worldbank.org/INTLAWJUSTICE/Resources/HumanRightsAndClimateChange.pdf.

15 Reconciliation and the Indian Residential Schools Settlement

Canada's coming of age?

Kathleen Mahoney

Introduction

At 3:00 pm on 11 June 2008, the prime minister of Canada rose in the House of Commons to make a historic apology. Part of that apology reads as follows:

> For more than a century, Indian Residential Schools separated over 150,000 Aboriginal children from their families and communities. In the 1870s, the federal government, partly in order to meet its obligation to educate Aboriginal children, began to play a role in the development and administration of these schools. Two primary objectives of the Residential Schools system were to remove and isolate children from the influence of their homes, families, traditions and cultures, and to assimilate them into the dominant culture. These objectives were based on the assumption Aboriginal cultures and spiritual beliefs were inferior and unequal. Indeed, some sought, as it was infamously said, 'to kill the Indian in the child'. Today, we recognize that this policy of assimilation was wrong, has caused great harm, and has no place in our country.[1]

'Coming of age' is a term usually associated with the transition from adolescence to adulthood. As a literary genre it focuses on the psychological and moral growth of the protagonist from youth to adulthood and in which character change is extremely important.[2] The Indian Residential Schools Settlement, in many ways, marks a 'coming of age' for Canada. With the Residential Schools Settlement Agreement, the country has finally been forced to come to grips with its dark and painful past and take full responsibility for the collective and individual devastation it caused to hundreds of thousands of its first citizens and provide remedies to compensate them. Whether or not the Settlement Agreement will result in a character change of the population has yet to be determined.

This chapter will bring the reader through a short history of the residential schools and the negotiations leading to settlement. I will discuss the challenges of achieving reconciliation between Canada, the first peoples, the churches and the non-aboriginal populations of the country – what has been done and what needs to be done to create the conditions for reconciliation.

History

Indian residential schools were created out of a very specific government policy of assimilation dating back to pre-confederation, but it was carried out most vigorously by a poet[3] who was also a civil servant. His name was Duncan Campbell Scott.[4] Scott never saw any reason to question the assumption that the indigenous peoples ought to become just like the colonizing peoples. Shortly after he became Deputy Superintendent, he wrote:

> The happiest future for the Indian race is absorption into the general population, and this is the object and policy of our government ... I want to get rid of the Indian problem. I do not think as a matter of fact, that the country ought to continuously protect a class of people who are able to stand alone ... Our objective is to continue until there is not a single Indian in Canada that has not been absorbed into the body politic and there is no Indian question, and no Indian Department, that is the whole object of this Bill.[5]

The creation of residential schools was an integral part of the assimilation project. Most of these schools were operated in conjunction with various church entities, which were responsible for the day-to-day operation of the schools. Aside from forcible removal from the family and community, there was a concerted effort to undermine and abolish the languages, spiritual and cultural traditions and way of life of the students. In addition, many students suffered inadequate living conditions including lack of food, clothing and shelter. Numerous students suffered serious emotional, physical and sexual abuse. In some schools, more than 50 per cent of the children died.

In the 1990s, former residential school students began suing Canada and the various church entities, individually and by way of class action. They sought compensation for the harm that they suffered at the residential schools. People began to speak out about their experiences to the media, not the least of which was Phil Fontaine, the first native leader to do so, and court dockets began to overflow with residential school claims. Without an alternative method of resolving disputes, it was estimated that the caseload would take another 53 years to conclude at a cost of $2.3 billion in 2002 dollars, not including the value of the actual settlement costs.

As a result of extensive negotiations to move the cases out of the courts into an alternative dispute resolution model, discussed below, a Canada-wide agreement was approved as a single class action in court orders from courts across the country. The agreement changed lives, exposed previously unknown truths and brought understanding and awareness to a population that had been kept in the dark about its own history for more than 150 years.

The negotiations

The negotiation process in this context was an experience for which no law school courses, years of teaching, or prior practice could have fully prepared me.

The search and commitment to find the best way to come to grips with a profound human rights violation perpetrated over generations put me and my team on a rollercoaster ride, oscillating between elation and despair more times than I can remember. The core mandate of my role as the chief negotiator for the Assembly of First Nations (AFN) was reconciliation. This ultimately led to the largest and most comprehensive Settlement Agreement in Canadian history, the architecture of which is an amalgam of human rights principles, Indigenous cultural values, restorative justice concepts, feminist legal theory, tort law and contract law.

Our team made it clear to the government that to succeed in creating conditions for reconciliation, any settlement agreement would have to be based on wide consultation with First Nation communities, elders, survivors, and church representatives.[6] There had to be formal acknowledgement by the government that the policy of residential schools and the deliberate attempt to destroy Aboriginal identity, culture, and family life was racist and legally wrong and that significant individual and collective compensation and an apology in the House of Commons would have to be part of the package.[7]

AFN insisted that the approach of the negotiations as well as any settlement reached would have to be underpinned by a set of values which included the following:

- be inclusive, fair, accessible and transparent;
- offer a holistic and comprehensive response recognizing and addressing all the harms committed in and resulting from residential schools;
- respect human dignity and racial and gender equality;
- contribute towards reconciliation and healing; and
- do no harm to survivors and their families.[8]

We believed that if the central element of the agreement was to be reconciliation, the survivor group and its perspectives had to be included at every step of the resolution process. This required the government to retreat from its initial attempt of an out-of-court ADR process which was based on the British-common-law torts model that was not only rigid and unresponsive to the notion of reconciliation, but was inherently biased against Aboriginal experience and perspectives.[9]

From a European-derived torts perspective, the only legal issues that needed to be resolved were the individual cases of physical and sexual abuse and unlawful confinement carried out by employees of the government and churches. The more general issue of the morality and legality of the residential school system *per se* and the collective harms – for example, reduced self-esteem; isolation from family; spiritual harm; and loss of language, culture, a reasonable quality of education, kinship, community, and traditional ways – through the exercise of government policy[10] were not seen as legal wrongs appropriate for consideration in the ADR model.[11] Secondary harms to parents, grandparents, spouses, and descendants were completely ignored.

It became obvious that the government's preoccupation with resolving individual tort claims[12] and the several class action suits that had been filed against it[13] was primarily a self-interested response to the building pressure on the courts.

I formed an 'expert group' to devise a new, holistic plan that would be more conducive to reconciliation, healing, truth telling, and fairness. The expert group was comprised of survivors, elders, national and international academics, lawyers, research students, and judges.[14] We conducted research, wrote briefs, and held workshops and study sessions for a three-month period in the summer of 2004 and then, with the results of the deliberations, I wrote the AFN Report that was made public in November 2004.[15]

The report's recommendations laid out a blueprint for a settlement agreement that included a lump sum payment for every former residential school student for loss of language and culture and loss of family life; improved, equal, context-sensitive, and inclusive compensation for individual abuse; provision for negotiated settlements, wider vicarious liability requirements, context-trained adjudicators, health supports, a truth commission, cultural provisions, and healing and commemoration reparations for survivors and their families, among many other suggestions.[16]

In a short period of six months, the Agreement in Principle, which collapsed all of the class actions and individual actions that had been started as well as all future actions into the Settlement Agreement, was signed in November 2005.[17]

The Truth and Reconciliation Commission (TRC)was a key element of the package. We were always of the view that the TRC would be the lasting legacy of the Settlement Agreement and its most important feature. We operated from the premise that, once a TRC has done its work and a history is known, it can never be unknown.

What was interesting about this phase of the negotiations was that only the AFN team participated in the negotiations for the TRC and for commemoration, education and healing funds. All the other claimants' lawyers avoided these negotiations, perhaps a function of their legal training and lack of comfort with non-traditional, restorative remedies. Canadian lawyers are trained to resolve disputes according to a set of 'common' rules and values that are not necessarily values common to those of their Aboriginal clients. As a result, their sole focus on compensation for individual claims was disappointing. Their attitude mirrored that of the government and some church representatives that restorative elements of the Settlement Agreement, so important to survivors, were not important. The failure to understand that healing and community-building remedies are antithetical to the adversarial, torts-based system that lawyers are used to in the Anglo-American legal system does not bode well for reconciliation going forward. When lawyers are not familiar with Aboriginal justice values, they shy away from what they would see as non-legal work requiring them to be culturally sensitive and aware of the deep, collective harms caused by the torts and crimes of people like the defendants. Unfortunately, this lack of cultural sensitivity and awareness can re-victimize the claimants they are retained to help.[18]

Bearing all of this in mind, while negotiating the mandate of the TRC, the AFN's position was that the TRC's mandate had to be built upon Aboriginal values with the understanding that reconciliation is an ongoing individual and collective process and requires commitment from all those affected.[19] The AFN specifically decided against subpoena powers for the Commission, for example, because we wanted to avoid the trap of litigation while creating a vehicle that would enable the survivors to tell their stories in a respectful and appropriate setting and ensure the preservation of the Indian residential school history in the most complete and accurate way possible.[20]

Through my ongoing involvement with survivors in the settlement process, I found that although most survivors are pleased with the Settlement Agreement and satisfied with the results of their claims, there is still a considerable amount of anger and mistrust of the process and of lawyers generally, and reconciliation for them is a remote possibility.

I think a lesson that we are all learning is that revealing the past, paying reparations, apologizing, and creating a unique, culturally sensitive and aware Settlement Agreement cannot, without more, accomplish reconciliation. Reconciliation is far too large a project for a settlement agreement to accomplish, even if it is the largest and most comprehensive one in Canadian history.

A member of the South Africa Truth Commission, Reverend Bogani Finca[21] told a story at the University of Calgary conference on truth commissions[22] that eloquently explains the challenge of reconciliation. It is the Cow Story:[23]

> It seems there was a Zulu gentlemen called Tabo, who had a cow. One day Mr White came along, overcame Tabo, took his cow, and went away. Tabo lived a miserable life without his cow. He lost his livelihood, he became poor, he became depressed, and he couldn't provide proper housing for his children or get them the things they needed to have a successful life. Many years later, a truth commission came along and Tabo and Mr White were brought together in a process of reconciliation. They were deeply moved. Mr White offered an apology. Tabo accepted the apology. They hugged, they kissed, and they had a cup of tea together, and even shared a few jokes. At the end of the day, Tabo was standing at the door of his shack as Mr White walked out to his car. Before they waved goodbye, Tabo had the courage to ask Mr White, 'What about the cow?' Mr White got very angry and said, 'Tabo, this is about reconciliation, it has nothing to do with the cow.'

I repeat this story because 'talking about the cow' is where the credibility of the Indian Residential Schools Settlement Agreement, the apology and reconciliation, will be tested and their ultimate success determined. 'Talking about the cow' would be asking the question, now that the Settlement Agreement has been implemented, compensation paid, apologies made and accepted, how can Canada move away from its past history of oppression, deprivation, and systemic discrimination of its first peoples?

To achieve reconciliation, it is not enough to merely stop doing the wrong things and say sorry. It must be understood that the Settlement Agreement deals with past wrongs, not future relationships. Systemic issues facing the survivors and their families and the majority of the Indigenous population before the Settlement Agreement was signed are still present. Their lives are still the same. The Indigenous communities continue to live lives that are less than they should be.

'Talking about the cow' can be expressed by being truthful about what divides us. It can be talking about power, understanding who has power over the past, so as to determine who has power over the present. It can be talking about what government policy should be – about how to best combat racism, deep prejudice, and discrimination – about how to advance the position of Aboriginal peoples in Canadian society. It can be about improving Aboriginal education, protecting and preserving Aboriginal languages, strengthening Aboriginal culture and traditions, sharing economic opportunities, and appointing Aboriginal Canadians to high positions of power. Aboriginal and non-Aboriginal Canadians have not lived the same history, and being truthful about that could go some distance towards achieving reconciliation.

Reconciliation also requires a meeting of the minds of Aboriginal and non-Aboriginal Canadians that justice has been done between us.

Rupert Ross[24] relates a story told to him by a Mohawk woman who said that, in her community, they had an expression for the meeting of the minds, that moment of justice, as they understand it – only it was called 'face cracking.'[25] It is the moment when the hard exterior of injustice cracks and the connection between people on the emotional level takes place. This place is the spiritual realm that comes into play when healing and reconciliation start to occur. But as the Mohawk woman explained, face cracking cannot take place unless there is an understanding of the victim's experience of being wronged. If we only look at wrongs in an abstract, objective, legal way, no connection will likely occur. Legal descriptions and solutions more often than not fail to describe things the way they really are and fail to take into account the actual experience of the victims.

To illustrate, Ross takes the example of a simple purse snatching on the street. In terms of the British common law followed in Canada, from an objective standpoint, all that happened is the loss of a purse. It can be fixed by a conviction, a return of the purse, and maybe a fine. That's all. But if we looked at the purse snatching from the victim's experience, Ross says, we would understand that the harm of the purse snatching is much more than the value of the purse. It harms the relationship the victim has with her community. He says unless it is recognized that her relationship with the community is altered, and unless that is talked about and dealt with, she will not feel safe in that community and it is likely that she will continue to feel that way indefinitely. In other words, the harm done by the experience will not heal and she will not be able to reconcile with the perpetrator; she will never have that face-cracking moment.

What the residential school policy did was to put two cultures into fundamental conflict with each other. The effects of that conflict altered the relationship between Aboriginal and non-Aboriginal people in a very profound and negative way. For reconciliation to occur, this damage must be understood from the victims' perspective and steps must be taken to repair it. When it comes to the Settlement Agreement, in order to experience a 'face-cracking' moment, those working with it, most of whom are lawyers, must strive to implement it as much as possible from the victims' perspective even though their inclination and legal training tells them otherwise. Trevor Farrow's argument that the dominant, traditional model of lawyering is unsustainable and that it is an ethical imperative for lawyers to be more sensitive to the diverse needs of the communities they serve challenges lawyers to take positive action.[26] Maintaining the *status quo* practice of abstract, decontextualized, exclusive professionalism not only violates the compact the legal profession has with society, it arguably ensures that justice for Aboriginal people will never be done.

Unfortunately, the Department of Justice lawyers and their superiors appear to be moving away from the approach Farrow encourages. There is a well-founded perception[27] among claimant's counsel and survivors that Canada has shifted from a more reconciliatory approach to IAP claims to a hyper adversarial one.[28] Counsel complain of delayed compensation payouts, arguments for increasingly narrow interpretations of the Settlement Agreement, refusal to provide documents, imposition of arbitrary new rules on the process, and shameless reliance on victim-blaming arguments.[29] These tactics not only defeat the purpose and intent of the Agreement and re-victimize the survivors, they have the effect of absolving Canadian society of its responsibility for the toxic legacy of the residential schools.[30] For example, the crumbling skull arguments routinely employed by Department of Justice lawyers in IAP hearings are based on the principle that tort law damages must only be awarded to put the plaintiff back into his original position before the tort was committed. In the abstract, this argument can be justified as supporting the corrective justice principle that underlies tort law generally.[31] However, as Kent Roach argues, when applied to residential school survivors in the context of residential school history, the crumbling skull argument is pernicious and manifestly unjust. It fails to consider that the original position of many survivors is diminished because of the harms visited on the parents, communities, and culture of Aboriginal peoples over many years of compulsory attendance at residential schools and because of other forms of discrimination and human rights abuses by the very defendants using the argument to try and defeat their abuse claims under the IAP.[32] Such tactics, I believe, bring the Canadian legal system and the legal profession into disrepute.

Recent litigation has seen judges weighing in to chastise the government for its attitude to its responsibilities under the Settlement Agreement.[33] In a case where Canada was refusing to provide documents regarding abuses at a residential school where children were put in an electric chair for punishment, Justice Paul Perell rebuked the government for its calculated and insensitive behaviour:

> Based on its unduly narrow interpretation of its obligations, Canada has not adequately complied with its disclosure obligations with respect to the St. Anne's narrative . . . If truth and reconciliation is to be achieved, if it is to be a genuine expression of Canada's request for forgiveness for failing our

Aboriginal people so profoundly, the justice of the system for the compensation for the victims must be protected.[34]

As explained above, the more legalistic the process of implementation, the more it moves away from the guiding principles and the more one-sided it becomes. This bias, in turn, diminishes the sense in the survivor community that justice or healing, or 'face cracking', will ever be achieved. The danger of this happening is before us now. The inevitable consequence of the adversarial climate that is being created will be a sense of betrayal and re-victimization, not justice or reconciliation.

At the 2004 conference, it was made very clear that reconciliation would be impossible under the government's ADR settlement plan. When the government of the day was presented with an alternative that embraced the guiding principles and it took the Aboriginal view of fairness and justice into account, the Settlement Agreement was achieved in record time. By agreeing to compensate survivors for loss of language and culture and loss of family life, to soften proof of causation requirements in individual abuse claims, to provide health supports, and to attend to the early compensation of the sick and elderly and to the truth telling, healing, and commemoration components, the government achieved something no other government in the world has been able to achieve. It stood alone as a government that understood that a victim-centred, inclusive, generous, holistic, and culturally-sensitive approach was the appropriate standard and methodology for dealing with mass violations of human rights.[35] Today, a different government is in place, and although the Harper government issued the apology, the words of the apology are not ringing true in terms of the strategies and tactics being employed by its servants in the IAP hearings and in the litigation with the TRC.

There is no question that the amounts of money awarded in CEP and IAP claims are tangible markers that go some way towards vindicating the wrongs by providing solace in what they can buy. By comparison with other countries and what they have awarded for similar abuses, the compensation is on the higher end of generous and that is something we can be proud of.[36] But it must always be borne in mind, and I have been told so very many times by survivors, that no amount of money can replace a lost childhood, replace the devastation of childhood sexual and physical abuse, the loss of language and culture, and most importantly, the loss of family life. That is why restorative justice principles and reconciliation are so important and why it is of great concern that the reconciliatory benefits of the Settlement Agreement are now in jeopardy.

Reconciliation is a multifaceted process that may take decades or generations to achieve. It will not happen on its own. Reconciliation will require much more than the Settlement Agreement, although it is a good start towards remedying the wrongs of the past. Healthier and safer relationships between Aboriginal and non-Aboriginal peoples in the future will require commitment and political leadership at all levels of government to end the inherited stigma of inferiority that corrupts the majority as much as it corrupts the minority. Political, social, and cultural change must take place because truth telling without substantive change will be a hollow and unproductive exercise.

This is why the final element of reconciliation, which is forgiveness, remains a question mark. For those who are able to forgive, it may complete the circle of their experience and bring some peace. For those who cannot forgive, their wounds will likely continue to fester and the burden of anger and bitterness may remain and be passed down, generation after generation. Perhaps the TRC will be able to provide the solace that money cannot and give comfort to those who still struggle with the aftermath of the residential school experience in their daily lives. Then again, perhaps it will not, if rancour and adversarial attitudes persist between Canada, the TRC, and claimants.

The longer-run success of the Settlement Agreement's objective of reconciliation will depend on attitudes, commitments, and circumstances hardly knowable at this moment – including attitudes of Aboriginal peoples about how much they wish to seek achievement within mainstream institutions and goals. What is clear is that the goal of reconciliation will not be achieved unless a climate conducive to reconciliation is nurtured and encouraged by all the parties.

Notes

1 For the full apology see www.aadnc-aandc.gc.ca/eng/1100100015644/11001000156 49.
2 In literary criticism, a novel of formation of character, novel of education, or coming-of-age story is called a *Bildungsroman*. For further details, see http://en.wikipedia.org/wiki/Bildungsroman.
3 Scott was honoured for his poetry during and after his lifetime. He was elected a Fellow of the Royal Society of Canada in 1899 and served as its president from 1921 to 1922. The Society awarded him the second-ever Lorne Pierce Medal in 1927 for his contributions to Canadian literature. Ironically, Scott's widely recognized and valued 'Indian poems' cemented his literary reputation. In these poems, the reader senses the conflict that Scott felt between his role as an administrator committed to an assimilation policy for Canada's Native peoples and his feelings as a poet, saddened by the encroachment of European civilization on the Indian way of life. See Canadian Poetry Archives, http://collectionscanada.gc.ca (accessed 30 March 2011).
4 Prior to taking up his position as head of the Department of Indian Affairs, in 1905 Scott was one of the Treaty Commissioners sent to negotiate Treaty No. 9 in Northern Ontario. Aside from his poetry, Scott was the head of the Department of Indian Affairs from 1913 to 1932.
5 John Leslie, *The Historical Development of the Indian Act*, 2nd edn (Ottawa: Department of Indian Affairs and Northern Development, Treaties and Historical Research Branch, 1978), 114.
6 See Phil Fontaine, 'National Chief's Message', *AFN Echo* 1(3) (November–December 2004), 2, available at http://64.26.129.156/cmslib/general/EchoVol1 No3.pdf.
7 'Canada Agrees to Reparations for All Residential School Students', *Cultural Survival* (17 June 2005), available at www.culturalsurvival.org/news/canada-agrees-reparations-all-residential-school-students.
8 *Ibid.*
9 See the AFN Report for a detailed critique of the ADR process as well as a map forward for a holistic and reconciliatory approach, available at http://epub.sub.uni-hamburg.de/epub/volltexte/2009/2889/pdf/Indian_Residential_Schools_Report. pdf.

10 Rosalyn Ing, Dealing with Shame and Unresolved Trauma: Residential School and Its Impact on the 2nd and 3rd Generation Adults, PhD thesis, Department of Educational Studies, University of British Columbia (2000).

11 See James Rodger Miller, *Shingwauk's Vision: A History of Native Residential Schools* (Toronto: University of Toronto Press, 1996); J. R. Miller, 'Troubled Legacy: A History of Native Residential Schools', *Saskatchewan Law Review* 66 (2003): 357; Basil Quewezance and Joe Nolan, *Indian Residential Schools: A Background Report* (Assembly of First Nations, 2004); Jennifer Koshan, 'Does the DR Model Violate the Charter?', in Kathleen Mahoney (ed.), *The Residential School Legacy: Is Reconciliation Possible?* (unpublished); Greg Hagen, 'The DR Model: Fair Treatment or Re-Victimization?' in Mahoney, *The Residential School Legacy*.

12 The Treasury Board Secretariat estimated that there would be approximately 18,000 claims, 14,000 of which would be proven valid.

13 *Baxter/Cloud* was an Ontario class action claiming loss of Indian language and culture in Indian residential schools for $76 billion. As of 1 May 2004, seventeen judgments had been issued; see *Cloud v Canada (Attorney General)* (2004), 73 OR (3d) 401; *Baxter v. Canada (Attorney General)* (2006), 80 OR (3d) 481.

14 The members of the team and their biographies are listed in *Assembly of First Nations Report on Canada's Dispute Resolution Plan to Compensate for Abuse in Indian Residential schools* (Ottawa: Assembly of First Nations, 2004), Appendix A at 28, available at http:// epub.sub.uni-hamburg.de/epub/volltexte/2009/2889/pdf/Indian_Residential_ Schools_Report.pdf.

15 *Ibid.*

16 The Canadian Bar Association as well as class-actions lawyers endorsed the AFN Report. See e.g. Canadian Bar Association, *The Logical Next Step: Reconciliation Payments for All Indian Residential School Survivors* (February 2005), available at www.cba.org/CBA/ Sections/pdf/residential.pdf.

17 *Agreement in Principle* (20 November 2005), available at www.residentialschoolsettlement. ca/aip.pdf.

18 Trevor C. W. Farrow, 'Residential Schools Litigation and the Legal Profession', *University of Toronto Law Journal* 64(4) (2014): 6596–619, available at http://muse. jhu.edu/login?auth=0&type=summary&url=/journals/university_of_toronto_law_ journal/v064/64.4.farrow.html.

19 *Ibid.*

20 Access to documents and other relevant information was dealt with in the Settlement Agreement, Schedule N in paragraph 2, 4, 10 part C and 11 (see www.residential schoolsettlement.ca/settlement.html).

21 In 1995, Reverend Bongani Finca was appointed a member of the South Africa Truth and Reconciliation Commission. He served on the Human Rights Violations Committee and became the commissioner responsible for the work of the TRC in the Province of the Eastern Cape.

22 Truth Commission: Sharing the Truth about Residential Schools – A Conference on Truth and Reconciliation as Restorative Justice, University of Calgary, June 2007.

23 Deon Snyman, 'Restitution Is about Returning the Cow' (5 July 2011), available at http://restitution.org.za/2011/07/restitution-is-about-returning-the-cow-dr-bongani-finca-former-trc-commissioner.

24 As a Crown Attorney working with First Nations in remote northwestern Ontario, Rupert Ross learned that he was routinely misinterpreting the behaviour of Aboriginal victims, witnesses, and offenders, both in and out of court. With the assistance of Aboriginal teachers, he began to see that behind such behaviour lay a complex web of coherent cultural commandments that he had never suspected, much less understood.

25 Rupert Ross, *Returning to the Teachings: Exploring Aboriginal Justice* (Toronto: Penguin, 1996).

26   Farrow, 'Residential Schools Litigation and the Legal Profession'.
27   This has been the author's personal experience; it is also this author's perception after talking with numerous claimants' counsels. No formal study has been done. But, for an example of a lawyer's experience in dealing with the federal government, see Dave Dean, 'The Government Has Been Forced to Hand Over Documents about Abuse of Native Children at St. Anne's Residential School', *Vice* (20 January 2014), available at www.vice.com/en_ca/read/the-government-has-been-forced-to-hand-over-documents-about-the-abuse-of-native-children-at-st-annes-residential-school. See also Angela Sterritt, 'Residential School Survivors Face "Adversarial" Government: Lawyers Say Government Attitude Has "Shifted" As Survivors Wait for Information,' *CBC News* (4 February 2014), available at www.cbc.ca/news/canada/north/residential-school-survivors-face-adversarial-government-1.2523520.
28   In the case of *Fontaine* et al. *v. Canada* et al., the government lawyers argued that reconciliation was only a value up until the Agreement was signed. After that point, if there is a problem with the interpretation of the Agreement, a 'no holds barred' adversarial battle is appropriate. See *Fontaine* et al. *v. Canada* 2014 MBCJ/93 page 20, para 48.
29   For a thorough discussion of the pernicious use of the crumbling skull principle of causation in cases involving Indian residential school survivors and their descendants, see Kent Roach, 'Blaming the Victim: Canadian Law, Causation and Residential Schools', *University of Toronto Law Journal* 64(4) (2014): 566–95, available at http://muse.jhu.edu/login?auth=0&type=summary&url=/journals/university_of_toronto_law_journal/v064/64.4.roach.pdf.
30   *Ibid.*, 24.
31   *Athey v. Leonati*, [1996] 3 SCR 458.
32   Roach, 'Blaming the Victim', *supra*, note 29.
33   In a recent decision, Justice Schulman of the Manitoba Court of Queen's bench stressed the importance of access-to-justice principles to questions of textual interpretation: *Fontaine* et al. *v. Canada (Attorney General)* et al., 2013 MBQB 27 at para 21. See also Frank Iacobucci, *Report of the Independent Review Conducted by the Honourable Frank Iacobucci* (February 2013), Ontario Ministry of the Attorney General, available at www.attorneygeneral.jus.gov.on.ca/english/about/pubs/iacobucci/pdf/First_Nations_Representation_Ontario_Juries.pdf.
34   *Fontaine v. Canada (Attorney General)*, 2014 ONSC 283 at para 226.
35   In 2010, the United Nations Permanent Forum on Indigenous Issues recognized the Canadian Truth and Reconciliation Commission as a model of best practices and an inspiration for other countries; see Truth and Reconciliation Commission of Canada, *Truth and Reconciliation Commission of Canada: Interim Report* (Winnipeg: Truth and Reconciliation Commission of Canada, 2012), available at www.attendancemarketing.com/~attmk/TRC_jd/Interim%20report%20English%20electronic%20copy.pdf.
36   The comments of Mr Tom Boland, Deputy Minister formerly in charge of the Irish Compensation Plan for Industrial School Survivors in Ireland, that the Canadian compensation for Indian residential schools was '*de minimus* and grudgingly given' were indicative of the Irish perception that the ADR Canadian model was not generous as it stood in 2004. See Kathleen Mahoney, *Report on Fact-Finding Mission to Ireland Regarding Compensation Scheme and Related Benefits for Industrial School Survivors in Ireland*, available at www3.telus.net/kmahoney/Documents/files/Irish%20report.pdf. Mr. Tom Boland, formerly the equivalent of a deputy Minister in the Department of Education of the Government of Ireland and recognized as the main architect of the redress scheme and the Government's policy there, told me that the decision was made early on in Ireland to compensate the survivors for 100 per cent of their injuries because it was the right thing to do whether or not the Church was willing to contribute. His view of the Canadian compensation scheme was that it is being given grudgingly and was *de*

minimus. He said that the attitude that guided the thinking behind the Irish scheme, both in amount of compensation offered as well as in the procedural requirements and the support services, was generosity. It was also understood that without some out-of-court resolution, the survivor's claims would 'gum up' the Courts so it was essential that they be kept out of the Courts as much as possible. He knew this was not possible unless the plan was generous. There was also the fear that the Courts might grant much higher awards than the current common law provides, given the unique nature of the harms suffered. By the time the ADR was revised into the IAP, the Canadian plan was considerably more generous than the Irish plan.

Part IV

Democracy, ecological concerns and the perils of environmental defence

16 Public health and environmental health risk assessment

Which paradigm and in whose best interests?

Colin L. Soskolne

Introduction to the role of epidemiology in public health risk governance and risk assessment

Epidemiology is the applied science central to public health. It provides the methods to study the distribution and determinants of diseases in populations. That is, epidemiologists are concerned about the geographic occurrence of diseases – both acute and chronic – as well as their causes. Because of its focus on populations, epidemiology is the science that bridges toxicological research on animals in laboratory settings to human health studies, critical to the formulation of rational, evidence-based health policy.

The focus of epidemiology is on community health; namely, on the prevention of disease, disability and premature death in communities of people as well as animals. In the chronic disease realm, epidemiologists study a health problem not only to contribute to knowledge on the relationship between exposure to putative causal agents (such as man-made pollutants) and disease occurrence, but – as an applied science – with a view to developing policy interventions to mitigate or correct the problem of exposure to harm.

The epidemiologic method is critical to environmental risk governance, which provides a framework that includes risk evaluation at the local level. Risk evaluation is one of the central steps in the classical health risk assessment paradigm to identify needs for risk management.

As with any scholarly discipline, there are approaches (or paradigms) that are more appropriate than others to address a problem. This chapter speaks to paradigmatic options that can be selected, with every intention of doing good as opposed to harm, or not. Paradigms may be used, or abused, to serve special/narrow interests rather than serving the public interest.

The challenge that emerged from a community's health complaints

Baytex, one of five oil companies operating in the bitumen-rich area of Peace River, Northern Alberta, Canada, used a method of heating sand–bitumen–water mixtures in above-ground tanks, to separate the components and to recover the bitumen. Vapours escape via leaks and during transferring, and notably the Baytex tanks had open vents to avoid freezing and tank rupture during the winter.

Residents in this small rural farming community, living in proximity to the tank 'pads', had complained of feeling sick and other health issues, sometimes in relation to odours. They left their homesteads because of fumes from the processing of bitumen that they claimed had caused their ill-health. On leaving the polluted environment, most of the more acute health concerns were relieved; however, some of the residents had developed severe chemical sensitivities to common exposures such that symptoms recurred when they were re-exposed upon returning to their homes.

Baytex contested the community's claims. Shell, another of the five oil companies operating in the area, decided in advance not to contest the community's health complaints.

Under the Alberta Energy Regulator (AER) Proceeding, two independent health risk assessment reports were commissioned in 2013 by the AER for the formal hearing, one from Dr M. Sears and the other from Dr D. Davies.[1] Each of the two reports was to shed light on the question of health effects as expressed through community complaints in relation to emissions from the bitumen recovery and transportation processes. The two reports delivered opposing messages. Essentially, one report (i.e. that of Davies) adopted a linear-reductionist paradigm; the other (i.e. that of Sears) used more of a post-normal science approach.

Davies concluded that the data he analysed provided no evidence of excessive risk, so the health complaints must have arisen as a result of mechanisms associated with odour annoyance. Sears highlighted that the additive risk assessment process was inappropriate given known synergies of components in the complex bitumen emissions mixtures; that problems with the analytical data before the panel resulted in an under-estimation of exposures; and that physician-diagnosed health effects (e.g. severe sinus inflammation resembling chemical burns) as well as the severe chemical sensitivities resulting in multi-system symptoms reported by the worst-affected individuals were the result of poisoning by the emissions that accumulated under the atmospheric inversions particularly common during the winter.

The author of this chapter was contracted to evaluate the two independent health risk assessment reports of Davies and Sears, and to present the merits of each. In this case study, some of the strengths and weaknesses of each approach are exposed, together with their implications for both science and public policy. A cautionary message is offered about the appropriateness of each approach in the context of serving special interests, as opposed to serving the interests of the local community. This discussion is offered as a contribution to advancing our collective understanding of how science, purportedly in pursuit of truth, can be misused and thus serve narrow, special interests as opposed to the public interest.

Collateral damage from linear reductionism in addressing issues of community health risk assessment negatively impacts the very direction that science itself takes. However, a fuller treatment of this latter dimension is beyond the scope of this chapter.

Problem definition

Heretofore, the mainstay of the modern chronic disease epidemiologist has been to rely almost exclusively on quantitative methods. Epidemiologists are now recognizing, however, the need for new methods and concepts in order to contribute usefully to studies relating to complex environmental interactions. To help in advancing this thinking, distinctions need to be made between Newtonian dualism and complexity paradigms.

The Newtonian approach relies on evidence based in reductionism, predictability, and linear thinking; its focus is on reducing uncertainties in the presence of assumed system equilibrium; it is deterministic in its approach. In contrast, the complexity paradigm embraces holism (i.e. systems thinking), and it accepts the unpredictability of nature by assuming system instabilities; it is decidedly non-linear, acknowledges uncertainty, and is non-deterministic in its approach.

The work of Funtowicz and Ravetz (1992) was seminal. It advanced the integration into quantitative methods of participatory, qualitative approaches for a post-normal scientific paradigm for addressing real-world complexity.

The risk governance-risk assessment paradigm

Since the 1960s, the United States Environmental Protection Agency (USEPA) has been instrumental in advancing thinking about frameworks for the conduct of health risk assessment. The approach has evolved with the release of its most recent report on the topic (USEPA 2014). In addition, analogous thinking was formally embraced in 2005 under the rubric of 'risk governance' by the International Risk Governance Council (IRGC). It serves as a useful checklist for engaging fully to address community concerns about a perceived health concern/risk.

The USEPA health risk assessment framework has been seen as one that is inherently linear-reductionist by virtue of the fact that it addresses single exposures and single health effects, one pair at a time. With growing recognition that the 'single exposure, single health effect' approach is not reflective of real-world experiences, the framework is being modified to take complex mixtures more into account. Nevertheless, the paradigm remains useful for grasping public health needs *vis à vis* policy approaches for risk governance in response to community concerns.

The framework provided in the box below is a five-step linear paradigm that makes transparent the needed considerations for addressing a community's concerns. It is adapted from that put forward more recently by both the IRGC (2005) and the USEPA (2014).

A five-step risk governance-health risk assessment framework

1   *Conduct a hazard assessment*: Are potentially harmful pollutants present in the area?

2   *Conduct a vulnerability assessment*: If potentially harmful pollutants are present in the area, could people/animals be exposed to them, and by what pathways (e.g. via the air, dust, water, soil, food)?

3   *Conduct a health risk evaluation*: If a pathway of exposure is established, is there epidemiologic evidence for health effects in the local population from research completed in populations having analogous exposures elsewhere in the world, or from experimental evidence of biological harms?

4   *Develop a risk management plan*: If health effects are evident/anticipated/demonstrated, how will the problem be managed to prevent further exposures from occurring?

5   *Develop a risk communication plan*: How will we communicate with the community of stakeholders?

Any health risk assessment process, from a risk governance perspective, needs to include consideration of how each of the five steps will be conducted and implemented in advance of attempts to address community health concerns. In so doing, a more rigorous, robust assessment and appropriate handling of a community's concerns is best assured. It should be clear that quantitative methods alone will not suffice for the participatory, integrative aspects that need to be addressed through the questions in the box above.

In the AER Proceeding, citizens had indicated in their submissions that they would like to talk to the 'health experts'. Consistent with the participatory method of the qualitative, post-normal science approach, Sears pursued this opportunity, and in the end Davies participated in two of the three sessions, having missed the first one.

Understanding influence and its impact

All members of society, including scientists, are influenced by their contextual narrative. The contextual narrative is the script by which we pursue our lives, comprising, among other things, social norms, environmental constraints, values of our peers and our community, and prevailing interests and focus. This script limits each of us in our world view. It influences each of us in the very questions that we ask of ourselves, of our community, and of our governments; and for scientists, in the very hypotheses we may pose and questions that we ask of our study participants/research subjects. Even in North America's pluralistic society, a dominant paradigm reigns, making somewhat predictable that in which each of us engages and how we approach it.

Several biases have been identified in epidemiology arising from the influence of dominance. These are repression and suppression/oppression biases (Porta 2008). The former occurs when scientists are not inclined to address any question that may go counter to dominant interests. The latter bias relates to the suppression/oppression of produced evidence that goes counter to prevailing dominant interests. In the pursuit of truth and of optimum public health, science and society are not well served when such biases operate.

A role for public interest science

Applied science in the public interest affords an impartial, structured process for dealing with complex societal issues. Scientists should strive to synthesize knowledge from various scientific disciplines and/or stakeholders such that integrated insights are made available to decision-makers. By committing to the integration of quantitative and qualitative methods and expertise in an interdisciplinary fashion, an understanding of cross-linkages and pathways, of complexity, is more likely to be revealed (Rotmans 1998). Indeed, participatory methods, one form of qualitative evaluation, provide a mechanism to broaden our understanding of complex issues.

The context at the time of the Alberta Energy Regulator decision on 16 April 2014

To apply the foregoing considerations to the Peace River Proceeding, the Alberta Energy Regulator (AER) ruling demonstrated wisdom despite some of its rationale not being especially cogent (in this author's view). Indeed, perhaps they had been influenced in their decision by Shell indicating that the elimination of emissions was possible. In addition, many dimensions were ignored and/or were not well reflected in the AER decision (e.g. toxic metals were not mentioned, the water and the soil were not addressed and, importantly, the laboratory data problems were not appreciated).

Regardless, the AER ruled that Baytex must install pollution protection equipment to stop the escape of gasses to the atmosphere. A four-month deadline was set. The displaced families would continue litigation against the company.

Context is provided by a local news radio (630 CHED) report by Jeremy Lye on the day of the decision:

> A statement from Baytex says the company 'has continually worked to lower its environmental impact and to reduce emissions from all its operations.'
>
> But in the meantime, the venting continues, and Keith Wilson, a lawyer who represents those families, says according to court documents, Baytex is making $45,000 a day net profit.
>
> 'While these families have been out of their homes, Baytex has generated a profit of forty-five million dollars,' Wilson says.
>
> 'The simple reality is they have a financial incentive to continue to open vent,' he says.
>
> Wilson says the fight for compensation after two years of health complaints will be a long one.

'This is probably going to be a five- to seven-year trial and it's a frightening thought for these families to have to continue to fight this out for that much more time, it's just wrong, it shouldn't be happening,' Wilson says.

Stakeholder and Community Relations director for Baytex, Andrew Loosley, says the company will defend itself against the litigation.

'They've filed a lawsuit against us and we're going to vigorously defend our interests,' Loosley says.

Loosley says Baytex has already begun installing vapour covers and says it will continue to make the changes the AER demands.

(Lye 2014)

To lightly deconstruct the above events and arguments, we observe corporate interests competing with community interests. In the AER decision, we see political expediency at play. Both sides are posturing for ongoing adversarial roles. After all, any decision from the Regulator could open the way for civil litigation.

To exemplify the real nature of corporate self-interest at work, how disingenuous and self-interested is Baytex when it stated during the hearing that they had installed some vapour recovery equipment, with Wilson replying that it was made of garbage bags and duct tape? What enforcement mechanism and monitoring will be needed to determine Baytex's compliance with the AER ruling and the purported intent to be responsive to community health concerns?

How useful was the AER Proceeding and what could scientists learn from the experience? The lessons for scientists are the focus of this chapter.

Lessons learned for health and risk scientists

Both professional and public education, including that of public servants in risk assessment and management as well as the regulatory enforcement arm of government, have a place in preventing harms that can arise through misapplication or misrepresentation of scientific methods. The more post-normal, quantitative and qualitative approach taken by Sears (as opposed to the linear-reductionist, quantitative approach of Davies) is demonstrably more credible, especially in light of the failure of Davies's predictive modelling to bear out that which happened in practice.

While peer review can be of benefit in reviewing how projects are done, those projects undertaken by consultants are not commonly subjected to peer review. One mechanism to avoid harm might be to require that consultants submit their proposed approach for review to an objective third party with the expertise to determine the appropriateness of what it is that is being proposed. More specific lessons for the training of epidemiologists would be as follows.

- Include in health sciences training improved basic ecological, biological, chemistry, toxicological, sociological, geopolitical, economic (i.e. full cost accounting), ethics and philosophy, as well as systems thinking.

- Insist on qualitative, as well as quantitative, approaches when any attempt is made to address complex problems that involve health and multiple stakeholder interests.
- Insist on high-quality, public reporting of scientific studies, to ensure transparency as to methods, data quality, and analyses.
- Develop and teach public health ethics in relation to the conduct of community health risk assessment; professional integrity must be emphasized over self-interest. Above all, the mission of epidemiology is to maintain, enhance, and promote health in communities worldwide. Professionals in this field therefore must work to protect the public health interest above any other interest, recognizing the realities of how relentless, tenacious and ruthless moneyed interests can be in derailing science (see 'Classical techniques that skew results' below) to serve their own short-term self-interest.

But what are we up against?

Several questions emerge, given that scientists are people with human frailties. While it is outside the scope of this chapter to address each of these questions, they do merit mention:

- What creates/drives misconduct in science?
- What tempts scientists away from the pursuit of truth?
- How does misconduct derail scientific discourse?
- How does misconduct influence public policy and hence population and global environmental health?

As professionals we have the moral duty, in the public interest, to call people on exhibited poor conduct. This ought to be done in a collegial way rather than one of confrontation. Sometimes, however, peer pressure is not effective to correct poor decisions before harm can arise, and calling people on what are seen as poor decisions has its challenges, especially in the context of speaking truth to power.

Classical techniques that skew results: from biased methods to junk science

Some scientists are engaged in the lucrative 'doubt science' industry, defending their work as supporting the notion that science advances through refutation (Epstein 1978; Davis 2007; Michaels 2008). However, it is now well known that standard techniques are used by such people as they contribute to 'junk science', the latter being produced usually through funding from moneyed interests. The latter is used to infiltrate the literature such that in court proceedings, doubt will work in favour of the defendant and make it unlikely that policy change will ensue. What are the standard techniques that these scientists use to foment uncertainty about cause and effect? They employ the following:

- statistically under-powered studies;
- inadequate follow-up methods;
- inadequate follow-up time;
- contaminated controls, and a broad range of degree and types of exposure among the presumed exposed group;
- ignoring known synergies among components of the mixture of chemicals to which people are exposed;
- inadequate laboratory practices that systematically under-estimate exposures;
- inappropriate analytical methods for calculations;
- unbalanced discussion;
- selective disclosure of competing interests; and/or
- linear-reductionist quantitative methods without post-normal qualitative approaches to complement them.

In addition, arguments that are used to delay action in support of maintaining the *status quo* include such classical techniques used to skew research results as:

- biased or selective interpretation;
- mechanistic information is ignored for inferring effects;
- exaggerated differences are made between human and toxicology studies, the insistence being on separating effects seen in animals from effects in humans, or the converse as is convenient; and/or
- the fact that molecular structures predict hazard potential is ignored.

Finally, techniques employed that skew and delay policy, and also create an unhelpful division among scientists, include:

- the insistence on demonstrating effects in local populations of exposed people despite demonstrated effects in humans elsewhere; and/or
- the failure to make explicit those value judgements that underlie decisions about selecting appropriate standards of evidence to draw policy-relevant conclusions.

A fuller exposition of the above techniques can be found in Cranor (2011).

Conclusions

Epidemiology is an applied science that bridges work from *in vitro* and animal systems (i.e. toxicology) all the way to public health and health policy. The scientific paradigm used in health risk assessment can pit industry against regulatory frameworks and the communities that regulation is intended to protect.

Both the traditional quantitative approaches, as well as qualitative, post-normal, integrative approaches are of value in applied science. The science is advancing, but most importantly, the dominant narrative and the nature of moneyed influence need to change if public health is to be protected through policy intervention. This applies both to retrospective reviews of current enterprises, and reviews in the course of permitting proposed extraction and industrial developments such as mines, pipelines and other developments with environmental and health impacts. Roleplay may help students to recognize when they are cast into conflict-of-interest situations and, if so, how to disclose and extricate themselves therefrom.

When governments elect to dismantle the infrastructure for environmental exposure monitoring, this creates data gaps for science; it also makes law enforcement impossible because, in the absence of evidence, a case cannot be contested. Public access to high-quality data is essential also under the public's right to know.

Acknowledgements

This chapter emerged from a report under contract to a participant in the Alberta Energy Regulator (AER) Proceeding No. 1769924 into Odours and Emissions in the Peace River Area, Alberta, Canada (2013–2014). The report was neither presented as evidence nor discussed at the hearing. Dr Soskolne's client has permitted publication of this case study in the interests of advancing his observations with respect to health science.

Note

1 All documentation and submissions relating to the Proceeding are accessible at the website of the AER at www.aer.ca/applications-and-notices/hearings/proceeding-1769924 (accessed 15 March 2015).

References

Cranor, C. F. (2011) *Legally Poisoned: How the Law Puts Us at Risk from Toxicants*. Boston, MA: Harvard University Press.

Davis, D. (2007) *The Secret History of the War on Cancer*. New York: Basic Books.

Epstein, S. S. (1978) *The Politics of Cancer*. San Francisco, CA: Sierra Club Books.

Funtowicz, S. and Ravetz, J. (1992) Three Types of Risk Assessment and the Emergence of Post-Normal Science. In S. Krimsky and D. Golding (eds), *Social Theories of Risk*. London: Praeger.

IRGC (2005) *Risk Governance: Towards an Integrative Approach*. White Paper no. 1. Geneva: International Risk Governance Council. Available at www.irgc.org/risk-governance/irgc-risk-governance-framework.

Lye, J. (2014) Displaced Families to Continue Baytex Litigation. 16 April. Available at www.630ched.com/2014/04/16/displaced-families-to-continue-baytex-litigation (accessed 15 March 2015).

Michaels, D. (2008) *Doubt is their Product: How Industry's Assault on Science Threatens Your Health.* Oxford: Oxford University Press.

Porta, M. (ed) (2008) *A Dictionary of Epidemiology*, 5th edn. New York: Oxford University Press.

Rotmans J. (1998) Methods for IA (Integrated Assessment): The Challenges and Opportunities Ahead. *Environmental Modelling and Assessment* 3: 155–79.

USEPA (2014) *Framework for Human Health Risk Assessment to Inform Decision Making.* Washington, DC: Office of the Science Advisor, US Environmental Protection Agency. Available at www.epa.gov/osa/framework-human-health-risk-assessment-inform-decision-making (accessed 15 March 2015).

17 Radical changes in Canadian democracy

For ecology and the 'public good'

Peter Venton

Introduction

Three of the major problems facing Canadian society today are the high degree of inequality in the distribution of income and wealth, the continuing degradation of the environment and the fiscal crisis in the government sector. All three problems stem from the failure of successive Canadian governments over the last 45 years to manage the capitalist market economy to maintain full employment and a fair distribution of income and wealth. In part, this failure arose from the corporate sector's capture of the political agenda. Corporations pressured governments into believing that lower taxes on corporations, their owners and their managers combined with a smaller government sector would lead to greater private investment and economic growth sufficient to ensure the achievement of the government's goal of full employment.

Today, none of the leaders of Canada's three major federal political parties agree with the majority of Canadians that taxes should be increased on corporations and high income individuals, despite the fact that surveys indicate that a majority of Canadians believe that greater taxation of high incomes and wealth should be implemented to correct the excessive degree of inequality in the distribution of income and wealth. Indeed, the views of these leaders are consistent with the corporate agenda of lower taxes. This observation has prompted the central argument of this essay – that the corporate capture of the political agenda was enabled by an erosion in Canadian democracy.

The erosion in Canadian democracy has led to the present 'democratic deficit' – a term defined by the gap between the ideal of democracy and actually existing democracy. In this chapter the ideal of democracy is described in terms of its definition and principles, its rationale and its prerequisites. The chapter also identifies the gaps between the ideal elements and actually existing democracy in Canada. Foremost among these gaps is the large proportion of the electorate who do not vote and whose interests would be better served by more progressive taxation and higher corporate income tax rates than prevail presently.

A related gap is the pursuit of political power by all major political parties for its own sake rather than for representing the majority of Canadians. In this pursuit, parties have focused on their base of supporters and the political centre.

Owing to the high degree of self-disenfranchisement, voters in the centre are not likely representative of the majority of Canadians. In addition, public opinion has been significantly influenced by the rise of conservative propaganda that has repeatedly promised Canadians that they would be better off with lower taxation of corporations and smaller, more efficient governments. At the same time, the growth in post-war consumerism that affected large numbers of Canadians spilled over into political life. As a consequence, many citizens looked to governments as purveyors of personal private benefits rather than institutions for serving the 'public good'. And politicians tended to confirm these citizen views and abandoned their leadership role of attempting to change citizens' minds about the benefits of government policies and services. In any case, the large majority of politicians had no formal education in political economy or public sector economics and were therefore ill-equipped to play such a leadership role. In this context of increasing citizen disengagement, rising conservative propaganda and consumerism among citizens and politicians, the Conservative party gained political power with Canada's 'first-past-the-post' electoral system and a split of votes between the opposing New Democratic Party (NDP) and Liberal Party to become the governing party with the support of a small base of the electorate; in the 2011 federal general election, less than 25 per cent of the electorate voted for a Conservative party candidate.

The last section of this chapter presents proposals for eliminating the gaps that comprise the democratic deficit. They include adopting compulsory voting, reforming the electoral system to reflect the objective of proportional representation, refocusing public education on civics and the principles of collective decision-making, training politicians in political science and public sector economics, increasing public financial support for political parties to enable politicians to engage citizens in democratic discourse, and establishing a royal commission on the public good and social progress.

Origins of problems facing Canadian society today

Since the late 1960s, successive federal and provincial governments in Canada have failed to heed John Maynard Keynes' revolutionary prescriptions in the 1930s for saving capitalist market economies from chronic tendencies to under-consumption and unemployment leading eventually to economic depressions. Briefly, Keynes's prescriptions were (a) to modify the market economy through progressive taxation to affect income and wealth distribution and (b) to socialize investment to ensure full employment (Minsky 2008: 147). Instead of adopting Keynes's prescriptions, they responded to pressure from corporations to reduce corporate income taxes to enable them to maintain after-tax profit rates in the face of the growing international and domestic competition that was putting downward pressure on their profits.

Starting in the late 1960s, governments in Canada dramatically reduced rates of taxation on corporations, the wealthy and high-income individuals. Federal estate taxes and Province of Ontario succession duties that prevailed in the 1960s were eliminated in the 1970s. In Canada, corporate and personal income taxes are shared between the federal and provincial governments. The combined federal and Ontario government tax rates on the income of Ontario corporations other than small businesses was reduced from 51 per cent in 1969 to 30 per cent in 2010 – a percentage reduction of 41.2 per cent (Cahill 2007; Weir 2012). Between 1969 and 2010, reductions in the combined federal and provincial government rates of taxation on personal incomes of Ontario taxpayers were greatly reduced. For the ten highest personal income brackets in 1969, tax rates were reduced by 57 per cent for annual earnings over $2.8 million in 2010 dollars and 19.6 per cent for the $81,941 to $88,491 income bracket. The average reduction for the ten highest income brackets was 39 per cent. The average reduction for the next eight highest income brackets was 26 per cent. These brackets were in the range of $37,106 to $81,940 in 2010 dollars (Department of National Revenue and Taxation 1969: 16–18; Canada Revenue Agency 2010).[1]

Fiscal crisis in the government sector

Reductions in tax rates over the past 45 years have lowered government revenues and led to increasing government deficits and rising government sector debts. The growth in debt was arrested starting in the mid-1990s with austerity programmes of the federal and provincial governments. However, after the world financial crisis of 2008, the growth rate in government-sector revenues in Canada declined while governments' massive fiscal stimulation initiatives increased their expenditures. The result was a series of annual budgetary deficits that increased government-sector debt over the period from March 2007 to March 2013. Federal government debt increased from $467 billion to $602 billion while Ontario government debt increased from $153 billion to $252 billion. The combined effect of these changes was to increase government sector debt attributed to Ontario households from $65,534 per household in 2007 to $96,801 per household in 2013 – a 45 per cent increase (Government of Canada 2008, 2013; Ontario Financing Authority 2014).[2]

During the same period, private consumer and mortgage debt of Canadian households increased from 137 per cent of personal income in 2007 to 163 per cent in 2013, although debt servicing costs fell from 9 per cent of disposable income to 7 per cent owing to falling interest rates (Parkinson 2014). Nevertheless, many Canadians are reported to be living from pay cheque to pay cheque and consequently appear to be in no mood to countenance tax increases. This situation represents a fiscal crisis for the government sector.

Environmental degradation

Successive federal and provincial governments hoped that reduced taxation of corporations would induce corporations to increase investment to create more jobs sufficient to reach government goals of full employment. These strategies called for a higher rate of economic growth than would have been needed with Keynes's prescriptions. However, not all of the greater private investment and related economic growth was accommodated by natural resources so that some degradation in the environment resulted. Meanwhile, starting in the 1980s, the scientific community made dire predictions of the consequences of not reducing the level of greenhouse gas emissions. Despite these warnings there was a general unwillingness of political leaders to change their strategy of economic growth at the expense of the environment. A notable exception was the leader of the Liberal Party of Canada, Stephane Dion, who proposed in 2008 a carbon tax aimed at reducing greenhouse gas emissions. Despite his assertion that the carbon tax would be offset by other tax reductions rendering it 'fiscally neutral', Mr Dion failed to win a majority of seats in the House of Commons in the 2008 federal election. In conclusion, the policies of most political leaders were for full employment by means of tax reduction policies at the expense of the environment.

Inequality in the distribution of income and wealth

Reductions in tax rates over the period from 1969 and 2010 were greater for high income earners and the wealthy than for the less affluent and this caused some increase in the inequality of income and wealth. However, there was another important factor at work. Economic growth enables corporations to realize economies of scale and greater specialization in production and distribution of their products and services, both of which increase their productivity and profits. However, not all of the resulting increase in profits is commensurate with increasing labour productivity owing to the strong bargaining power of corporations with respect to workers in circumstances where changes in the total supply of labour in the economy due to demographic, cultural and institutional factors exceeds changes in total demand. For example, labour productivity in Canada increased by 37 per cent between 1980 and 2005 but virtually all of the increase went to owners, investors and managers of businesses who are generally more affluent than workers. If workers had received a fair share of the profits generated by their productivity, their median income would have gone up by $15,000 a year by 2005. Instead, the annual median earnings adjusted for inflation for full-time full-year employees in Canada went up by $53 (Dobbin 2013: 22).

A failure of democracy

Inequality of income and wealth became a popular and controversial issue following the Occupy Wall Street movement in 2011 – a movement which spilled over into Canada. The current degree of inequality of income and wealth in Canada is misaligned with the views of the majority of Canadians. Andrew Moran reported the results of a Broadbent Institute 'Equality Project' survey of a sample of Canadians on the issue of income and wealth inequality. It revealed that over three-quarters of the respondents believe that income inequality is a serious issue and that many were willing to do something about it. Specifically,

83 per cent of respondents said that they supported higher income taxes for the affluent in Canada and nearly three-quarters said they wanted corporations to pay higher tax rates than 2008 levels. Some 69 per cent supported the introduction of a 35 per cent inheritance tax on any estate that is estimated to be valued at $5 million or more (Moran 2012). If the survey were a random representative sample, the foregoing results could be said to apply to the majority of Canadians.

Canada has a Westminster style of parliamentary government modelled on the British system. In addition, Canada has a federal system comprised of a federal government and ten provinces and two territories. Major tax bases such as corporate income, personal income and value added taxes are shared between the federal government and the provinces. Further public services are divided between the federal and provincial and territorial governments.

From 2006 to 2011, Mr Harper was Prime Minister as leader of the Conservative Party, which held a minority of seats in the House of Commons. But in the federal election in the spring of 2011, the Conservative party won a majority of 166 of the 308 seats and formed the government. In that election the NDP led by Mr Jack Layton won 103 seats and became the official opposition party, replacing the Liberal party, which had been the official opposition party prior to the 2011 election. The Liberal Party, led by Mr Michael Ignatieff, won only 34 seats and Mr Ignatieff resigned immediately following the election. He was replaced by Mr Bob Rae, who acted as an interim leader pending the search for a new leader. In April 2013, Mr Justin Trudeau, son of the former Prime Minister, Pierre Elliot Trudeau, was elected leader of the Liberal party. Some four months after the 2011 federal election, Mr Layton resigned as leader of the NDP for health reasons and Mr Thomas Mulcaire subsequently became leader of the NDP.

The leaders of the three parties following the 2011 federal election had markedly different views from the majority of Canadians about income inequality. Mr Harper had publicly stated that 'there is no such thing as a good tax'. Furthermore, the Harper-led government reduced the tax rate on corporate income from the 21 per cent that prevailed in the 2004 to 2007 period to 18 per cent by 2010 and to 15 per cent in 2012. These changes represented a 28 per cent reduction in tax rates over a five-year period (Stanford 2013: 71, 72). This suggests that Mr Harper would not support any increase in tax rates on wealthy or high-income earners.

Under NDP leader Jack Layton, the more traditional social democratic positions, such as increased taxation of wealthy individuals (i.e. an inheritance tax), the acceptance of deficit spending, the rapid creation of new universal social programmes and references to expanding public ownership were gradually eliminated from the party's discourse (Martin 2012).

Following Mr Layton's resignation as NDP leader, one of the leadership candidates, Mathew Cullen, proposed increased taxation on higher income earners. However, he was defeated by Thomas Mulcaire who became leader. Mr Mulcaire does not support an increase in taxation of higher-income earners, although he does support an increase in the rate of taxation on corporate income.

Meanwhile, in a speech in 2011, interim Liberal party leader Bob Rae stated that 'Liberals believe the greatest respect we can show people is to talk candidly about the problems we face and how we might deal with them, not pander to every whim and prejudice . . . A simple bumper sticker slogan of "tax the rich" will not provide answers to the real issues we face as a country' (Rae 2011).

In summary, the views of these leaders are inconsistent with the preferences of the democratic majority and are closer to the interests of the corporate sector. This inconsistency can be explained by the hypothesis that corporate capture of the political agenda has been enabled by an erosion in Canada's democracy. The erosion has led to a substantial 'democratic deficit' – that is, a gap between the ideal of democracy and the way democracy actually functions in Canada today.

The nature of democracy

The first step in the analysis of the democratic deficit is to describe the ideal of democracy in terms of its definition, its principles, its rationale and its prerequisites. Briefly, democracy is defined as a system for organizing collective human life in a society that involves government by all the people through elected representatives for the benefit of at least a majority of citizens. It also involves the acceptance of the principles of equality of rights, equal treatment and social justice.

The principles of equality of rights and equal treatment mean that every citizen has a right to vote and every vote is equal. They also mean that all citizens have equal access to, and participation in, the legislative process of developing, monitoring and administering of laws, policies and programmes governing the behaviour of all citizens. Furthermore, all groups, whether religious, cultural, gender, ethnic or racial, are treated equally. Embodied in the principle of equal treatment is the idea of fraternity, which means a group joined by common interest, fellowship or ideas. Fraternity implies treatment of the common people and elites as equals.

Social justice principles are about two aspects of fairness. The first principle is equality of opportunity, which requires that all should have an equal opportunity to participate in the economic and social life of society. The second principle is that inequality of wealth and authority are just only if they result in compensating benefits for everyone and, in particular, the least advantaged members of society. This second principle is referred to as the 'difference principle', as enunciated in philosopher John Rawls's 1971 book, *Theory of Justice* (Kloppenberg 2011: 90–92).

A social contract

Democracy represents an implied social contract between citizens whereby citizens give up some of their individual liberty in exchange for collective benefits. This means that freedom is not absolute in democracy but rather competes with citizens' preferences for collective benefits. To put it another way, every citizen in a democracy gives up a degree of freedom by accepting a degree of responsibility for the welfare of his/her fellow citizens.

The 'public good' is democracy's rationale

The public good is a collection of society's ends. These ends are numerous, diverse and shared. For example, they include the objectives of peace, security, health, limits on the inequality of private income and wealth, a balance between work and leisure time for social, cultural, recreational, physical activity and other pursuits, democratic engagement, and the level and quality of environmental and ecological resources. They also include the absence of 'social bads' as measured by the incidence of crime and family breakdown.

Over the last thirty years, some social scientists have developed indexes composed of indicators of the wellbeing of citizens in nation states. These indexes are positioned by some public choice theorists as summary indicators of the values of all citizens if, hypothetically, all citizens had an opportunity to vote on their preferences for each and every one of the elements of the index. The indexes are variously labelled as 'social welfare', 'quality of life', 'happiness' or 'genuine progress'. Whatever the label, the indexes comprise statistics that can be considered as measures of elements of the public good. One example is the Genuine Progress Indicator developed in the 1980s for the USA. Another is the array of indicators published each year by the Australian government – indicators selected after a public consultation (Coyle 2011: 202).

In Canada, the only such index for the whole country of which I am aware is the University of Waterloo's Canadian Index of Well Being. Table 17.1 shows percentage changes in eight broad indicators of wellbeing of Canadians over the period 1994 to 2010. Each of the percentage changes in the eight broad indicators is weighted by a fraction of one and the weighted percentages are aggregated to form a single overall percentage. As well each broad indicator is an index of several component detailed indicators that are also weighted by a faction of one and aggregated to form the broad indicator. Some examples of these detailed indicators are presented in the notes to Table 17.1. The results of the process of weighting all of the detailed and broad indicators is the overall index of wellbeing that has increased by 5.7 per cent over the 16-year period from 1994 to 2010. By contrast, the median income of Canadian families, adjusted for inflation, increased by 28.6 per cent over this time period. Insofar as the increases in family income are viewed as an indicator of success of Canada's efficient economic system, the gap between the two suggests that Canada's economic system is not nearly as effective as it could be in serving the public good.

Citizen responsibilities are the prerequisites for democracy

Democracy demands certain responsibilities. Unlike the logic of consumerism, these demands are not matters of choice but matters of duty. In a democracy, only

Table 17.1 Trends in elements of the Canadian Index of Well Being, 1994–2010.

Category	Change
Living standards[a]	14.3%
Education	21.8%
Community vitality	10.3%
Democratic engagement	7.0%
Healthy populations	4.9%
Time use	1.3%
Leisure & culture[b]	−7.8%
Environment[c]	−10.8%
Overall Index	**5.7%**

Source: University of Waterloo (2014)

a. The index of living standards increased by only 14.3 per cent despite the fact that after-tax median income of families in 2010 increased by 28.6 per cent (in inflation adjusted dollars). The gap between the two figures is due to the presence of negative factors of wellbeing. These include an 11.4 per cent increase in the inequality of income as measured by the ratio of average after-tax family income of the top quintile of family income earners to the bottom quintile of family income earners. Also, there was a 2.8 per cent reduction in employment quality as measured by the CIBC index of employment quality. Finally, there was a decline in the index of economic security produced by the Centre for the Study of Living Standards.
b. Leisure and culture indexes declined as annual visitations to national parks and historic sites declined by 28.7 per cent and annual hours spent in volunteering for culture and recreation organization declined by 21.3 per cent. Also, time spent in social leisure activities declined by 19.7 per cent. Finally, average attendance per performance fell 10.7 per cent. These negative changes more than offset positive increases: a 24 per cent increase in the frequency of participation in physical activity over 15 minutes and a 7.2 per cent increase in the average number of nights away per trip in the past year on vacations over 80 km from home.
c. The quality of the environment decreased substantially as a result of an increase in the ecological footprint (17.2%) and a reduction in the Canadian Living Planet Index of 23.8 per cent and a 9.8 per cent increase in annual megatons of greenhouse gas emissions. The Viable Metal Reserves Index declined by 40.3 per cent. These negative changes outweighed positive changes: 11.7 per cent increase in primary energy production and a 3.9 per cent increase in water yield in southern Canada.

citizens determine the nature of the public good and discuss the choices among the various societal ends that make up the public good. In this regard, citizens have three responsibilities:

•   a duty to vote in elections;
•   a duty to engage in discourse with their fellow citizens and elected representatives; and
•   a duty to spend time to become informed about issues related to the public good.

These are the requirements of good citizenship. Without good citizenship, power will be ceded to corporations, the military or oligarchies and their interests. The former premier of Ontario and subsequent leader of the Liberal Party of Canada, Bob Rae, stated, 'Without good citizenship the whole game is given over to spin doctors' (comment recorded by the author at the Reflections on the Public Good conference at the University of Toronto, 7 November 1998).

In a democracy, all eligible citizens rule. Using the metaphor of business organization this means that every citizen is, in effect, 'the boss'. The citizen (i.e. the boss) delegates his/her elected representative (i.e., subordinate or direct report) the responsibility for achieving the public good for all citizens. So the organization of democratic government is analogous to an inverted private organization with many bosses and few subordinates where all bosses have equal power in the organization. In the language of the theory of governance of organizations, the citizens are 'principals' and their political representatives are 'agents' who are contracted to serve the interests of the principals. Oliver Williamson, a Nobel Laureate in economics, pointed out that all modes of organization are flawed (Coyle 2011: 214). The flaw is due to the fact that principals and agents have different interests and agents will tend to act in their own self-interest rather than in the interests of the principals. Owing to the existence of asymmetry of information between the principal and the agent, there is a lack of transparency in the relations between the two and consequently the differences between them are unlikely to be fully reconciled in favour of the principal – despite the fact that the principal has power over the agent. In these circumstances the best that can be expected for good governance in a democracy is a responsible citizenry that has a sense of empowerment for insisting on accountability and performance from their political representatives. To this end the citizenry need to have a modicum of knowledge about the public good and a broad knowledge of the effectiveness of alternative feasible policies or approaches for achieving the public good. These are the prerequisites of democracy.

Canada's democratic deficit

This section analyzes the gaps between the foregoing ideal democracy and actually existing democracy in Canada – the gaps that make up the democratic deficit.

Citizen disengagement is rampant

The first problem of democracy starts at the top with the citizens/bosses. A majority of citizens are neither interested in, nor knowledgeable about, public issues that affect their own wellbeing, much less the wellbeing of their fellow citizens and the country. The following are rather startling revelations about democracy in the US, Britain and Canada.

In 1933, Clem Whitaker and Leone Baxter formed Campaign Inc., the first political consulting firm in the US – and indeed the world. Over a successful career of 20 years they developed a number of techniques for running campaigns. One of these was:

> Make it personal because candidates are easier to sell than issues. A wall goes up when you try to make Mr. and Mrs. Average American work or think.

The average American doesn't want to be educated; he doesn't want to improve his mind; he doesn't even want to work, consciously at being a good citizen.

(Lepore 2012: 55)

Meanwhile, British historian Tony Judt argued that the generation of the 1960s, as manifested in the new left, adopted a more selfish focus on individual rights than the old left. A by-product of this shift was a decline in the shared sense of purpose that had characterized the old left. The propensity of a younger generation to look exclusively to their own needs resulted in a steady falling away from civic engagement in public decision making (Judt 2010: 90, 91).

Only a minority of Canadians are actively involved in politics and public policy issues. A Samara Canada survey of a sample of 2,287 Canadians revealed that, in the last 12 months, 40 per cent of Canadians had discussed a societal/political issue face to face or on the phone, 31 per cent had contacted an elected official about an issue that concerned them, 30 per cent had worked with others on a community issue, 20 per cent had attended a political meeting and 10 per cent said that they were, or had been, a member of a political party in the past five years (Samara Canada 2013: 1, 3, 11, 12). In the federal general election of 2011, almost 40 per cent of Canadians eligible to vote did not vote. Fifty years ago the comparable figure was only 20 per cent.

Low voter turnout may not be a problem if the interests of the non-voters are similar to the interests of the voters. However, with respect to the issue of inequality of income and wealth, interests vary among different socioeconomic groups. In this regard there is some evidence to suggest that there may be wide differences in voter participation among different income groups. For example, Table 17.2 shows that the estimated US voter participation rate in the 2010 congressional elections for the affluent who earn more than $50,000 per annum was 68 per cent – far higher than the 30 per cent participation rate for the less affluent who earn less than $50,000 per annum. The result is that an affluent minority of 40 per cent of the US electorate account for 60 per cent of votes cast in the 2010 congressional elections.

For Canada, data on voter participation by income group are unavailable. However, in the 2011 federal election, 71 per cent of Canadian homeowners voted while only 54 per cent of non-homeowners, who are generally less wealthy, voted (Sharanjit and LaRochelle-Cote 2012: 8).

It should not be surprising that a greater proportion of high income earners and the wealthy participate in politics because they have an apparent, clear stake in policies of lower taxes and government austerity. For the highest 20 per cent of Canadian family income earners, the proportion of taxes to income is around 48 per cent (Lee 2007: tables 1, 16, 17). By contrast, the portion of benefits from government-sector spending is only 24 per cent (Mackenzie and Shillington 2009: tables 33–38). Therefore, it is in the economic interests of the highest 20 per cent of income earners to support a political party that promises reduced taxes and reduced government expenditures and, conversely, to resist tax increases that

Table 17.2 Voter participation rates in the November 2010 US elections.

Annual income range	Percentage of population	Percentage of voters	Percentage of population voting in the 2010 congressional elections[a]	Implied voter participation rate[b]
Less than $50,000	60%	40%	18.2%	30.3%
More than $50,000	40%	60%	27.3%	68.3%
Total	100%	100%	45.5%	45.5%

Source: percentages of population and voters are from Krugman (2009: 193), from which other data are derived (as explained in notes a and b below).

a. 45.5% of the citizen population over 18 reported having voted in the November 2010 elections (US Census 2014). Multiplying the percentage of voters in column 2 by 45.5% yields the results in column 3.
b. The implied voter participation rate is derived by dividing the figure in column 3 by the figure in column 1.

would fund government spending on infrastructure or operating programmes such as health care and public education. The reverse applies to the middle-income and, particularly, lower-income groups.

Consumerism has infected politics

Over the last 50 years, businesses have developed new products and services and innovative ways to convince Canadians to buy goods and services that they never knew they needed. As well, businesses have made greater efforts to increase customer satisfaction with more convenient products and friendlier service. Advertising has expanded to focus on the more impressionable (and affluent) young. All of this infused the Canadian public with a consumer mentality. Marketing and advertising are built on the concept that the customer is always right. The reality is that the customer is often not right in the sense that he/she has not spent the time to think about his or her needs or the transitory utility of the product or service, or their capability of affording the product.

Consumerism in Canada has spilled over into political life. In a 'consumer democracy', many citizens are keen to know what tangible specific things the government has done for them lately. They are often uninterested in or unwilling to sacrifice for the public good and believe taxes are solely a cost as opposed to a price for collective benefits. 'The consumer–citizen cannot be easily sold on anything that increases his/her taxes or reduces his/her consumption of material goods and services' (Delacourt 2013). For consumer citizens, political issues may be messy and complex and therefore inconvenient to address.

In a consumer democracy, the behaviour of politicians and political parties also changes. Politicians no longer attempt to change citizens' minds about the collective benefits of taxation. Instead they tend to confirm citizens' views in order to gain their support. Citizens' views that tax increases are just plain bad or that any tax increase will certainly be wasted through government inefficiencies are not to be challenged (*ibid.*). Political slogans replace comments on broad public issues related to the public good. Where such issues are controversial, political parties avoid them for fear of upsetting their political base or prospective recruits outside their base. This mindset of many politicians may explain Tony Judt's observation:

> The men and women who dominate western politics today are over-whelmingly products or by-products of the 60s. They do not seem to believe very firmly in any coherent set of principles or policies . . . They convey neither conviction nor authority . . . They fail to inspire their electorates . . . Convinced that there is little they can do, they do little.
>
> (Judt 2010: 133–4)

Conservative propaganda has influenced public opinion to support business interests

Economic propaganda is defined as statements unsupported by the facts or statements comprising economic fallacies. They are used by conservatives to serve three purposes:

- to reduce taxes and regulations on businesses;
- to increase government subsidies to businesses; and
- to increase demand for business products and services by privatizing government operations.

The following are some examples of conservative propaganda.

- *'Corporate productivity increases generate wealth and employment in the economy.'* In fact, overall productivity increases in the economy combined with offshore production of manufactured goods have reduced jobs in the last ten years. For example, Statistics Canada reports that the Province of Ontario lost 255,000 manufacturing jobs over the period January 2003 to September 2012 as employment fell from 908,900 to 654,200 (Beltrame 2013). The Ontario Ministry of Finance attributes this loss to improvements in technology and the movement of manufacturing jobs to lower-cost jurisdictions (Ontario Ministry of Finance 2014: ch. 3).
- *'Wealth is a reward for hard work and should not be taxed.'* A substantial portion of wealth is inherited rather than earned from lifetime savings. French economist Thomas Piketty estimates that in the US, inherited wealth probably accounted for at least 50–60 per cent of total private capital in the period 1970–1980 (Piketty 2014: 428).
- *'Tax cuts increase jobs.'* Tax cuts have only one fifth the effect on GDP that government expenditure on public infrastructure does according to the Federal Department of Finance estimates contained in report number six of *Canada's Economic Action Plan*. By inference they have only one fifth the effect on job creation (Stanford 2011).

- *'Tax rates over 50 per cent on high incomes reduce incentives to work hard.'* Economists Peter Diamond and Emmanuel Saez estimated that marginal tax rates for high-income earners could be increased up to 70 per cent without any material reduction in business profitability or productivity (Diamond and Saez 2011).
- *'Government is generally less efficient than private businesses.'* These generalized statements ignore many facts. For example, according to OECD reports, per capita expenditures for health in Canada, where most of the costs are publicly funded, are $4,445 per capita – far lower than the $8,233 per capita costs of health in the US, where most costs are privately funded (Kane 2012). Average per pupil costs of Ontario's public elementary and secondary schools were $11,207 in 2011–12 – substantially lower than average per pupil costs of many private elementary and secondary schools in the province where average tuition for K–12 pupils can be as high as $25,000 (Ontario Ministry of Education 2011: 1). From 1996 to 2011, public sector funds in Canada generated annual investment returns of 7 per cent, compared with 5 per cent in private sector funds (Leech and McNish 2013: 29).

Politicians are untrained for their main job in parliament

In Westminster-style governments such as Canada's, the paramount role of members of parliament (MPs) when in parliament is to establish policy and pass laws for achieving the public good. Their second role is to hold the government to account for the effective administration of laws. And their third role is to authorize expenditures ensuring that expenditures are cost effective in terms of the stated programme objectives. In this context, the main functions of MPs are to coordinate the roles and activities of the 'institutions' of businesses, free markets, families, unions and non-profit organizations, for example, to maximize the public good. To this end, they need to have an understanding of how the institutions of free markets and businesses, work with other institutions in society to achieve the public good. In this regard, formal education in political science and public-sector economics is an advantage but the vast majority of MPs do not have it (Samara Canada 2010: 7).

Political parties' aims are corrupted by the quest for power

In a 'consumer democracy' the main objective of the political party is to achieve political power for power's sake. To this end, parties appeal primarily to their 'base supporters'. Governing parties tend to initiate legislation that serves the interests of their base. Since the bases of parties often represent a minority of the electorate, these legislative initiatives do not tend to represent the interests of the majority of Canadians. The governing party's focus solely on achieving political power thus represents a corruption of the democratic ideal that government policies should represent the interests of all, or at least a majority, of the electorate.

The electoral system is partly undemocratic

In Canada's 'first-past-the-post' (FPTP) electoral system, the number of political candidates elected as MPs who are members of a particular party is based on how many MPs from that party received more votes in their electoral district than any of the candidates from other parties. When results such as these are aggregated for all electoral districts in the country, a party may win a majority of seats in the parliament with a minority of total votes cast. For example, after the election of 2011 the Conservative Party formed the government with a 54 per cent majority of seats in the House of Commons, although its candidates received only 39 per cent of the popular vote. As previously mentioned, the governing Conservative Party has tended to introduce and pass legislation by the force of its majority, which represents the interests of only a minority of the electorate at the expense of the majority of the electorate.

Proposals for restoring Canada's democracy

Adopt compulsory voting

If Canada adopted compulsory voting along the lines of Australia for federal, provincial and local governments, participation levels would rise. It is anticipated than compulsory voting along Australian lines would raise voter participation rates to close to 90 per cent.

Reform the electoral systems to reflect the objective of proportional representation

Federal and provincial electoral systems need to be reformed to reflect the principle of proportional representation (PR). PR refers to an electoral system in which the proportion of MPs in the House of Commons who are members of a particular party is the same as that party's proportion of the popular vote in the previous election. PR is aimed at significantly reducing the level of corruption of political parties that tend to represent only the interests of the minority of the electorate that make up their base. PR electoral systems reduce the political power of the governing party because they often lead to minority governments. Governing parties in minority governments are forced to take more collaborative and cooperative initiatives with other parties in order to pass legislation. As a result, minority governments tend to represent a greater diversity among the electorate and therefore a higher proportion of the electorate.

Refocus public education on civics and the principles of collective decision making

By definition, democracies do not work with uninformed electorates. For this reason Ancient Greece limited the franchise to the few who were presumed to be knowledgeable and engaged. This is not to suggest that the franchise should be limited in Canada but rather that all citizens need to be educated so that they can be fully engaged in political discourse and participate in the political system. This education should include precepts in economics and political science that are related to the successful operation of economic systems that governments are expected to establish and manage. Further, in an ideal democracy, the variations in educational outcomes among socioeconomic classes and regions ought to be minimal in order to realize the democratic principle of equal participation in legislative processes for developing policies.

Provide politicians with formal education in political science and public-sector economics

Although time constraints would make it difficult, formal training programmes for politicians would assist the operation of democracy. Politicians would be educated in areas that are relevant to performing their main roles in Parliament. This education would also enable them in an informed way to lead the public in discourse on the public good and economic systems based on the principles of democracy. As well, it would enable them to counter conservative propaganda with rational evidence-based discourse.

Knowledge that citizens should be able to expect of their political representatives includes an understanding of the principles of democracy, the mechanics of civics, the principles of economic systems and public-sector economics. This knowledge would help politicians to serve the goal of governments in a democratic state to coordinate the various 'institutions' of families, businesses, markets, and monetary and financial systems for the optimal achievement of elements of the public good. Effective, accountable government requires engaged and informed citizens who demand knowledgeable political representatives with the integrity to serve the public good on their behalf.

Increase public support for political parties to engage in democratic discourse

There is a massive imbalance between societal resources devoted to the promotion and manufacture of material wants and the resources available to politicians for engaging with citizens in democratic discourse. This imbalance needs to be reduced. To this end, funding of political parties from general tax revenues should be substantially increased. Moreover, the funding should be distributed democratically through government per vote subsidies to parties so that political representatives of the less affluent electorate are on a more level playing field with political representatives of the more affluent electorate.

Establish a royal commission on the public good and social progress

The indexes of wellbeing that have been developed by social scientists appear as plausible indicators of the elements of the public good. However, we cannot be sure. Only a process of democratic deliberation and political confrontation among Canadians can determine the public good. Such a process should lead to some broad definition of the nature and magnitude of the desirable elements of the public good that reflect the values, interests and priorities of the majority of citizens. To this end, the government of Canada should establish a royal commission on the public good and social progress to engage as many Canadians as possible in that discourse.

Conclusions

The implementation of the foregoing proposals for restoring Canada's democracy and reducing its democratic deficit will take a considerable time. However, they are expected to transform Canadian democracy in several respects.

With compulsory voting, political parties will have incentives to represent the interests of many more Canadians than they currently do. Electoral systems based on the principal of proportional representation are expected to prevent the domination of one party that chooses to implement policies for the benefit of a small minority of the electorate at the expense of the majority.

The refocus of education on the public and formal training of politicians is expected to change the current content of political discourse that features ideology, rhetoric, emotional appeals, slogans, and character assassination to one that features discussion and debate about alternative evidence-based pragmatic solutions to issues that are relevant to the needs and values of the majority of Canadians.

Citizen knowledge of civics and the principles of collective decision making are expected to give many more citizens a sense of empowerment. With that sense citizens will feel comfortable about insisting on the integrity and competence of their political representatives. As well, they will feel comfortable about insisting on accountability for results and insisting on real explanations where there has been a failure to achieve results. However, it is recognized that this sense of empowerment will not materialize without hard work of citizens to become informed and to actively engage in political discourse. For the role of the citizen is not that of a customer but rather that of an employer. This means that citizens are responsible for ensuring that their elected political representatives are honest, ethical and competent. As essayist John Ralston Saul explains, 'If democracy fails then it is ultimately the citizen who has failed, not the politician' (Saul 1997: 80).

Formal training of politicians combined with increased funding of political parties to engage in political discourse is expected to substantially increase Canadians' understanding and respect for politicians. In an ideal democratic state, politicians are viewed as the country's most important citizens by virtue of the fact that they represent the interests of millions of citizens.

Finally, the outcome of a royal commission on the public good and social progress is expected to increase citizens' awareness of the elements of the public good as well as inform political representatives of the degree of importance that they attach to each of the elements. On the basis of the evidence in this chapter – albeit circumstantial – it is hypothesized that the majority of the electorate will opt for a combination of slower economic growth, a more equal distribution of income and wealth, a healthier environment and more leisure time for a better work–life balance and greater security of employment. This shift in citizen priorities is expected to result in the acceptance of a reduction of somewhat lower levels of material consumption. Only a fraction of the lower consumption of the top 20 per cent of income earners resulting from tax increases and fewer hours worked would be offset by increased consumption of those in the lowest 20 per cent of income earners. Much of the reduced consumption might be for uneconomic luxury and status goods and services. Slower economic growth will enable the achievement of environmental health and sustainability and the maintenance of ecological integrity, both of which are now held hostage by the present version of a capitalist market economy that – while efficient – is relatively ineffective in achieving an optimal public good.

Notes

1 Annual income brackets in 1969 were multiplied by 5.8996 to reflect cumulative inflation between 1969 and 2010 as indicated by the Bank of Canada's inflation calculator. The 1969 tax rates were then compared with the 2010 tax rates for each of the augmented 1969 income brackets and the percentage changes in the tax rates were calculated. Note that the top ten income brackets in 1969 were eliminated subsequent to 1969; by 2010 the top ten were included in one wider income bracket. The next eight highest 1969 income brackets were, by 2010, merged into two wider income brackets.
2 Federal debt attributed to Ontarians increased from $35,078 per Canadian household in 2007 to $45,223 in 2013, while Ontario government debt increased from $31,450 per Ontario household in 2007 to $51,578 in 2013. Thus debt from the two levels of government attributable to Ontario households increased from $66,534 to $98,346 – an increase of 45 per cent. Note that data of 13.320 million Canadian households and 4.888 million Ontario households used in these calculations were taken from the 2011 census of Canada. Consequently, the percentage increase of 45 per cent is overstated somewhat by the percentage increase in households between 2007 and 2013.

References

Beltrame, J. (2013) Manufacturing Jobs: Ontario's Decade Long Job Decline Led by Factory Sector. *Canadian Press* (28 January). Available at www.huffingtonpost.ca/news/manufacturing-Canada.

Cahill, S. (2007) Corporate Income Tax Rate Database: Canada and the Provinces, 1960–2005. Available at www4.agr.gc/resources/prod/doc/pol/pu.

Canada Revenue Agency (2010) *T1 General 2010 Ontario, Schedule 1*. Ottawa: Canada Revenue Agency.

Coyle, D. (2011) *The Economics of Enough: How to Run the Economy as if the Future Matters.* Princeton, NJ: Princeton University Press.

Delacourt, S. (2013) Why Politicians Always Have Time for Tim Hortons. *Toronto Star* (28 September).

Department of National Revenue and Taxation (1969) *1969 T1 General Tax Guide.* Ottawa: Department of National Revenue and Taxation.

Diamond, P., and E. Saez (2011) The Case for a Progressive Tax: From Basic Research to Policy Recommendations. *Journal of Economic Perspectives* 25.

Dobbin, M. (2013) Free Market CEOs Have Freed Themselves from Consumers. *CCPA Monitor* (September): 22–3.

Government of Canada (2008) *Annual Financial Report of the Government of Canada for 2007–08.* Available at www.fin.gc.ca/toc/2008/afr2008_-eng.asp.

—— (2013) *Annual Financial Reports of the Government of Canada for 2012–13.* Available at www.fin.gc.ca/afr-rfa/2013/report-rapport-eng.asp.

Judt, T. (2010) *Ill Fares the Land.* New York: Penguin Books.

Kane, J. (2012) Health Costs: How the US Compares with Other Countries. 22 October. Available at www.pbs.org/news hour rundown.

Kloppenberg, J. (2011) *Reading Obama: Dreams, Hope and the American Political Tradition.* Princeton, NJ: Princeton University Press.

Krugman, P. (2009) *The Conscience of a Liberal.* New York: W. W. Norton.

Lee, M. (2007) *Eroding Tax Fairness: Tax Incidence in Canada, 1990 to 2005.* November. Ottawa: Canadian Centre for Policy Alternatives.

Leech, J, and J. McNish (2013) *The Third Rail: Saving Canada's Pension System.* Toronto: McClelland & Stewart.

Lepore, J. (2012) The Lie Factory: How Politics Became a Business. *The New Yorker* (24 September): 50–59.

Mackenzie, H., and R. Shillington (2009) *Canada's Quiet Bargain: The Benefits of Public Spending.* April. Ottawa: Canadian Centre for Policy Alternatives.

Martin, L. (2012) The Eternal Leavening of the Canadian Left. *Globe and Mail* (10 April).

Minsky, H. P. (2008) *John Maynard Keynes.* New York: McGraw Hill.

Moran, A. (2012) Ed Broadbent Calls for More Taxes to Battle Income Gaps in Canada. Available at www.broadbentinstitute.ca/en/blog/ed-broadbent-calls-more-taxes-battle-income-gaps-canada.

Ontario Financing Authority (2014) 2014–15 Borrowing Program: Province and Ontario Electricity Financial Corporation. Available at www.ofina.on.ca/borrowing_debt/borrowing.htm.

Ontario Ministry of Education (2011) Memo to Directors of Education from Assistant Deputy Minister Gabriel F. Sekaly Re: Education Funding for 2011–12. 31 March. Available at www.edu.gov.on.ca/eng/funding.

Ontario Ministry of Finance (2014) *Ontario's Long Term Report on the Economy.* Available at www.fin.gov.on.ca/en/economy/ltr.

Parkinson, D. (2014) Canadians' Household Debt Burden Edges Higher in Second Quarter. *Globe and Mail* (15 September).

Piketty, T. (2014) *Capital in the Twenty-First Century.* Cambridge, MA: Belknap Press of Harvard University Press.

Rae, B. (2011) Speech to the Economic Club, Toronto, 9 November.

Samara Canada (2010) *The Accidental Citizen?* MP Exit Interview Report number 1. Toronto: Samara Canada.

Samara Canada (2013) *Lightweights? Political Participation Beyond the Ballot Box.* Democracy Report number 6. Toronto: Samara Canada.

Saul, J. R. (1997) *The Unconscious Civilization.* New York: Free Press.

Sharanjit, U., and S. LaRochelle-Cote (2012) Factors Associated with Voting. Available at www.statcan.gc.ca/pub/75-001-x/2012001/article.

Stanford, J. (2011) How Corporate Tax Cuts Can Actually Destroy Jobs. 27 January. Available at www.progressive-economics.ca/2011/01/27/how-corporate-tax-cuts-can-actually-destroy-jobs.

Stanford, J. (2013) The Failure of Corporate Tax Cuts to Stimulate Business Investment Spending. In Richard Swift (ed.), *The Great Revenue Robbery*, 66–83. Toronto: Between the Lines.

University of Waterloo (2014) *Canadian Index of Well-Being*. Available at https://uwaterloo.ca/canadian-index-wellbeing/resources/infographics (accessed 21 April).

Weir, E. (2012) Corporate Taxes and Investment in Ontario. 23 January. Available at www.progressive-economics.ca/2012/01/23/ontario-corporate-tax-investment.

18 Corporate media, ecological challenges and social upheaval

Rose A. Dyson

In recent years, telecommunications and the internet have underscored the extent to which we now live in a global village. Issues have become more diffuse and interconnected so that change in one sector or region affects others, irrespective of national boundaries, much like an ecosystem. These trends are having a huge impact on policy making and demand attention on an international basis. Public health issues now merge with formerly unrelated fields, such as education, the environment, culture, gender studies, security and the economy. Holistic principles based on interconnectedness have never been more relevant. New media offer enormous potential for mobilizing for political action, limiting climate change, maintaining the biosphere and securing energy needs. But they can also be mobilized for harmful, destructive purposes.

Introduction

Surging income inequality is unfolding throughout the world. The resulting gap is becoming so wide and so apparent that even the political right, traditionally allergic to discussions of class and inequality, has started to take notice. 'Crony capitalism' became the rallying cry, not only of Occupy Wall Street protests but also of American Tea Party stalwarts (Freeland 2013: 190). The US and Canada, with just 5 per cent of the world's population, control almost one-third of the world's private consumption expenditures. Western Europe, with 6.4 per cent of the population, controls almost 29 per cent. Essentially, this means that almost 11.4 per cent of the world's population controls 60 per cent of the world's consumer spending (Barber 2007: 10). But inequality leaves capitalism with a dilemma because the overproducing capitalist market must either grow or expire. If the poor cannot be enriched enough to become consumers, then those with vast disposable incomes but few needs must be enticed into shopping through massive advertising campaigns.

Homer Dixon, Chair of Global Systems at the Balsillie School, University of Waterloo, points out that as low economic growth globally becomes the new normal, political instability is rising. We first saw it in the Middle East with the Arab Spring in 2010 and 2011. In the spring of 2014, we saw it in Thailand, Turkey, Venezuela and Ukraine. Key precursors to unrest tend to be endemic corruption, economic cronyism and rising middle-class grievances (Homer-Dixon 2014: A11). Adding to the list, although seldom acknowledged, is growing evidence of fall-out from climate change in droughts, threatened food and water supplies, unemployment and, I argue in this chapter, a cultural climate that encourages the use of violence as a conflict resolution strategy.

In his book *Climate Wars*, Gwynne Dyer (2009) points out that, to prevent runaway heating and keep the temperature rise to two degrees Celsius, only massive mobilization and wartime-style controls in every major industrialized and industrializing country would stop the rise in greenhouse-gas emissions by 2015. Clearly, to avoid 'cooking our goose' we are going to have to adopt a much more ambitious and radical agenda for change. In his commencement address to the class of 2014 at the West Point Military Academy, US President Barack Obama spoke of global warming as a 'creeping national-security crisis' that would shape the soldiers' time in uniform as they are called upon to respond to refugee flows, natural disasters, and conflicts over water and food (Koring 2014).

The perils and promises of digital technology

Two exceptionally optimistic people about the future are Google big thinkers Eric Schmidt and Jared Cohen, co-authors of *The New Digital Age: Reshaping the Future of People, Nations and Business* (2013). With unbridled enthusiasm for the promise of all things digital, the authors believe that great potential will be unleashed for everyone once the number of the world's population with an internet connection of some kind rises to around 5 billion. Cyberspace will, they concede, make things more complicated and states will find they need domestic and foreign policies to deal with both the physical and virtual worlds. But the authors gloss over privacy issues, terrorism and the darker side to the digital revolution as minor impediments to the coming new age. It is anticipated that the networking power of the Internet will, in itself, serve as a deterrent to potential perpetrators of brutality and corruption as ex-combatants are 'persuaded' to change their ways and resort to entertainment and other less harmful pastimes.

Overlooked entirely is growing evidence of low-level radiation adding to the toxic carbon emissions in our air pollution, from cell phone towers and other wifi devices, research demonstrating how emails contribute to rising stress levels, and evidence of digital dementia (Roser 2014; Davis 2010; Anderssen 2014).

American media scholar Robert McChesney examines the Internet through a lens of political economy and free-market capitalism. Separating the celebrants from the sceptics, he says '[a]t the center of political debate will be economics: what sort of economy can best promote democratic values and structures and self-governance while nurturing the environment? And at the center of everything will be the Internet' (McChesney 2013: 22). The economy and the Internet will rise and fall together. So far, trends are not encouraging. As he points out, many successful giants such as Apple and Google were begun by idealists who may not have wanted to be old-fashioned capitalists, but the system demands it. Any qualms about privacy, commercialism, avoiding taxes or paying low wages to

Third World factory workers had to be set aside. It is not that the managers are particularly bad and greedy people – indeed, their individual moral makeup is mostly irrelevant – but rather the system sharply rewards some types of behavior and penalizes others so that people either get with the programme and internalize the necessary values, or they fail.

McChesney is a rare media analyst who points to ways in which non-renewable sources of cheap energy that fuel climate change, drive up health costs and resist policy development for a sustainable future are an integral part of the new digital age. Most pundits are inclined to overlook the extent to which these technologies themselves are voracious energy users and toxic polluters. But even ecologically modernizing digital production cannot avoid energy and resource use. Each advance depends on a further cradle-to-grave cycle of extraction and transport from manufacture to market – then from consumption to waste pit.

These electronic cast-offs contain a witch's brew of heavy metals and toxic substances, such as lead, cadmium and mercury. In Canada, some of this is recycled but most of this waste gets shipped off to developing countries willing to trade cash for trash. According to a report released by Greenpeace in April 2014, digital technology products and services accounted for about 2 per cent of worldwide emissions in 2012, about the same as the airline industry. Some of the biggest electricity demands come from huge data centers offering 'cloud computing'. Collectively these services consume more electricity than all but five countries – China, the US, Japan, India and Russia (Liedtke 2014: FP2).

A look at the content offered by these high-tech gadgets sheds light on another dimension to the problem, in addition to their insatiable appetite for energy. In December 2011, *The Economist* predicted that video games will be the fastest-growing and most exciting form of mass media over the coming decade, estimated at around $82 billion by 2015 (Cross 2011). The action-packed video game *Call of Duty: Black Ops*, for example, had in one month taken in more than $1 billion in sales as fans in countries around the world queued for blocks on the first day of its release in 2010, to purchase a coveted early copy. In his book *Consumed: How Markets corrupt Children, Infantilize Adults and Swallow Citizens Whole*, Benjamin Barber (2007) offers other examples of a global economy that overproduces goods and targets children as consumers in a market where there are never enough shoppers and where the primary goal is no longer to manufacture required goods but to create needs and desires.

On the whole, it is impossible to discuss media violence, either systemic or as action-packed entertainment, without taking into account some of the deeper and broader manifestations of dominant media interests in every aspect of social, political, economic and cultural life. Throughout the literature on mass media, dating back over half a century, there is a recurring theme of indignant campaigns and suppressions, of the shading of certain messages to the exclusion of others, and a general agenda which is highly functional for established power (Dyson 1995). Following the release of the Ontario government-funded Royal Commission on Violence in the Communications Industry Report in 1977, which built on the findings of similar American studies, all of which indicated that the overwhelming weight of evidence pointed towards harmful effects, Mr Justice Lucien Beaulieu, one of its three commissioners, argued that, as with other toxic products, the principle of reverse onus should apply to potentially harmful media as well as other products. Harmful effects from media violence should be ruled out *before* products are released onto the market, not after.

For decades, industry opposition has quickly surfaced in response to any hint of new legislation restricting production and distribution of harmful material. An example arose in 1986 in the form of a 'leaked' memo from North American public relations affiliate, Argyle Communications Inc., based in Toronto, Canada:

> A way must be found to discredit the organizations and individuals who have begun to disrupt the legitimate business activities of publishers. This can be accomplished by creation of a broad coalition of individuals and organizations opposed to the Commission's findings and recommendations . . . these new groups would include academics, civil libertarians, religious leaders, civic and community leaders, politicians, columnists, commentators and entertainers. It might be called 'Americans for the Right to Read' or 'The First Amendment Coalition.'
>
> (Johnson 1986)

Over the years many Americans have pointed out that in drafting the First Amendment, Thomas Jefferson emphasized that their legal system must be responsive to changing historical conditions. In his book *Amusing Ourselves to Death*, Neil Postman (1986) called for a ban on political commercials such as those for cigarettes and alcohol. To counter objections on the basis of First Amendment violations he suggested a compromise not unlike the one proposed by Joyce Nelson for the Canadian Radio-Television and Telecommunications Commission two years later: 'Require all political commercials to be preceded by a short statement to the effect that common sense has determined that watching political commercials is hazardous to the intellectual health of the community' (Nelson 1988). Throughout the world the rights of tobacco companies to advertise their products under the guise of freedom of expression continue to be challenged and these trends need to be broadened to encompass other toxic cultural messages as well.

On the eve of launching the Cultural Environment Movement (CEM) at Webster University in St. Louis in 1996, founding director George Gerbner referred to American Attorney General Janet Reno's national consultation on youth and violence in 1993 as the first high level examination of media policy. The conclusion was that the issue of media violence was really just the first phase of a major cultural debate about life in the twenty-first century and that citizens should take lessons from the wider environment movement to form a 'cultural environment movement' (Dyson 2000). By August, 1994, Gerbner, then based at Temple University in Philadelphia, had gathered more than 2,700 names in his database with representatives of 88 different organizations from various states and countries around the world.

At the Smart Data International Symposium held at the University of Toronto in 2012, Kirk Jordan, who spoke on high performance computing innovations, was from the IBM Watson Research Center in Cambridge, Massachusetts, and concluded his remarks with a call for new approaches to how ever-increasing amounts of data and new systems are applied, pointing out that too often these innovations end up fuelling the gaming industry, most of it violent, and little else (Jordan 2012).

In the aftermath of countless tragic shootings such as that at the Columbine high school in Littleton, Colorado in April 1999, or the bombs set off during the 2013 Boston Marathon, security specialists weigh in with calls for better legislation but these quickly get mired in debates over privacy concerns (Ward 2014: A9). While digital technology has greatly assisted law enforcement in tracking down and apprehending perpetrators, intelligence officers caution that the ways in which the Internet fuels such attacks cannot continue to be ignored.

But despite decades of research findings showing harmful effects from violent media entertainment, there is still nothing to suggest that public safety will take precedence over corporate profitability. When new gun control legislation was signed into law by New York State Governor Mario Cuomo in 2013, following the massacre at a Connecticut Kindergarten School in 2012, criticism from Hollywood quickly surfaced. Officials in the movie and television industry issued warnings that 'the new laws could prevent them from using the lifelike assault weapons and high-capacity magazines that they have employed in shows like *Law & Order: Special Victims Unit* and *Dark Knight Rises*' (DiManno 2013: pA2). Gun rights activists, on the other hand, mocked the idea of a Hollywood exception. But Cuomo quickly fell into line, promising exemptions despite firearms permits having already been granted to the industry by the New York Police Department, as they have also been in the State of California (Kaplan 2013: A18).

The lure of the Internet has created some perilous spaces for young people in particular. One has only to examine the missing discourse in the search for solutions to cyberbullying which now focuses exclusively on the need for exemplary behavior on the part of parents, teachers and students. Remarkably absent is any recognition or acknowledgment of the impact of role models which glorify violence, power differentials and abuse of women and girls in popular culture in a hyper real world of virtual entertainment and networking. Students often use more than one medium at a time, and are further seduced into hours of engagement in cyberspace, increasingly fraught with the dangers of cyber bullying and sexual predators. Other perils present themselves on the issue of teenaged 'sexting', commonly defined as the production and exchange of nude or semi-nude images or videos via mobile phone. There is widespread confusion over how these differ, if at all, from child pornography. Given the youthful inclination towards a sense of invincibility, the dangers to the potential loss of control of personal information and associated damage to one's reputation are usually overlooked entirely:

> Lurking throughout the web are social networking spaces designed to encourage surveillance, harassment, and denigration of individuals. Many of

these sites are crafted so their web designers and those who communicate on them assume limited (or no) liability for their actions.

(Parsons 2013)

Pornography is now mainstream. This is underscored by a recent reprimand from the CRTC to three Toronto-based erotica TV channels for failing to broadcast sufficient levels of Canadian-made pornography to meet established Canadian content quotas (Hopper 2014: A1).

Exploiting children for profit is big business

Merchants of digital technology point to 'gamification' as a way of 'revolutioniz-ing' education. This not only leads to the marginalization of the role of teachers and classes such as physical education, art and music. Programmes that require students to bring their own digital devices to the classroom pose great risks as well as rewards. These include the potential for distraction, gaming addictions, unwanted video and audio recording, digitally-enabled academic dishonesty, pornography and cyber bullying.

In addition, we now have well-documented evidence of physical health problems such as obesity, diabetes and heart problems resulting from too much screen time. And the list keeps growing. As Joel Bakan points out in his book *Childhood Under Siege*, by tapping into teen emotions like love and fear, addictinggames.com has become one of the Internet's premier casual gaming sites. Every month, over ten million players log on to play games such as 'Whack Your Soul Mate', involving excessive amounts of blood and gore. Gaming expert and psychologist Douglas Gentile says the video game industry considers addictiveness a game's 'main indicator of success', certainly not as a problem for mental health or well-being (Bakan 2011). Ninety per cent of Canadian children over the age of 5 years, according to a recent global study, are not meeting a target of an hour a day of moderate to vigorous exercise (McGinn 2014: A6). On the basis of overall physical activity Canada received a D–, below Mozambique and Mexico. And although 66 per cent of Canadian parents think their kids spend too much time watching TV or on the computer, recommended remedies focused, once again, almost entirely on families 'to get out and play'. Other solutions focused on school boards and municipalities 'to revisit policies, bylaws and playground rules' for outdoor activity (*ibid.*).

Eruptions of violence, serial murder and senseless deaths continue to dominate news coverage with the usual speculation on why or how these occurred. Was there a connection between Norwegian mass murderer, Anders Breivik's rampage that killed 77 people in July 2011 and police reports that he was a computer game abuser, having played 'World of Warcraft' for a total of 500 hours over a period of three months (Pidd 2012)? According to US lieutenant colonel and professor of military science David Grossman, operant conditioning, a technique widely used in military training to teach soldiers to kill with impunity, is initiated by the rewards aspect of points received for brutal action in video games. When talking of conditioned responses, he explains, 'we must also talk about violent video games, because to understand how we can make killing a conditioned reflex – stimulus-response-stimulus-response-stimulus-response – it is important to understand how the average opponent has been trained' (Grossman 2004: 77). In the military, these techniques are reluctantly employed as a trade-off for national security reasons, but to indiscriminately reward children for repeatedly blowing off the heads of their virtual playmates is to encourage dangerous and dysfunctional play. In the past, children who beat up their friends or classmates were punished. Today, not only are we teaching them that such violence can be entertaining, we are rewarding them with points for being good at it.

In response to the London riots in the summer of 2011, retired prison psychiatrist Theodore Dalrymple wrote about how deeply embedded in everyday life criminality has become (Dalrymple 2011: A18). Pioneering media theorist George Gerbner called the normalization of anti-social, criminal behavior resulting from violence in the media the 'mean world syndrome' (cited in Morgan 2002). Predictable consequences usually result in calls for tougher law and order measures such as more prisons and longer sentences. The hidden curriculum in video games and complementary forms of social media constantly reinforce the notion that boys and men should be brutally violent, girls and women are sexual objects, buying and accumulating material things leads to happiness, parents and other authority figures are uncool and that obsessive–compulsive behavior is normal.

Gerbner argued, 'We distinguish the long-term cultivation of assumptions about life and values from short-term "effects" that are usually assessed by measuring change as a consequence of exposure to certain messages' (*ibid.*: 296). In other words, the accumulation of exposure to negative media imagery over time can have deeply corrosive effects on how we think and the values we espouse, even if those effects appear only to be experienced in the short term. Chris Hedges is much more direct in his assessment of unfolding instability, reinforced by irresponsible media trends. In his book *The Empire of Illusion*, he argues that we are now immersed in a culture that has 'passively given up the linguistic and intellectual tools to cope with complexity, to separate illusion from reality' (Hedges 2009: 44).

On 8 June 2011, it was reported in *The Toronto Star* that a Statistics Canada study found that police reported incidents of hate crimes had increased 42 per cent nationwide between 2008 and 2009. Two years later it was reported that, based on a briefing released in Washington DC by the Simon Wiesenthal Center, over a period of one year Twitter Inc., the web-based social-messaging service, helped spur a 30 per cent growth in online forums for hate and terror, bringing the number of sites being tracked by the centre up to about 20,000 (MacMillan 2013: B1). Everywhere in the developed world it seems that hate and intolerance resulting in crime is on the rise. The full impact of the four killings at a Jewish school in France by a radicalized Islamic citizen and lone gunman on a motorcycle on 19 March 2012 has yet to be determined, but that they were hate-tinged and add to a culture war influencing French politics there can be little doubt (Saunders 2012: A9). We ignore at our own peril the connections between popular culture that encourages antisocial behavior and rising evidence of intolerance in society at large. To continue to examine these issues in isolation from one another is counterproductive.

Institutional responses to media issues

The Toulouse shootings in France spawned reflections on how increasingly elusive and dangerous those that counter terrorism agents now hunt, are becoming. Today, terrorists are as likely as not to be citizens radicalized via the Internet in any country, including Canada. Prevention is getting harder. Federal agencies have had to come up with creative strategies – including what those on the front lines of fighting cyber-crime refer to as 'disruption' techniques (Freeze 2012: A4).

In Canada, debate on the issue was initiated with bill C-30, introduced in the House of Commons in February 2012. Although vetted by every attorney general serving provincially and in the territories, the opposition quickly branded it a draconian attempt at privacy invasion. Huge advances in the uses and capabilities of information technologies and how these provide safe havens for criminal activity such as organized crime, sexual exploitation, fraud and identity theft were ignored. It was argued that a warrant should be required before any data is gathered. Now, neither telecommunications nor Internet providers are required to preserve data but, as pointed out by the Ontario Provincial Police, by the time law enforcement officers obtain a warrant, which could take weeks or even months, the desired content may not be available. Ironically, as the attention of opposition critics shifted to the alleged wrongdoing during the previous federal election involving a robo-call scandal, a call was made for the CRTC to get involved in the investigation, because it does *not* require a warrant to gather data (Chase and Leblanc 2012: A6). This begs the question of why the telecommunications federal regulator enjoys a privilege denied to officers working on the front lines of cyber-crime. That public indignation focuses entirely on privacy encroachment by government institutions, while similar encroachment from commercial interests on our buying habits and psychological vulnerabilities to addiction are ignored, belies common sense.

Following the Boston terrorist attacks in April 2013, two Muslim immigrants in their thirties were charged for an alleged plot to derail a Via Rail passenger train. *The Globe and Mail*, in its editorial on 27 April, decided to revisit the proposals inherent in bill C-30 and endorsed University of Ottawa law professor Craig Forcese's call for more informed and serious debate on tools for communication interception (Globe and Mail 2013). A legal regime, said Forcese, 'designed for wired technology in the 1970s is probably not suitable for wireless technologies, mobile phone numbers, and disposable communications tools'.

Occasionally, attempts are made to restrict gratuitously violent media for entertainment purposes from being marketed to children. In 2010, California Senator Leland Yee proposed a ban on violent video games backed by the state chapter of the American Academy of Pediatrics, other health organizations, and supported by eleven other states. Nevertheless, the US Supreme Court ultimately ruled in favor of the software industry (Bascaramurty 2011: A13). This case demonstrates how the court system has leaned in favor of big business in recent years, protecting political speech, pornography, violent media entertainment and marketing of junk food under the First Amendment. In other words, no distinctions are being made between individual freedom of expression and corporate freedom of enterprise. Similarly, distinctions between news media and entertainment media have been blurred if not obliterated. The rise of public relations advocacy campaigns warning against the dangers of government interference in the market place has conditioned the public at large to accept the gravitational pull in social codes of conduct involving all forms of media towards the consensus of those in power. But time is running out. In 2002, McChesney, recognizing the urgent need to generate popular awareness of the issues and organize it as a political force, co-founded the public interest group Free Press. Although it has grown and had some success, he laments what has become a chronic problem for media scholars. It remains too isolated from other organized popular groups that still fail to understand the importance of media policy making.

Calls for restrictions on marketing violence to children as harmless entertainment along with other harmful products such as junk food, over the years, have come from the American Psychological Association, American Psychiatric Association, American and Canadian Paediatric societies, the US Center for Disease Control and Prevention, the Canadian Center for Science in the Public Interest, EDUPAX and Canadians Concerned About Violence in Entertainment (Dyson 1995, 2000). As psychologist Craig Anderson (2003) points out, 'After 40 years of research, one would think that the debate about media violence effects would be over.' Many criticisms in the new debate are recycled myths from earlier objections to restrictions on media violence that have repeatedly been debunked on theoretical and empirical grounds. Valid weaknesses have been identified and often corrected by media violence researchers themselves.

Legislation in Quebec, the UK, Scandinavia, France, Switzerland, Italy, Malta, New Zealand, Greece and other countries banning advertising to children based on research showing harmful effects has yet to be adopted in the rest of Canada, despite several bills introduced at both the federal level and provincial level in Ontario in recent years, calls from Toronto and other boards of health in Ontario, EDUPAX, the Center for Policy Alternatives, the Center for Science in the Public Interest and Canadians Concerned About Violence in Entertainment (Dyson 1995, 2000; Jeffery 2007; Linn 2010). McChesney repeats the call in his book, *Digital Disconnect: How Capitalism is turning the Internet Against Democracy* (2013). Similar demands have been made and bills introduced to eliminate taxes and funding for audio–visual productions deemed to be contrary to the public interest. In Canada, bill C-10, introduced by the minority Harper government in 2008, although initially passed in the House of Commons, was stalled in the Liberal-dominated Senate and ultimately died on the order paper when the next election was called.

In 2011, it was reported in *The New York Times* that wasteful government spending in both Canada and the US involves generous tax breaks for video game producers regardless of the content. According to tax professor, Calvin H. Johnson at the University of Texas in Austin, a collection of deductions, write-offs, and credits now makes production of video games such as the gory *Dead Space 2* one of the most highly subsidized businesses in the US. In 2008, Ontario paid one game company more than $321,000 for each job to relocate on the north side of the border. Montreal has also been generous with a rich package of incentives for relocation (Kocieniewski 2011: A1). In Ontario and other provinces in Canada, much of this generous tax money is discreetly packaged under subsidies for 'electronic arts'.

It is important to recognize that all advertising is condoned and encouraged by government policies and regulations. Allowing businesses to write off their advertising expenditures as a business expense on their tax returns not only costs the government tens of billions annually in revenues, but also encourages ever-greater commercialism in our culture: not a healthy outlook for long-term environmental sustainability. In other words, the capitalist system as we have known it in the past three centuries has outlived its usefulness. As McChesney explains, 'It means one recognizes that a system that promotes poverty, inequality, waste, and destruction – to the point of making the planet uninhabitable – deserves no free pass from democratic interrogation in the present, whatever its past achievements' (McChesney 2013: 228). And despite endless claims that a great new capitalism system is just around the corner thanks to digital technologies, there is little evidence to back them up. There is clearly plenty of money for those at the top – who want to keep things the way they are – but little evidence that the benefits are reaching any of the rest of us.

Conclusion

Growing concern over how the Internet fuels terrorism, cyber bullying and other excesses antithetical to social health and well-being is just the tip of the iceberg in the challenges that lie ahead for elected officials. The debate over the harmful effects of violent entertainment in popular culture, its sedentary nature and the need to protect children from commercial exploitation is not new and predates the arrival of the Internet. And while countless youth amuse themselves with action-filled electronic toys of one kind or another and still manage to grow up to be responsible adults, the reality is that the overwhelming weight of research points towards harmful effects. At the very least it must be acknowledged that the proliferation of these seductive and often energy-intensive pastimes serve as weapons of mass distraction from the urgency of looming problems posing serious threats to the long term survival of humankind as a whole. The need to address them crosses over party lines and surpasses our institutions' current abilities to manage them. Clearly we must learn to better regulate powerful new information technologies. But who should be in charge of new breakthroughs in scientific laboratories? The market – which translates into large, profit-driven, corporate interests – or those who are elected to act in the public interest? The older challenges of how to govern our economy are still with us but now, we are increasingly being overtaken by the need, in the twenty-first century, to decide how we should govern technology. It offers us enormous potential, in mobilizing for political action, limiting climate change, maintaining the biosphere, and securing energy needs, but governments and educators must engage these issues in a serious way soon or risk losing entirely the ability to influence outcomes.

References

Anderson, C. (2003) Violent Video Games: Myths, Facts and Unanswered Questions. Available at www.apa.org/science/about/psa/2003/10/anderson.aspx.

Anderssen, E. (2014). Inbox of Burden. *Globe and Mail* (31 March): L1.

Bakan, J. (2011) *Childhood Under Siege: How Big Business Targets Children*. Toronto: Allen Lane.

Barber, B. (2007) *Consumed: How Markets Corrupt Children, Infantilize Adults, and Swallow Citizens Whole*. New York: W. W. Norton & Co.

Bascaramurty, D. (2011) U.S. high court rejects video game ban. *Globe and Mail* (28 June): A13.

Chase, S., and D. Leblanc (2012). Elections Canada turns to CRTC for help. *Globe and Mail* (3 March): A6.

Cross, T. (2011) All the World's a Game. *The Economist* (10 December). Available at www.economist.com/node/21541164.

Dalrymple, T. (2011) Lenient justice begets jobs – and London burns. *Globe and Mail* (11 August): A18.

Davis, D. (2010) *Disconnect: The Truth About Cell Phone Radiation, What the Industry Has Done to Hide It, and How to Protect Your Family*. New York: Dutton.

DiManno, R. (2013) Newtown marks a grim anniversary. *The Toronto Star* (30 December): A2.

Dyer, G. (2009). *Climate Wars. New Material From Copenhagen and Beyond*. Oxford: Oneworld.

Dyson, R. A. (1995) The Treatment of Media Violence in Canada Since Publication of the LaMarsh Commission Report in 1977. Doctoral thesis, OISE/UT, Toronto.

—— (2000) *Mind Abuse: Media Violence in an Information Age*. Montreal: Black Rose Books/ UT Press.

Freeland, C. (2013) *Plutocrats: The Rise of the New Global Super-Rich and the Fall of Everyone Else*. Toronto: Doubleday.

Freeze, C. (2012) The Silent Threat of the Lone Wolf. *Globe and Mail* (22 March): A4.

Globe and Mail (2013) Keeping the Gloves On. *Globe and Mail* (27 April): F9.

Grossman, D. (2004) *On Combat: The Psychology and Physiology of Deadly Conflict in War and Peace*. Millstadt, IL: Human Factor Research Group.

Hedges, C. (2009) *Empire of Illusion: The End of Literacy and the Triumph of Spectacle*. New York: Alfred A. Knopf.

Homer-Dixon, T. (2014) Why These Fractures? Stalled Economies. *Globe and Mail* (11 April): A11.

Hopper, T. (2014) Porn Not Canadian Enough, CRTC Warns. *The National Post* (6 March): A1.

Jeffery, B. (2007) *Hitting the Easy Mark: The Law on Marketing to Children is Ripe for Reform*. Ottawa: Canadian Center for Policy Alternatives.

Johnson, S. (1986) Pro-pornography 'leaked' memorandum. 5 June. In *Pornography and Free Trade: A Time for Action*, Appendix A, pp. 1–6. Toronto: METRAC.

Jordan, K. E. (2012) IBM's Blue Gene/Q System and Implications for Simulation and Data Analysis. Available at www.ipsi.utoronto.ca/sdis/pdf/Kirk-Jordan-SmartData-May16.2012.pdf.

Kaplan, T. (2013) In New York's Gun Law, Hollywood Sees Obstacle to Filming. *New York Times* (2 May): A18.

Kocieniewski, D (2011) Tax Breaks Bolster Video Game Industry. *New York Times* (11 September): A1.

Koring, P. (2014) Obama Sees More Restrained Role for US. *Globe and Mail* (29 May): A13.

Liedtke, M. (2014) Greener Apple Unveils Free Recycling Program. *The National Post* (22 April): FP2.

Linn, S. (2010) The Commercialization of Childhood and Children's Well-Being: What is the Role of Health Care Providers? *Paediatrics and Child Health* 15(4).

MacMillan, D. (2013) Twitter Fuels Rise in Web-Based Hate Forums, Report Says. *The Toronto Star* (8 May): B1.

McChesney, R.W. (2013) *Digital Disconnect: How Capitalism is Turning the Internet Against Democracy.* New York: New Press.

McGinn, D. (2014) Why Kids Over 5 Aren't Doing Enough Physically. *Globe and Mail* (21 May): A6.

Morgan, M. (2002) *Against the Mainstream.* New York: Peter Lang.

Nelson, J. (1988) *The Colonized Eye. Rethinking the Grierson Legend.* Toronto: Between the Lines.

Parsons, C. (2013) We are Big Brother: Sex, Lies and Digital Memory: How Surveillance Threatens Communities. In J. Greenberg and C. Elliot (eds), *Communications in Question: Competing Perspectives on Controversial Issues in Communications Studies*, 2nd edn. Toronto: Nelson.

Pidd, H. (2012) Anders Behring Breivik Spent Years Training and Plotting for Massacre. *The Guardian* (24 August). Available at www.theguardian.com/world/2012/aug/24/anders-behring-breivik-profile-oslo.

Postman, N. (1986) *Amusing Ourselves to Death.* New York: Penguin.

Roser, M. (2014). Too much screen time for kids seen as fuel for 'digital dementia'. *Globe and Mail* (10 March): L5.

Saunders, D. (2012) Hate-tinged killings during culture war set to shift French politics. *Globe and Mail* (20 March): A9.

Schmidt, E., and J. Cohen (2013) *The New Digital Age: Reshaping the Future of People, Nations and Business.* London: John Murray.

Ward, O. (2014) Big Brother – bad or good? Debate will decide. *The Toronto Star* (2 May): A9.

19 A complex adaptive legal system for the challenges of the Anthropocene

Geoffrey Garver

Law and legal systems are inevitably embedded in social, political, cultural and ecological contexts that continually adapt and evolve. In turn, law itself is dynamic and always evolving, despite the consistency, predictability and order that law generally is meant to instil. Several scholars theorize that legal systems are complex adaptive systems that evolve according to the same fundamental systems dynamics that underlie the evolution of other complex adaptive systems (Hornstein 2005; Ruhl 2008; Craig 2013; Garmestani and Benson 2013). According to this theory, legal systems co-evolve with other complex adaptive social and ecological systems – collectively, an evolving web of systems called 'panarchy' (Holling *et al.* 2002) – through interactions and feedbacks at and across multiple temporal and spatial scales. This co-evolution involves a constant 'interplay between change and persistence, between the predictable and unpredictable' (*ibid.*: 5).

How legal systems interact and evolve with other subsystems of the human–Earth system takes on heightened importance when considered in the context of the assertion that Earth has entered a new, human-dominated epoch called the Anthropocene. The Earth's life systems, and humanity itself, face a predicament without historical precedent due to expanding aggregated impacts of human activity that have already overshot global ecological limits (Rockström *et al.* 2009; Carpenter and Bennett 2011; Steffen *et al.* 2011; WWF 2014). Human beings have influenced the global ecosystem to such an extent that many scientists have called for formal geological recognition of the Anthropocene – an era most often associated with the onset of the industrial revolution in the late eighteenth century (Crutzen and Stoermer 2000; Rockström *et al.* 2009; Steffen *et al.* 2011).[1] The Anthropocene puts in perspective a new way to consider the past, present and future of the human–Earth relationship.

This chapter analyzes law as a complex adaptive system in the Anthropocene. Maintaining ecological integrity and fostering a mutually enhancing human–Earth relationship (Berry 1999) are taken as overarching objectives for humanity that are more desirable than the current dominant commitment to perpetual economic growth. Following an elaboration of these more desirable societal goals, this chapter considers three central inquiries. First, how does the concept of the Anthropocene influence analysis of the role of the legal system in the evolution of the ecosphere in the past, present and future? Second, how do legal systems function as complex adaptive systems that interact with other complex adaptive systems at and across multiple temporal and spatial scales? Third, how can human societies orient law and legal systems so as to maintain and enhance ecological integrity and promote an enduring, mutually enhancing human–Earth relationship in the Anthropocene?

Ecological integrity and a mutually enhancing human–Earth relationship as core objectives

The growing role of humans in shaping the global ecosystem is a central factor in determining what the goals of society should be in the Anthropocene. The declarations of the Group of Twenty (G20) nations and the international community's prevailing approach to sustainable development reflect these dominant goals, which give primacy to sustained economic growth. Although those goals are at the root of long-term trends of impairment or destruction of the ecological foundations for human and other life, they provide the contextual foundations of most legal systems from the local to the global scale. Aiming for a mutually enhancing human–Earth relationship with a core commitment to ecological integrity provides a more hopeful trajectory for humanity and the ecosystems in which human beings are embedded.

G20 leaders declared in 2013 that '[s]trengthening growth and creating jobs is our top priority and we are fully committed to taking decisive actions to return to a job-rich, sustainable and balanced growth path' (G20 2013: 3), maintaining their earlier commitments to sustained economic growth (G20 2010, 2011). The G20's declarations are consistent with the international community's prevailing conception of sustainable development 'that meets the needs of the present without compromising the ability of future generations to meet their own needs' (United Nations 1987: ch. 2, para. 1). The principles set out in the 1992 Rio Declaration on Environment and Development indicate a broad international commitment to environmental protection, but the sustainable development they connote requires an overriding and intractable commitment to economic growth. More recently, the outcome document of the Rio+20 Summit in 2012, *The Future We Want*, refers to nations' shared commitment to economic growth, or 'sustained, inclusive and equitable economic growth' ('sustainable' has morphed into 'sustained'!), more than twenty times (United Nations 2012).

UNEP's 'green economy' strategy (UNEP 2011) maintains this growth-insistent interpretation of sustainability. Greening of the global economy is cast as a 'new engine of growth' that builds, rather than runs down, social and natural capital (*ibid.*: 16). In this vision of sustainability, decoupling sustained economic growth from ecological impacts is fundamental (*ibid.*: 629). Sustainable Development Goals (SDGs), under development at the United Nations at this writing, are likewise growth-insistent. The initial draft includes seventeen SDGs for 2030 on a range of environmental, social and economic themes, including a goal to 'promote sustained, inclusive and equitable growth, full and productive employment and decent work for all' (United Nations 2014: 3).

Under any definition of sustainability, action to reduce material and energy demands of human society is essential (Jackson 2009; UNEP 2011). However, economic growth has a strong historical correlation with ecological degradation and persistent social injustice, and at some point sufficient decoupling of ecological degradation from growth becomes impossible (Daly 1996; Jackson 2009; Krausmann *et al.* 2009; WWF 2014). Because true long-term sustainability cannot be growth-insistent, the current overriding commitment to economic growth and the subsidiary goal to raise living standards throughout the world to those in wealthy countries must eventually give way to a more hopeful alternative.

The concept of 'right relationship' offers such an alternative (Brown and Garver 2009). Right relationship calls for fostering ecological integrity and a mutually enhancing human–Earth relationship (Berry 1999), by which a prosperous future for all includes prosperity for non-human life and encloses prosperity within the ecological limits of 'our diverse and finite earth' (Brown and Garver 2009: 10). Ensuring prosperity in an ecologically finite world means providing all present and future members of life's commonwealth 'bounded capabilities' to thrive, contingent on Earth's limited capacity to support life and on fair intragenerational, intergenerational and interspecies sharing of that capacity (Jackson 2009: 45–7; Brown and Garver 2009). The emerging concept of planetary boundaries of safe operating space for humanity (Rockström *et al.* 2009) is a leading approach to defining those limits.

To foster ecological integrity and a mutually enhancing human–Earth relationship, a hierarchy of societal goals for the Anthropocene is appropriate. Instead of perpetual economic growth, the primary goal should be to respect the Earth's ecological limits (Daly 1996) according to planetary boundaries or other appropriate metrics, with pursuit of prosperity for present and future generations within the social and economic spheres contained within those aggregate limits.

The challenges of the Anthropocene

The concept of the Anthropocene underscores the notion that humanity has reached a point of no return: stratigraphical markers of new geological epochs must be more or less permanent, and of the same magnitude as markers of earlier Epoch boundaries (Zalasiewcz *et al.* 2011a; Waters *et al.* 2014). Thus, the Anthropocene cannot be undone (Ellis 2011); it can only frame the current human–Earth dilemma and potential responses to it.

Whether humanity's impacts have left the kind of unique, enduring marker or set of markers that is required for recognition of a new geological Epoch is for geologists to determine (Zalasiewicz *et al.* 2011b; Waters *et al.* 2014). However, the term 'Anthropocene' emerged outside of geology, as a way to highlight evidence that humanity 'now rivals some of the great forces of Nature in its impacts on the functioning of the Earth system' (Steffen *et al.* 2011: 843). For the social sciences and humanities, the geological inquiry is tangential; the Anthropocene has more to do with considering how ethical, social, cultural, political, economic and legal systems have contributed to dangerous accumulation of human impacts on

Earth's life systems, and how human society might adjust those systems to ward off risks of catastrophic evolution of Earth systems and promote a mutually enhancing human–Earth relationship (Zalasiewicz *et al.* 2011a; Kellie-Smith and Cox 2011; Robinson 2014).

The Anthropocene concept has implications for the meaning of ecological integrity, which has been defined with reference to 'wild nature . . . that is virtually unchanged by human presence or activities' (Westra *et al.* 2000: 20). Ecological integrity can be assessed by comparing an ecosystem's structure and function with benchmark conditions at a comparable ecosystem that is relatively free of human impacts, taking into account natural variability (Angermeier and Karr 1994; Westra *et al.* 2000). The Anthropocene underscores the possibility that the global reach of human impacts on Earth's life systems is such that ecosystems completely free of anthropogenic impacts no longer exist, and that widespread human impacts on 'wild nature' have occurred since Palaeolithic times (Ellis *et al.* 2013; Foley *et al.* 2013). In that case, full ecological integrity would no longer technically be possible.

The metrics and indices that have been developed to assess ecological integrity suggest this possibility, in that benchmarks of ecological integrity have been tied to ecosystems that are 'virtually' pristine or 'relatively' free of anthropogenic impacts (Westra *et al.* 2000). In addition, researchers seeking to define ecological integrity implicitly acknowledge that, scaled to the global level, full ecological integrity for the global ecosystem would only be possible without humans – or at least modern humans (*ibid.*). The most pristine ecosystems are part of higher-scale landscapes and ecosystems, and ultimately the entire ecosphere (*ibid.*) and the anthropogenic impacts or alterations in those larger landscapes and ecosystems at some point become significant, compromising ecological integrity at least to some degree. Thus, although humanity should seek to maintain or restore the highest levels of ecological integrity possible, inevitable human impacts make full ecological integrity everywhere on Earth impossible even in the best of circumstances.

A mutually enhancing human–Earth relationship must allow for some degree of human impacts on the ecosystems in which people are embedded. Studies showing human impacts on Earth systems dating back to Neolithic or Palaeolithic times (Ellis *et al.* 2013; Ruddiman 2013; Foley *et al.* 2013) suggest that for much of human history, such a relationship may well have existed – and therefore, that environmental history can help reveal factors that either help or impede it (McNeill 2003; Diamond 2005). As well, planetary boundaries research is framed around maintaining the global ecosystem within an envelope of parameters that appears to correlate with a mutually enhancing human–Earth relationship. These studies strongly imply that preserving and restoring ecological integrity as much as possible is consistent with a mutually enhancing human–Earth relationship.

However, fostering this mutually enhancing relationship is not yet a prevalent societal goal. Instead, the Anthropocene highlights the extent to which human impacts have impaired ecological integrity and the long-term capacity of human-ity and Earth's life systems to be mutually enhancing. Thus, the Anthropocene brings into sharp focus the need to move past competing but flawed narratives of the ideal human–Earth relationship. One is an anthropocentric narrative of humanity '"mastering" nature to reclaim Eden' (Merchant 2003: 3). The other is a purely ecocentric narrative that laments the 'decline from a pristine earth to a paved, scorched, endangered world' (*ibid.*: 2) and calls for the restoration of wild nature on Earth. The first is flawed because, at the least, taken to the extreme, the construction of artificial nature destroys the capacity of the ecosphere to support human and other life. The second is flawed because at its extreme it leaves no meaningful place for humans in Earth's life systems. Both fall short because they rely too much on 'the linear approach of modern scientific thinking' (*ibid.*: 4). Systems-based, eco-anthropocentric approaches to law and other normative regimes will be needed to yield '[n]ew kinds of stories, new ways of thinking, and new ethics' (*ibid.*) for the Anthropocene, ones that promote ecological integrity and a mutually enhancing human–Earth relationship.

Legal systems as complex adaptive systems

Panarchy theory was developed not only to describe how complex adaptive systems interact and change at multiple temporal and spatial scales, but also 'to improve the design of policy responses [to declines in environmental quality and the growing scale of human activities]' (Ruhl 2012: 1). Law is necessary for applying panarchy theory so as to implement desired policy responses (*ibid.*). Moreover, several scholars have been building the case that legal systems themselves are complex adaptive systems that must be accounted for in panarchy theory (Ruhl 1996, 2008, 2012; Hornstein 2005; Craig 2013). Because 'law is where humans write the rules for other social systems' (Ruhl 2008: 897), one such scholar asserts that applying panarchy theory to address the current ecological crisis 'will require managing the complex adaptive legal system to adaptively manage other complex adaptive natural and social systems, all in a way that maintains some level of social order' (Ruhl 2012: 2).

An adaptive system is one in which the agents or components of the system make adjustments based on what occurs in their interactions. Complexity occurs in an adaptive system when the agents and elements develop non-linear interdependencies that also evolve as the agents and elements adapt according to their interactions and external inputs, giving the system as a whole characteristics that transcend the characteristics of its individual parts (Hornstein 2005; Ruhl 2008; Craig 2013). Complex adaptive systems are emergent, which means that although their components interact according to deterministic rules, the system-wide results of the interactions are continually novel, not entirely predictable and characteristic of the system as a whole rather than of its individual components (Hornstein 2005; Ruhl 2008). Emergence explains in part why attempts to control complex adaptive systems typically lead to unintended consequences. Complex adaptive systems are also path dependent, such that the past trajectory of the system affects the range of its possible futures (Hornstein 2005; Ruhl 2008). Emergence, non-linearity and path dependence have been identified as particularly important for considering how legal systems adapt along with the systems with which they interact (Hornstein 2005).

Incorporating human intention and design complicates the extension of complexity theory from natural ecosystems to social constructs such as legal systems (Ruhl 1996, 2008). Nonetheless, legal scholars have developed a strong case that legal systems behave like complex adaptive systems, oriented around a core tension among rights, freedoms and constraints (Ruhl 1996). Ruhl (2008) provides detailed examples of how legal systems exhibit both the agent properties (heterogeneity, deterministic rules, non-linearity, feedback mechanisms) and system properties (emergence, path dependence, self-organization, critical states, 'power law event distribution, adaptive resistance and resilience, phase transitions) of complex adaptive systems. For example, the actors and institutions involving legal systems – lawyers, clients, voters, elected officials, legislators, enforcers, judges, courts, legislatures, administrative bodies and so on – are diverse and heterogeneous; judicial decisions and new statutes or constitutions can bring about sudden, major – that is, non-linear – shifts in the legal landscape; the full meaning of a particular law or legal rule emerges as the legal system evolves, through the not entirely predictable interplay of the actors and processes involved; and the existence of a particular constitution or set of laws has a strong effect on the possible futures states of the legal system (Ruhl 2008).

A key defining characteristic of legal systems and the ecological, social and institutional systems with which they interact is the extent to which they lock in features or behaviours that either impede or improve the prospects for attaining desired objectives. Resilience, adaptiveness and resistance are key systems concepts in this regard, and all are value-neutral, in that they tend to keep in place system structures or functions that may impede or advance progress toward a desired goal (Ruhl 2011). To assess the role of complex adaptive legal systems in the build up to the Anthropocene and in potential responses to the dangerous ecological disruption that marks this new epoch, consideration must extend to the lock-in features of both the legal system and the other complex adaptive systems with which the law interacts – and for which the existence or lack of legal rules might itself be a strong lock-in feature (Ruhl 2011, 2012). For example, the global economic system has features with a significant degree of lock-in, such as a strong commitment to growth, capitalism and relatively unrestricted markets, that have historically maintained varying degrees of inequality in wealth and income. In turn, unequal income and wealth can lead to unequal access of individuals and organizations to the political process at different scales, and the political system may have its own locked-in features that help maintain those arrangements. Further, the result of this special access may be strong protection of the interests of wealthy individuals or organizations to procure rights and freedoms to engage in activities with detrimental ecological impacts, which may be of global significance as with climate change. The economic, political and ecological systems at stake interact as well with the legal system, which may contain several features with varying degrees of lock-in that reinforce wealth inequalities and unequal access to the political process. If, for example, constitutional provisions are invoked to maintain such arrangements, as in the case of the US Supreme Court's decision in *Citizen United v. Federal Election Commission* (558 US 310 [2010]), the degree of lock-in is very strong, given the difficulty of amending the US Constitution or overturning Supreme Court precedents.

A framework for law in the Anthropocene

A core challenge for law in the Anthropocene is to assess and respond to lock-in features that affect the potential of the legal system, as well as of the social, ecological and institutional systems with which it interacts, to promote ecological integrity and a mutually enhancing human–Earth relationship. Such an assessment should include three main categories of lock-in features of those systems: existing lock-in features that promote ecological integrity and a mutually enhancing human–Earth relationship, existing lock-in features that undermine those objectives, and new lock-in features that would advance progress towards those objectives. Each of those categories can be further subdivided according to the spatial or institutional scale (from local to global) at which the lock-in operates. For each of those categories and sub-categories, the assessment should account for the degree of lock-in, the nature and importance of the ecological and social stakes the lock-in affects, and strategies and reasonable timeframes for responding – by removing or preventing undesirable lock-in and by enhancing or integrating lock-in that is favourable. Such an assessment should take a systems-based approach, with ongoing consideration of how various lock-in features interrelate across systems and across temporal and spatial scales.

A comprehensive assessment of this kind would clearly require a major, multi-year transdisciplinary undertaking, but could enhance the development of coordinated strategies to prioritize responses to lock-in challenges according to their ecological or social stakes and the ease with which the lock-in challenge can be addressed. The desired result of this process of assessment and action would be a complex adaptive legal system that measures up to humanity's current ecological and social dilemma. Adaptive management and precaution against undermining ecological integrity have been identified as key features of such a legal system (Hornstein 2005; Ruhl 2012). Adaptive management approaches give much greater emphasis to back-end adaptation to the results of monitoring, as opposed to front-end process and impact analysis, than is common in contemporary environmental law (Ruhl 2012; Garver 2013). Precaution against undermining ecological integrity or crossing critical ecological thresholds such as those associated with planetary boundaries is appropriate in light of the irreducible uncertainty and inevitable potential for unintended consequences that exist in complex adaptive systems. A precautionary approach implies a need to focus adaptive management on the management of humans and their social constructs and physical undertakings, rather than the structure and function of the ecosystems in which they are embedded – although some direct ecosystem management is inevitable to remedy human disruptions of ecosystems and to promote ecological restoration and regeneration.

Adaptive management requires much more robust systems monitoring and response than exist in typical environmental and natural resources law regimes (Ruhl 2008; Craig 2013; Garver 2013). In this regard, historic or other types of baselines play an important role. A key challenge is to identify times and places when the human–Earth relationship appeared to be in a desirable, mutually enhancing state (Ruhl and Salzman 2011). In some instances, relatively static baselines may be useful, as in the case of referring to greenhouse gas emissions in 1990 as a historic baseline for climate change mitigation (*ibid.*). However, the 'stationarity' that is assumed in many aspects of environmental law is problematic given the dynamic and non-linear evolution of ecosystems (Craig 2013). Thus, the baselines for policy development should be more broadly based on reference conditions that incorporate the evolutionary capacity of systems to adapt and maintain resilience. For example, ecological restoration involves intentional intervention to return to a desired evolutionary trajectory based on objectives that integrate historical information – both socio-cultural and ecological – and criteria of ecological integrity (Higgs 2003).

Overall, the proposed prescriptions for law that have emerged from complexity theory scholars are in line with features that have been identified for 'the rule of ecological law' (Garver 2013). In addition to adaptiveness, enhanced monitoring and precaution about transgressing ecological and adaptiveness, those features include treating humans as embedded in Earth's ecosystems; giving primacy to ecological limits; fully integrating the rule of ecological law into all areas of law and policy; radically reducing humanity's use of materials and energy; establishing global and supranational rules distributed according to the principle of subsidiarity; and committing to intragenerational, intergenerational and interspecies fairness (*ibid.*). Much of the scholarship on law as a complex adaptive system focuses solely on environmental law as the domain in which solutions based on adaptive management and precaution should be developed. This focus is too limited. A transformative approach that brings law in line with promoting ecological integrity and a mutually enhancing human–Earth relationship must also fully integrate the legal infrastructure that supports practices in finance, trade, property transactions, contracts, commercial enterprises, taxation and so on, all of which play important roles in the human–Earth relationship.

Conclusion

In the dominant growth-insistent paradigm, technological solutions to environmental challenges are emphasized. The focus for innovation is on technology that will decouple economic growth from negative ecological and social impacts. The role of law is to maintain the reliance of the economic, social and political systems on economic growth and technological innovation. With societal goals shifted to fostering ecological integrity and a mutually enhancing human–Earth relationship, the focus for innovation broadens to include innovations in normative regimes such as ethics, economics and law. Although technology that reduces material and energy throughput in the economy and preserves or enhances ecological integrity remains important, much more emphasis is placed on developing alternatives that challenge or reject the orthodoxies of human constructs like economics and law that underlie the growth-insistent paradigm. A systems-based assessment of the potential of existing or innovative features of the legal systems and others with which it interacts to lock-in systems behaviours that work to advance or impede ecological integrity and a mutually enhancing human–Earth relationship would help illuminate a more hopeful pathway for law in the Anthropocene.

Note

1 Some researchers assert the Anthropocene began well before the industrial era with the extensive land use changes that emerged in the Neolithic Transition from hunter-gatherer societies to agricultural societies (Ruddiman 2013; Smith and Zeder 2013) or possibly even with human-driven extinction of megafauna in the Palaeolithic era (Foley *et al.* 2013).

References

Angermeier, P. L., and J. R. Karr (1994) Biological Integrity versus Biological Diversity as Policy Directives. *BioScience* 44: 690–97.

Berry, T. (1999) *The Great Work: Our Way into the Future*. New York: Three Rivers Press.

Brown, P. G., and G. Garver (2009) *Right Relationship: Building a Whole Earth Economy*. San Francisco, CA: Berrett Koehler.

Carpenter, S.R., and E. M. Bennett (2011) Reconsideration of the Planetary Boundary for Phosphorus. *Environmental Research Letters* 6: 1–12.

Craig, R. K. (2013) Learning to Think About Complex Environmental Systems in Environmental and Natural Resources Law and Legal Scholarship: A Twenty-Year Retrospective. *Fordham Environmental Law Review* 24: 87–102.

Crutzen, P. J., and E. F. Stoermer (2000) The Anthropocene. *Global Change Newsletter* 41: 17–18.

Daly, H. E. (1996) *Beyond Growth*. Boston, MA: Beacon Press.

Diamond, J. (2005) *Collapse: How Societies Choose to Fail or Succeed*. New York: Viking Press.

Ellis, E. (2011) Anthropogenic Transformation of the Terrestrial Biosphere. *Philosophical Transactions of the Royal Society A* 369: 1010–35.

Ellis, E. C., J. O. Kaplan, D. Q. Fuller, S. Vavrus, K. K. Goldwijk, and P. H. Verburg (2013) Used Planet: A Global History. *Proceedings of the National Academy of Science* 110(20): 7978–85.

Foley, S. F., D. Gronenborn, M. O. Andreae, J. W. Kadereit, J. Esper, D. Scholz, U. Pöschl, D. E. Jacob, B. R. Schöne, R. Schreg, A. Vött, D. Jordan, J. Lelieveld, C. G. Weller, K. W. Alt, S. Gaudzinski-Windheuser, K.-C. Bruhn, H. Tost, F. Sirocko, and P. J. Crutzen (2013) The Palaeoanthropocene – The Beginnings of Anthropogenic Environmental Change. *Anthropocene* 3: 83–8.

G20 (2010) The G20 Seoul Summit Leaders' Declaration. Available at www.g20.org/Documents2010/11/seoulsummit_declaration.pdf (accessed 28 October 2014).

G20 (2011) Cannes Summit Final Declaration. Available at www.g20.org/Documents2011/11/Cannes%20Declaration%204%20November%202011.pdf (accessed 28 October 2014).

G20 (2013) G20 Leaders' Declaration, Saint Petersburg Summit. Available at https://www. g20.org/sites/default/files/g20_resources/library/Saint_Petersburg_Declaration_ ENG_0.pdf (accessed 28 October 2014).

Garmestani, A. S., and M. H. Benson (2013) A Framework for Resilience-Based Governance of Social-Ecological Systems. *Ecology and Society* 18(1): article 9. Available at www.ecologyandsociety.org/vol18/iss1/art9 (accessed 20 November 2014).

Garver, G. (2013) The Rule of Ecological Law: The Legal Complement to Degrowth Economics. *Sustainability* 5(1): 316–37. Available at www.mdpi.com/2071-1050/ 5/1/316/htm (accessed 2 November 2014)

Higgs, E. (2003) *Nature by Design*. Cambridge, MA: MIT Press.

Holling, C. S., L. H. Gunderson, and D. Ludwig (2002) In Quest of a Theory of Adaptive Change. In L. H. Gunderson and C. S. Holling (eds), *Panarchy: Understanding Transformations in Human and Natural Systems*, 3–22. Washington, DC: Island Press.

Hornstein, D. T. (2005) Complexity Theory, Adaptation, and Administrative Law. *Duke Law Journal* 54: 913–60.

Jackson, T. (2009) *Prosperity Without Growth: Economics for a Finite Planet*. London: Earthscan.

Kellie-Smith, O., and P. M. Cox (2011) Emergent Dynamics of the Climate-Economy System in the Anthropocene. *Philosophical Transactions of the Royal Society A* 369: 868–86.

Krausmann, F., S. Gingrich, N. Eisenmenger, K-H. Erb, and M. Fischer-Kowalski (2009) Growth in Global Materials Use, GDP and Population during the 20th Century. *Ecological Economics* 68: 2696–705.

McNeill, J. R. (2003) Observations on the Nature and Culture of Environmental History. *History and Theory* 42(4): 5–43.

Merchant, C. (2003) *Reinventing Eden: The Fate of Nature in Western Culture*. New York: Routledge.

Robinson, N. (2014) Fundamental Principles of Law for the Anthropocene? *Environmental Law and Policy* 44(1–2): 13–27.

Rockström, J., W. Steffen, K. Noone, Å. Persson, F. S. Chapin, E. Lambin, T. M. Lenton, M. Scheffer, C. Folke, H. Schellnhuber, B. Nykvist, C. A. De Wit, T. Hughes, S. van der Leeuw, H. Rodhe, S. Sörlin, P. K. Snyder, R. Costanza, U. Svedin, M. Falkenmark, L. Karlberg, R. W. Corell, V. J. Fabry, J. Hansen, B. Walker, D. Liverman, K. Richardson, P. Crutzen, and J. Foley (2009) Planetary Boundaries: Exploring the Safe Operating Space for Humanity. *Ecology and Society* 14(2): article 32. Available at www. ecologyandsociety.org/vol14/iss2/art32 (accessed 29 August 2011).

Ruddiman, W. F. (2013) The Anthropocene. *Annual Review of Earth and Planetary Sciences* 41: 45–68.

Ruhl, J. B. (1996) Complexity Theory as a Paradigm for the Dynamical Law-and-Socity System: A Wake-up Call for Legal Reductionism and the Modern Adminsitrative State. *Duke Law Journal* 45(5): 849–928.

—— (2008) Law's Complexity: A Primer. *Georgia State University Law Review* 24(4): 885–911.

—— (2011) General Design Principles for Resilience and Adaptive Capacity in Legal Systems – With Applications to Climate Change Adaptation. *North Carolina Law Review* 89: 1373–1403.

—— (2012) Panarchy and the Law. *Ecology and Society* 17(3): article 31. Available at http:// dx.doi.org/10.5751/ES-05109-170331 (accessed 28 October 2014).

Ruhl, J. B., and J. Salzman (2011) Gaming the Past: The Theory and Practice of Historic Baselines in the Administrative State. *Vanderbuilt Law Review* 64(1): 1–57.

Smith, B. D., and M. S. Zeder (2013) The Onset of the Anthropocene. Available at www.academia.edu/7162531/The_Onset_of_the_Anthropocene (accessed 26 January 2015).

Steffen, W., J. Grinevald, P. Crutzen, and J. McNeill (2011) The Anthropocene: Conceptual and Historical Perspectives. *Philosophical Transactions of the Royal Society A* 369: 842–67.

UNEP (2011) *Towards a Green Economy: Pathways to Sustainable Development and Poverty Eradication – A Synthesis for Policymakers*. Nairobi: United Nations Environment Programme. Available at www.unep.org/greeneconomy/GreenEconomyReport/tabid/29846/language/en-US/Default.aspx (accessed 2 November 2014).

United Nations (1987) *Report of the World Commission on Environment and Development: Our Common Future*. New York: United Nations.

—— (2012) *The Future We Want: Outcome Document Adopted at Rio+20*. New York: United Nations. Available at www.un.org/en/sustainablefuture.

—— (2014) *Introduction and Proposed Goals and Targets on Sustainable Development for the Post 2015 Development Agenda*. New York United Nations. Available at http://sustainabledevelopment.un.org/content/documents/4523zerodraft.pdf (accessed 28 October 2014).

Waters, C. N., J. A. Zalasiewicz, M. Williams, M. A. Ellis, and A. M. Snelling (2014) *A Stratigraphical Basis for the Anthropocene?* London: Geological Society.

Westra, L., P. Miller, J. R. Karr, W. E. Rees, and R. E. Ulanowicz (2000) Ecological Integrity and the Aims of the Global Integrity Project. In D. Pimentel, L. Westra, and R. F. Ness (eds), *Ecological Integrity: Integrating Environment, Conservation, and Health*, 19–41. Washington, DC: Island Press.

WWF (2014) *Living Planet Report 2014: Species and Spaces, People and Places* (ed. R. McLellan, L. Iyengar, B. Jeffries, and N. Oerlemans). Gland: World Wide Fund for Nature.

Zalasiewicz, J., M. Williams, A. Haywood, and M. Ellis (2011a) The Anthropocene: A New Epoch of Geological Time? *Philosophical Transactions of the Royal Society A* 369: 835–41.

Zalasiewicz, J., M. Williams, R. Fortey, A. Smith, T. L. Barry, A. L. Coe, P. R. Bown, P. F. Rawson, A. Gale, P. Gibbard, F. J. Gregory, M. W. Hounslow, A. C. Kerr, P. Pearson, R. Knox, J. Powell, C. Waters, J. Marshall, M. Oates, and P. Stone (2011b) Stratigraphy of the Anthropocene. *Philosophical Transactions of the Royal Society A* 369: 1036–55.

20 Seeking justice in a land without justice

The application of anti-corruption principles to environmental law

Kathryn Gwiazdon

Introduction

> Across Africa, oil, gas and minerals are being discovered more often than ever before ... Used wisely, these natural resource revenues could lead to sustainable economic growth, new jobs and investments in health, education and infrastructure. But sadly, history teaches us that a more destructive path is likely – conflict, spiraling inequality, corruption and environmental disasters are far more common consequences of resource bonanzas. The cliché remains true: striking oil is as much a curse as a blessing.[1]

Generally, corruption can be understood as a spiritual or moral impunity or deviation from an ideal, a breaking of ethical values.[2] It has existed since the beginning of recorded history and is one of the most widespread forms of human behaviour. Even with this pervasiveness, however, it has been difficult to address at the social, political and legal levels.

To begin, there is no precise definition that is universally applicable to all forms, types and degrees of corruption. Simply because an act is legal does not mean it is free of corruption, nor does an illegal act mean that it is corrupt. Also, what is considered corrupt is ever-changing. It has a historical and cultural narrative that allows something to be corrupt at one time or to one people, and yet not corrupt at another time or to another people. The scale of the subject matter and the potential parties involved can also be overwhelming. Rarely is it one person acting alone, and accountability can go to the highest rungs of power.

What is clear, however, is its potential for destruction: corruption 'eats away at the soul of a nation'.[3] Whether ignored or participated in, it has the potential to destroy systems and societies, and all the culture, language and heritage connected to them. Corruption damages democratic integrity and threatens the rule of law. It has the power to create a non-participating, disillusioned civil society where voters do not vote, money wins campaigns, and more unjust laws are created, enforced and defended. It has the potential to create shadow economies and parallel authorities, where organized crime rules the resources and distorts development priorities, or governance is determined by bribes, cronyism and nepotism.

Anti-corruption laws today focus on actions that go against, or corrupt, the proper functioning of public, private and government activities. Although an individual crime, the body of law is not necessarily concerned with the integrity of the offending individual, but rather the potential harm to the integrity of the governance system in place, and its fair and equitable administration of justice. Does it (the arguably corrupt act) harm the rule of law? Does it harm the system?

Although corruption may be difficult to define, the principles to combat it are not: integrity, solidarity, democracy, justice, transparency, accountability and courage.[4] There must be a culture of integrity that embraces ethical individuals and actions, and promotes an honest, transparent and informed democracy, where those harmed can seek justice, and those who harm can be held accountable. People must have the courage to 'act without fear or favor'.[5] This courage should not be undervalued, as there is great personal and professional risk in shedding light on corruption.

So, what breeds corruption? What are its conditions? What types of corruption most affect the rule of law? And what are society's responses? For something so pervasive and permanent, is there a way forward? This chapter seeks to address those questions and, due to the high incidence of corruption within the natural resource sector and the extractive industry, within the realm of environmental law.

Conditions for corruption

> It is the world's vulnerable who suffer first and worst by corruption such as the theft of public money or foreign aid for private gain. Corruption is not some vast impersonal force but the result of personal decisions, most often motivated by greed.[6]

In order to have a better understanding of how to manage corruption, it is important to have a better understanding of the causes of corruption. If corruption is linked to a predatory yearning for power, whether political and/or financial, one only has to look for the seemingly powerless to find the corrupt. So what factors create a powerless people, or a vulnerable people? How can society and the rule of law better empower people to deter corruption?

Quite simply, countries that are rich in resources and weak in the rule of law are magnets for corruption. The countries it affects seem to be in an endless cycle, what came first: the corruption or the [discovery of] natural resources? This has been termed, 'the resource curse'.[7] The resource curse recognizes that one of the most prevalent conditions for corruption is the mere existence of the extractive industry,[8] and that the income generated from development projects is not trickling down to everyday citizens or GDP. The transnational nature of the industry also makes transparency and accountability difficult to maintain.

Wars stem from and create vulnerable peoples, and wars continue to be fueled by the profits attached to natural resources, including but not limited to water, oil, lumber and minerals. The physical act of extraction itself can cause civil unrest, such as when indigenous peoples are forcibly removed or their homeland made unliveable. In Botswana, Bushmen are being forcibly removed from the Kalahari to allow for diamond mines.[9] Tribes in India are being threatened with police action and 'being labeled Maoists' if they do not move.[10] Contamination caused by the oil spills in the Niger Delta, the Alberta tar sands, and generally, fracking and mountain-top removals have also caused legal, political and social unrest. Resource extraction has also been linked to the spread of disease, such as with the recent outbreaks of Ebola originating in West Africa.[11] And trade in illegal wildlife and illegal logging are billion-dollar businesses, where poachers and illegal loggers even turn to violence and murder.[12]

Corruption follows power and money, and vast amounts of money are given to foreign aid and development projects. Foreign aid is meant to help victims of war, humanitarian crises, diseases, etc. that often result directly or indirectly from resource extraction, and although some of it does reach the intended parties, it can also find its way to the same people who caused the conditions. Development projects, particularly water management projects, such as dams, canals, drains, groundwater access (e.g. large agriculture taking the majority of the water resources) and sanitation, are also prey to the corrupt, both in terms of financial gain and power over access.

A weak rule of law is an unreliable system of justice, where public officials are self-serving and the citizenry is kept in the dark. It can be a state that fosters nepotism and elitism, or one weakened by war. The system does not provide accountability to individuals who are causing harm, justice to those who have been harmed, or protection to those who seek to shed light on corruption. This may be due to a lack of actionable laws, or weak or corrupt law enforcement officers, prosecutors and judges. The state may not have whistleblower protections, or it may manipulate government licensing and contracts, embezzle funds intended for environmental protection programs, or wrongfully allow the clearing of customs for illegal goods and trade.

When there is a weak rule of law, there is also greater opportunity to distort or hide information that would otherwise empower citizens. The corrupt are attracted to and work to create an information deficit, where people have a lack of understanding from a lack of information, or a lack of understanding because they are purposely misinformed. This may be due to a lack of transparency in government processes, or the result of restrictions on the media, such as in Canada where government meteorologists are 'forbidden from publicly discussing climate change'.[13] With a lack of transparency, the role of the citizen in the democratic process can be more easily manipulated, and democracy more easily eroded.

The first case study addresses the resource curse and involves one of the most important anti-corruption principles: transparency. The global economic crash in 2008 was due in large part to corporate and government malfeasance, and part of the US response was to draft a transparency law particular to extractive industries.

Case 1: The resource curse

> Publicity is justly commended as a remedy for social and industrial diseases. Sunlight is said to be the best of disinfectants; electric light the most efficient policeman.[14]

In October 2012, American Petroleum Institute (API), the principal lobbying firm in the United States for the oil industry, filed suit against the US Securities and Exchange Commission, seeking to strike down The Dodd Frank Wall Street Reform and Consumer Protection Act, section 1504.[15] The regulations required oil, gas and mining companies to disclose payments over US$100,000 that are made to national and foreign governments for extractive projects.[16]

Section 13(q) of the provision addresses the phenomenon known as the 'resource curse', whereby 'oil, gas reserves, and minerals . . . can be a bane, not a blessing, for poor countries, leading to corruption, wasteful spending, military adventurism, and instability' when 'oil money intended for a nation's poor ends up lining the pockets of the rich or is squandered on showcase projects instead of productive investments'.[17] As a result of this corruption, many of the world's 'most wealthy mineral countries are the poorest countries in terms of their citizens' quality of life.[18]

One of the most outstanding arguments made by API was that mandatory disclosures were unconstitutional violations of companies' First Amendment rights and that they would amount to compelling speech, 'the right not to speak is just as paramount as the right to speak.'[19] API also argued that they would violate certain criminal laws where the host law prohibits disclosure of payments. The SEC found no evidence that any such laws even existed, and that to provide exemptions would undermine Congress' intent to promote international transparency. API also claimed that irreparable harm of billions of dollars would be caused due to the disclosure of competitive information. The SEC continued to reinforce the importance of transparency in combatting corruption, and that the rule was the result of a thorough, democratic process.

The Honourable John D. Bates vacated the rule and remanded it back to the SEC, 'the agency's interpretation was arbitrary and capricious.' The SEC will not be appealing the decision and so must rewrite the rule.

It is not difficult to see the corruption that can arise when industry and states are not transparent and are not being held accountable. However, even if transparency rules are required, data can be manipulated, and permits and contracts can be awarded due to bribes. Constant vigilance, oversight and accountability must follow transparency. The case below provides a good example of the important role of the media in bringing light to corruption, the international nature of the crime, and how corruption by a private entity can tarnish a nation's reputation.

Case 2: A flawed consultancy agreement

In September 2003, a scandal was uncovered by Norwegian paper *Dagens Næringsliv* involving a national oil company, Statoil, and an Iranian consultancy firm, Horton

Investments.[20] Statoil had paid Horton Investments US$15.2 million for the purpose of '[influencing] decision-makers in the Iranian oil and gas industry' to grant oil contracts to Statoil. Horton Investments was owned by Mehdi Hashemi Rafsanjani, Director of the National Iranian Oil Company, and son of former Iranian President Hashemi Rafsanjani.

In June 2004, Statoil was found guilty of bribery by the Norwegian courts and ordered to pay approximately US$3 million in fines. The uncovering of the corruption led to the resignation of the Statoil Chairman, the Statoil CEO, and their Director for International Operations. Statoil agreed to pay the fines, but insisted that this did not imply any admittance of criminal culpability, although an internal investigation had shown 'ethical lapses'.[21]

In October 2006, Statoil settled with the US Securities and Exchange Commission for violations of the US Foreign Corrupt Practices Act (FCPA). In addition to fines, Statoil had to agree to the following counts:

• Statoil had entered a 'flawed consultancy agreement' with an Iranian public servant;
• these bribes were paid to secure contracts in the country, and to get hold of confidential information; and
• Statoil had used inaccurate accounting procedures in order to hide the bribes from its records.[22]

Types of corruption that affect the rule of law

Several types of corruption affect the rule of law, including political corruption, judicial corruption, legal corruption and institutional corruption.

Political corruption

The World Bank relies on a straightforward definition of corruption: 'the abuse of public office for private gain'.[23] Political corruption includes political nepotism and cronyism, such as the hiring of friends, siblings or donors, which creates a general patronage system of governance. Political corruption also includes excessive lobbying. It is important to look at the laws being drafted and adopted, and by whom. Bribery of and by public officials also falls within political corruption. Bribes are sometimes masked as campaign contributions, or as with Statoil, 'consultancy fees'. Bribes can also be offered to be awarded valuable contracts, as seen in Statoil. It is important to follow the money going to candidates, and to see how that money influences decisions.

This manipulation of public and private funds is occurring to such a great extent in some countries that kleptocracies are being created. A kleptocracy is a form of political corruption where the government exists to increase the personal wealth and political power of its officials and the ruling class at the expense of the wider population, and often with the pretence of honest service.

Case 3: dissolution of an anti-corruption body

> PNG's governments are notorious for corruption, and ever run the risk of turning the state into a fully-fledged kleptocracy.[24]

Papua New Guinea (PNG) is a victim of the resource curse: they are a resource-rich, but citizen-poor, country. A recent report by the Pacific Research Institute found that 'revenues from past extraction of PNG's natural resources have failed to impact the lives of the average citizen', and that 'the average citizen has become poorer as a result of poor management and accountability'.[25]

In this politician-turned-businessman culture, public funds are regularly being co-mingled with private accounts. These public servants are earning vast wealth, in part due to their involvement in the extractive industries, and they are accused of using their executive authority to manipulate the other branches of government.

In 2011, Papua New Guinea Prime Minister, Peter O'Neill, created Investigation Task Force Sweep as part of an anti-corruption campaign. In June 2014, an arrest warrant was issued for O'Neill after the watchdog organization alleged he had illegally wired US\$30 million of government funds to a law firm. This same firm was charged a few years earlier with fraud and money laundering. O'Neill denied the charges, claiming that they were politically motivated. He then disbanded the anti-corruption body, ousted its chair, and among other high-level dismissals, replaced the Deputy Police Commissioner and Attorney General.[26] The PNG is now calling upon help from Australia, as Australia has invested billions of dollars in establishing the rule of law in the PNG.[27]

Case 4: The drought industry and ghost projects

Transparency International (TI) is the leading international organization in highlighting and combatting corruption.[28] They work with governments and other NGOs to foster a culture of integrity. TI publishes annual Global Reports that focus on a particular subject matter; for example, in 2007, they focused on judicial corruption, and in 2011, they focused on climate change. They also publish an Annual Corruption Perceptions Index and the National Integrity System Assessments.

Through their work, they have shown how political corruption can be found even in the exploitation of scarce resources, as well as with the money given to combat it. Ghost projects are public development projects that never occur, even if they were reported as completed.

Where water is scarce, access is power. Where water is scarce, people are vulnerable. According to TI, Piauí, Brazil is one of the country's poorest and driest states. Federal money had been allocated to improve the drought situation, but development was being delayed or not occurring at all. Arimatéia Dantas, coordinator of A Forca Tarefa Popular, called this exploitation the 'drought industry.' He said, 'The places supplied by the trucks are determined by politicians', and 'at times, water supply is used as electoral currency'.[29] Dantas, along with other citizen action groups, seeks to highlight this predatory business and hold them accountable.[30]

A Forca Tarefa Popular also discovered a ghost project in a community that had been awarded a US$7,000 grant for plumbing infrastructure. The records showed that the money was already allocated and that the project was complete. Transparency International helped them report the case to the public prosecutor, and a few months later, the community received access to water in their homes.[31]

Case 5: National and international responsibility

In 2005, the UN Convention Against Corruption (UNCAC) was drafted.[32] It is the only legally binding, universal anti-corruption instrument. The Convention covers many different forms of corruption, such as trading in influence, the abuse of power and various acts of corruption in the private sector.

The UNCAC directly addresses the threats to democracy and stability that corruption causes, and particularly its relation to sustainable development:

> *Concerned* about the seriousness of problems and threats posed by corruption to the stability and security of societies, undermining the institutions and values of democracy, ethical values and justice and jeopardizing sustainable development and the rule of law . . .[33]

And:

> *Recalling also* the Johannesburg Declaration on Sustainable Development, adopted by the World Summit on Sustainable Development, held in Johannesburg, South Africa, from 26 August to 4 September 2002, in particular paragraph 19 thereof, in which corruption was declared a threat to the sustainable development of people . . .[34]

In 2014, France indicted the eldest son of President Teodoro Obiang Nguema Mbasogo, Teodoro Nguema Obiang, on money-laundering charges and raided his Parisian apartment. In the raid, police seized his US$500 million estate and millions of dollars in luxury items. 'France's pursuit of cases like this sends a strong message that foreign governments can do more to ensure financial integrity of resource-rich countries', said Lisa Misol of Human Rights Watch.

Equatorial Guinea is an oil-rich West African country with widespread high-level corruption. In a country where the average citizen lacks reliable access to safe drinking water and electricity, the political elite are known for their exorbitant personal global spending.

In what would be the first UNCAC case brought before the International Court of Justice, Equatorial Guinea is now suing France, claiming that the son had diplomatic immunity as a UNESCO delegate. Immunities of convenience are common in domestic laws, as they seek to ease international relations. It can be argued, however, that they are by definition an example of legal corruption, as they legally overlook illegal behaviour in order to allow for more convenient international relations.

In October 2014, the son also settled a major anti-corruption case with the United States Department of Justice, and agreed to sell more than US$30 million worth of assets. Leslie Caldwell, US Assistant Attorney-General, stated, 'After raking in millions in bribes and kickbacks, Nguema Obiang embarked on a corruption-fuelled spending spree in the US.' Court documents showed that 'while he received a salary of less than $100,000 a year, he used his influence in government to amass more than 300 million USD worth of assets'.[35]

Judicial corruption

Judicial corruption directly threatens the rule of law and the governance systems in place. The courtroom is the centre of justice, where those who have been harmed can be made whole again, and those who harm can be held accountable. When corruption poisons the judiciary, it poisons the entire rule of law and stability of the state.

For judicial corruption, it is important to look at the entire justice process, from prosecution to final decision, and then enforcement of that decision. Who is being prosecuted? Are the judges or attorneys pressuring the litigants to drop or settle (and thereby taking the proceedings off public record)? Does the very existence of a private settlement process harm or help corruption? What is being admitted or heard by the court? What is being decided? What is being enforced? Who are, and who is paying, the experts in expert testimony? What evidence is being admitted? Has there been unnecessary delay?

Judicial electioneering raises many opportunities for corruption, as well. While many countries do not allow for the elections of their judges, it is a common practice in the United States. Of course, judicial appointments are not free from corruption, either.

Due to the potential of serious harm to the rule of law, judges and attorneys are held to a higher standard of integrity and must abide by a professional code of ethics. They must make all efforts to avoid any appearance of impropriety, and recuse themselves from cases if necessary.

Case 6: Judicial removal – And For The Sake Of The Kids

In *Caperton v. A. T. Massey Coal Co.*, the US Supreme Court overturned the West Virginia Supreme Court of Appeal's dismissal of the Trial Court's US$50 million damage award against the major coal company.[36] The Court stated that the Due Process Clause required the Appellate Judge to recuse himself from the case, due to 'serious risk of actual bias'. It was shown that the CEO of the defendant coal company created a shell organization, And For The Sake Of The Kids, which funnelled nearly US$3 million to the judge's election campaign.[37] Another judge had previously recused himself as photos surfaced of him with the Company's CEO in the French Riviera.[38]

Both Justice Roberts and Justice Scalia disagreed with the Court's decision. In Justice Roberts' dissent, he stated that 'the majority decision would have dire consequences for public confidence in judicial impartiality.'[39] Scalia, who joined him in his dissent, stated that this 'would permit . . . Due Process claims asserting judicial bias in all litigated cases in (at least) those 39 States that elect their judges'.

The disagreement among the Supreme Court Justices is not uncommon in many professions, where people seek to protect the integrity of the institution over the accountability of the individual. It is arguable which undermines the integrity of the rule of law more: the corrupt act itself, or hiding the corrupt act. If hiding corruption is meant to protect the system, what kind of system is being protected?

Legal corruption

Legal corruption exists when the laws in place protect corruption. Lobbying is as old as legislating itself. The technical role of the lobbyist is to provide a legislator with information on a particular subject matter in order to influence a decision, similar to a court expert providing information to a judge. The lobbyist, however, has a known bias to their employer and seeks to advance legislation to protect the special interests of that employer.

In the case below, not only was there no hearing on a controversial provision, but the provision was buried deep within a large and unrelated spending bill. This aspect of legal drafting, called riding, could be considered legal corruption, as the process itself prevents provisions from being appropriately read, discussed and voted upon. These are key aspects to democracy being manipulated, and yet this deceptive practice is legal and occurs with numerous bills.

Case 6: The Monsanto Protection Act

In early 2013, HR 933, a spending bill that included Section 735, the Farmer Assurance Provision, was passed 'under the cover of darkness' by the US Congress.[40] The provision later became known as the Monsanto Protection Act, as it blocked the US Department of Agriculture's review of genetically-modified crops and effectively gave companies such as Monsanto immunity in federal courts, even if harm to human health was found.

In addition to the manipulation of the legal process, the author of the bill, US Senator Roy Blunt, had worked directly with Monsanto in drafting the provision.[41] Monsanto was also the largest donor to Blunt's campaign, although he had received massive donations from agribusinesses as a whole.[42]

The Monsanto Protection Act had a term of six months which ended in December 2013. It was not renewed by Congress.

Institutional corruption

A common concern for any donor-reliant entity is undue influence by a benefactor. Institutional corruption occurs when an organization accepts money from a donor, or a donor gives money to an organization, with the intent to corrupt its purpose. In other words, the donor is giving money on expectations of change, whether it is to take a certain action, or to not take a certain action. Institutional corruption could also include financial support from people or organizations that have interests that *prima facie* conflict with the primary purpose of their institution. Similar to the role of judges, or legislators, organizations have a responsibility to avoid the appearance of impropriety and to protect the interests of their members. As they may also receive tax benefits as non-profit organizations, they have a particular legal and ethical duty to civil society.

Environmental organizations, and their members, need to be particularly wary when the organization accepts money from sources that historically work in opposition to their interests. Is the company seeking to greenwash its image? Is the company seeking to silence dissent on particular issues? It is important to bring all interested parties to the decision-making table, but the terms of the relationship must be set. Where is the line drawn between donation and influence? There can be difficulty with labelling institutional corruption, however, due to the amorphous nature of organizations: it may be a corrupted organization, or just a changing organization.

A way forward

There is a way forward, and it begins first and foremost by acknowledging that corruption exists. The topic has historically been difficult to address because of the fear of retaliation. Whistle-blower statutes are integral to combatting corruption, but there must be more. Shedding light on corrupt, and powerful, individuals can be at great personal and professional risk. This is why it is so important to have rules in place to protect those individuals courageous enough to address it. However, there must also be a culture of integrity: a culture that shuns corrupt behaviour, embraces those that defend democracy and justice, and fosters individuals and actions of integrity.

It is crucial for public and private entities to continue to advance anti-corruption policies and laws, and to broaden and particularize the topics they encompass. Environmental laws need to incorporate anti-corruption principles and address corruption specifically, as well as highlight the industries with the most potential for corruption. Anti-corruption laws, on the other hand, need to incorporate principles of environmental law such as ecological integrity, the right to a healthy environment, the rights of future generations, common but differentiated responsibilities, etc.

Indeed, the entire governance structure needs to support anti-corruption measures. A good example is South Africa's Office of the Public Protector, whose mandate is to strengthen constitutional democracy through its anti-corruption efforts,[43] and 'be the conscience of the state and ensure that it acts with integrity and justice at all times.'[44] In addition, more countries need to adopt lifestyle clauses for public officials (are they living beyond their means?) and more banks need to follow politically prominent individuals.

Another important legal avenue is to increase citizen suits, instead of relying on government bodies that may be weighed down by legislators, prosecutors, or judges, who may be purposely delaying or preventing justice. In this same vein, parties have to be creative and diverse when seeking redress, and should look to higher authorities and advocacy organizations, such as Transparency International, for assistance.

Corruption feeds off of the complexity of foreign relations, and so international efforts must be strengthened. For example, there is a movement for the International Criminal Court to include corruption in their actionable crimes. And, at the very least, G20 countries, who are expected to have more stable governments and economies, should ensure that corrupt public officials are not able to travel to other G20 countries to spend or hide stolen public money.

It is important, in particular, to foster judicial integrity and a strong sense of judicial honour. Any judicial review body needs to be transparent: the process, as well as the background of those that sit on the panel. The legal settlement process as a whole also needs to be revisited as it has become a tool used to shadow corrupt behaviour. It should be reviewed through a human rights lens, and take into consideration the public's right to know, particularly if there is potential for future harm.

The fight to combat corruption must go beyond the political and legal sectors. The media, for example, has a crucial role to play. Not only can investigative journalism bring attention to corruption, which has been shown to lead to legal action, but the very act, or even threat, of naming and shaming can be a deterrent.

In conclusion, corruption will always exist. But so will the people who fight it. Ethical behaviour is rooted in our human nature: we know it when we see it, and we know when it is absent. Humans have come together to form societies, governed by a rule of law. Corrupt individuals seek to manipulate these societies and these laws for personal gain. Only by having the appropriate safeguards, targeting the individuals and the organizations most prone to corruption, and fostering integrity, solidarity, democracy, justice, transparency, accountability and courage, can corruption be managed and effectively deterred.

Notes

1 Kofi Annan, Momentum Arises to Lift Resource Curse, *New York Times*, 14 September 2012, available at www.nytimes.com/2012/09/14/opinion/kofi-annan-momentum-rises-to-lift-africas-resource-curse.html. This article was written in response to the implementation of the transparency rules in The Dodd Frank Wall Street Reform and Consumer Protection Act (2010). Unfortunately, the rules that gave Mr Annan so much hope were later vacated by a federal district court, as was discussed in the first case study.

2 Anthony Wanjohi, Prevalence of Corruption in Kenya, *Kenya Projects Organization*, 17 December 2012, available at www.kenpro.org/corruption-in-kenya.

3 Nicky Rehbock, Thuli Madonsela: Corruption Eats Away at the Soul of the Nation, *Transparency International*, 17 October 2014, available at https://blog.transparency.org/2014/10/17/thuli-madonsela-corruption-eats-away-at-the-soul-of-the-nation. Thuli Madonsela, Public Protector of South Africa, was awarded Transparency International's annual Integrity Award.

4   Transparency International, Mission, Vision, and Values, available at www.transparency. org/whoweare/organisation/mission_vision_and_values. Transparency International is the leading international organization in combatting corruption. Their core values provide a thorough list of anti-corruption principles that states and organizations should seek to implement.

5   See Rehbock, Thuli Madonsela.

6   Ban Ki-Moon, On the Occasion of International Anti-Corruption Day. 9 December 2009, available at www.unodc.org/unodc/en/corruption.

7   The resource curse has been analyzed extensively in books and journals. A concise explanation with examples can be found in Nicholas Shaxson, The Resource Curse, or the Paradox of Poverty from Plenty, available at www.opendemocracy.net/ ourkingdom/nicholas-shaxson/resource-curse-or-paradox-of-poverty-from-plenty.

8   See generally, Ann Hollingshead, Why are Extractive Industries Prone to Corruption? *Financial Transparency Coalition*, 19 September 2013, available at www. financialtransparency.org/2013/09/19/why-are-extractive-industries-prone-to-corruption-part-ii.

9   Oliver Tickell, Botswana Government Lies Exposed as $5bn Diamond Mine Opens on Bushman Land, *The Ecologist*, 4 September 2014, available at www.theecologist. org/News/news_analysis/2542508/botswana_government_lies_exposed_as_5bn_ diamond_mine_opens_on_bushman_land.html.

10 Survival International, India: Tiger Reserve Tribes Face Illegal Eviction, *The Ecologist*, 14 October 2014, available at www.theecologist.org/News/news_round_up/2594211/ india_tiger_reserve_tribes_face_illegal_eviction.html.

11 IUCN, Ebola Outbreak Highlights Critical Links between Biodiversity Loss and Human Health Says IUCN's Wildlife Health Specialist Group, 8 October 2014, available at www.iucn.org/about/union/donors/?18439/Ebola-outbreak-highlights-critical-links-between-biodiversity-loss-and-human-health-says-IUCNs-Wildlife-Health-Specialist-Group.

12 Al Jazeera, Opponent of Peru Illegal Logging Murdered, 9 September 2014, *Al Jazeera*, available at www.aljazeera.com/news/americas/2014/09/opponent-peru-illegal-logging-murdered-2014995222652634.html.

13 Alex Greig, Canada Bans Government Meteorologists from Talking about Climate Change, *Daily Mail*, 3 June 2014, available at www.dailymail.co.uk/news/article-2646897/Canada-bans-government-meteorologists-talking-climate-change.html.

14 Justice Louis D. Brandeis, *Other People's Money and How the Bankers Use It* (1914).

15 *Am. Petroleum Inst. v. SEC*, No. 12–1668 (DDC 2 July 2013). Opinion available at www.sec.gov/rules/final/2013/34-67717-court-decision-vacating-rule.pdf. A helpful article with links to the parties' briefs available at www.earthrights.org/legal/american-petroleum-institute-v-sec-revenue-transparency-litigation.

16 Dodd-Frank Wall Street Reform and Consumer Protection Act, §1504, Pub. L. No. 111–203, 124 Stat. 1376, 2220 (2010).

17 *Am. Petroleum Inst. v. SEC*, No. 12–1668 (DDC 2 July 2013), citing 156 Cong. Rec. S3816 (May 17, 2010) (statement of Sen. Lugar).

18 *Am. Petroleum Inst. v. SEC,* No. 12–1668 (DDC 2 July 2013), citing 156 Cong. Rec. S5872 (July 15, 2010) (statement of Sen. Cardin).

19 *Am. Petroleum Inst. v. SEC*, No. 12–1668 (DDC 2 July 2013), citing *West Virginia State Board of Education v. Barnette*, 319 US 624 (1943).

20 BBC News, Statoil Fined Over Iranian Bribes, available at http://news.bbc.co.uk/2/ hi/business/3849147.stm.

21 *Ibid.*

22 Settlement terms available at www.sec.gov/news/press/2006/2006-174.htm.

23 The World Bank, Helping Countries Combat Corruption: The Role of the World Bank: Corruption and Economic Development, available at www1.worldbank.org/ publicsector/anticorrupt/corruptn/cor02.htm.

24 J. F. Canberra, Near Neighbors, Worlds Apart, *The Economist*, 8 August 2011, available at www.economist.com/blogs/banyan/2011/08/papua-new-guinea-and-australia.

25 Freddy Mou, People Poorer Despite Big Mines: NRI, *PNG Loop*, 21 October 2014, available at www.pngloop.com/2014/10/21/png-people-poorer-despite-big-mines-nri/§hash.VLOLfkWI.dpuf.

26 Helen Davidson, PNG Aid Under Scrutiny as Corruption Taskforce Head Arrives in Australia, *The Guardian*, 24 June 2014, available at www.theguardian.com/world/2014/jun/24/australia-well-placed-sanctions-png-corruption.

27 *Ibid.*

28 See generally www.transparency.org.

29 Transparency International, The Drought Industry, available at www.transparency.org/news/story/the_drought_industry.

30 *Ibid.*

31 *Ibid.*

32 Text available at www.unodc.org/documents/treaties/UNCAC/Publications/Convention/08-50026_E.pdf.

33 *Ibid.*, Appendix, Preamble, page 5.

34 *Ibid.*, page 2.

35 Javier Blas, Equatorial Guinea Leader's Son Settles US Corruption Case, *Financial Times*, 12 October 2014, available at www.ft.com/cms/s/0/700a81d6-51f0-11e4-b55e-00144feab7de.html#axzz3I7eVtDAQ.

36 *Caperton v. A. T. Massey Coal Co.*, 556 US 868 (2009), available at www.supremecourt.gov/opinions/08pdf/08-22.pdf.

37 *Ibid.*

38 *Ibid.*

39 *Ibid.*

40 Connor Adams Sheets, Monsanto Protection Act: 5 Terrifying Things To Know About The HR 933 Provision, *International Business Times*, 27 March 2013, available at www.ibtimes.com/monsanto-protection-act-5-terrifying-things-know-about-hr-933-provision-1156079.

41 *Ibid.*

42 Tom Philpott, Sen. Roy Blunt: Monsanto's Man in Washington, *Mother Jones*, 4 April 2013, available at www.motherjones.com/tom-philpott/2013/04/sen-roy-blunt-monsantos-man-washington.

43 South Africa Office of the Protector. See generally www.pprotect.org.

44 See Rehbock, Thuli Madonsela.

21 Environmental defenders

The green peaceful resistance

Antoni Pigrau and Susana Borràs

Introduction: who are environmental defenders? And why defend them?

In the framework of the United Nations, 'human rights defender' is a term used to describe people who, individually or with others, act to promote or protect human rights.[1] These include all sorts of rights, including the rights to health and food, or the rights to land and natural resources, so the defenders of the environment are a specific subgroup of the defenders of human rights in general.

The frequency of killings and threats against those who defend the environment and the rights of their people, and the subsequent impunity enjoyed by the majority of perpetrators of such serious violations affect the protection of the environment as well as our most fundamental human rights.[2] Human rights violations committed against these environmental defenders or activists are generally directly related to their activities of claiming, defending, and protecting territories and natural resources, or their defence of the right to autonomy and the right to cultural identity:

> Between December 2006 and May 2011, a large number of communications sent during the reporting period (106) concerned alleged violations against defenders and activists working on land and environmental issues. According to the information received, this group is thoroughly heterogeneous. It includes defenders carrying out a vast range of activities related to land and environmental rights, including those working on issues related to extractive industries, and construction and development projects; those working for the rights of indigenous and minority communities; women human rights defenders; and journalists.[3]

The right to participation and the right to information constitute two fundamental pillars for the actions undertaken by environmental activists and defenders. Environmental defenders have been the targets of violence because they have challenged the environmental impacts of a variety of activities, especially those of the extractive industries and their effects on fundamental human rights, such as the right to life and to housing, the right to water and food, the right to property, and the right to free, prior, and informed consent.

Thus, the clear relationship between environmental protection and the effective exercise of human rights cannot be denied. It therefore follows that the work of environmental defenders is a key component in the protection of human rights. These people provide information to society, and it falls upon the state to assume its obligation to guarantee the rights of its people. Environmental defenders also organize communities to fight for their rights and for environmental justice. However, during recent years, environmental defenders have been subjected to threats and violence, and this does not involve only isolated cases.

National and multinational companies, especially those dedicated to the extractive industries, are primarily responsible for stigmatizing environmental defenders, often with the complicity of state authorities, who prosecute the defenders and levy unjustified civil and criminal penalties for the purpose of shutting down social protest and forcing the environmental movement to focus on trying to free its leaders from incarceration.

The failure of the state to investigate aggressions and crimes committed against environmental defenders often complements such business activity and helps to weaken their defensive roles. Various practices have been identified: putting obstacles in the way of reporting companies, not recognizing the right to challenge and demand the revocation of state concessions or pursuing environmental claims with a clear lack of diligence, arguing that budgeting does not allow inspections to be performed. Time and time again, victims lose trials for such reasons as their lack of legitimacy, the destruction or deterioration of evidence or the simple unjustified delay of the authorities in solving the case, leading to damages being incurred and the resolutions being issued too late. Added to all of this, even when resolution occurs in favour of those affected, there is often a lack of authority to enforce the legal rulings. In particular, a judicial system that is not impartial can favour impunity and become a mechanism for covering up human rights violations. Claims frequently go uninvestigated, even when evidence is submitted. Furthermore, in cases involving environmental claims, prosecutors and judges who have tried to diligently perform their duties end up being thwarted, transferred, or even removed from their positions.

All of these obstacles create circumstances in which those who oppose environmental pollution, the environmental defenders, must live in a state characterized by defencelessness and personal and legal insecurity. A case like the one of Greenpeace activists on the *Arctic Sunrise* – protesting against oil exploration in the Barents Sea, they were arrested on September 2013, and incarcerated in Russia, but supported by a powerful organization capable of immediately making the facts visible worldwide – are, unfortunately, the exception.[4]

One clear example is that of Mr Alexander Nikitin, who was awarded the Goldman Environmental Prize in 1997.[5] Until 1985 Alexander Nikitin was a naval captain in the Soviet Northern Fleet, where he served on nuclear-powered submarines. From 1987 to 1992 he worked for the Department of Defence as the senior inspector for its Nuclear and Radiation Safety Inspection Department. Nikitin joined the Bellona Foundation, a Norwegian non-governmental organization addressing north-western Russia's environmental problems. In 1996 Nikitin co-authored a report entitled The Russian Northern Fleet – Sources of Radioactive Contamination, on the contamination caused by the facilities and submarines of the Russian nuclear fleet, especially in the Kola peninsula, which is one of the regions in the world with the highest levels of radioactive pollution. In 1995 Bellona's Russian office was ransacked by the Federal Security Police (FSB) and all references to the report were confiscated. Nikitin was trying to reconstruct the report when he was suddenly arrested on 6 February 1996. He was imprisoned on charges of high treason and divulging state secrets.

Mr Nikitin was sent to the FSB prison in St. Petersburg. During his first six weeks in prison, Nikitin was denied the opportunity to choose his own lawyer. He was held in solitary confinement and denied bail.

The case generated protests the world over. For this reason, Amnesty International, the UN's High Commissioner for Human Rights, and the Helsinki Committee declared Alexander Nikitin to be a prisoner of conscience.[6] The European Parliament also issued a resolution about the case, asking the Russian authorities to provide a full and detailed explanation of the charges against Alexander Nikitin and a clear schedule for his trial, which must be a public, just trial before a civil court, and requested that he be set free immediately in advance of the trial. This request was made because of the general concern that Nikitin may be judged by a military court, which would impede public access to the proceedings and to information, which would eliminate the possibility of a fair trial.[7]

On 14 December 1996, the Attorney General released Nikitin from prison and his case was sent back to the FSB for further investigation. After his release, he was not permitted to leave St. Petersburg. After several months, the FSB completed its investigation and filed additional charges of treason – seven in total – against Nikitin.

In June 1998, Nikitin's case was transferred to the City Court of St. Petersburg. In the first trial against Nikitin, in October 1998, the City Court of St. Petersburg sent the case back to the FSB for additional investigation. The Supreme Court confirmed this decision in February 1999, and the FSB filed new charges in July 1999. The second trial started at the St. Petersburg City Court in November 1999, and ended on 29 December with a full acquittal. The prosecution appealed to the Supreme Court, but the acquittal was confirmed and reached legal force on 17 April 2000.

The Nikitin case illustrates the problem perfectly. This chapter focuses on some aspects of the problem, which affects numerous areas in a wide range of countries and which will require further development in future work: the international legal framework to protect environmental defenders, the protection of environmental defenders through the Inter-American System of human rights, other mechanisms of protection and some final considerations.

The international legal framework to protect human rights and environmental defenders

The cases presented are just few examples of a pattern of a large and systematic criminalization against environmental defenders, as a means to discipline and do away with social protest. Moreover, environmental defenders and their families are defenceless: lack of recognition and legal protection increases their vulnerability. A part of the general instruments of human rights, which could provide a certain level of protection and means to denounce the violations of their human rights, the international society still remains passive in the face of human rights violations and environmental degradation.

According to the UN's Universal Declaration of Human Rights[8] of 10 December 1948, 'everyone has the right to an effective remedy by the competent national tribunals for acts violating the fundamental rights granted him by the constitution or by law' (Art. 8). Therefore, 'no one shall be subjected to arbitrary arrest, detention or exile' (Art. 9). Also, according to Article 10, 'everyone is entitled in full equality to a fair and public hearing by an independent and impartial tribunal, in the determination of his [or her] rights and obligations and of any criminal charge against him [or her]'. Furthermore, 'everyone charged with a penal offence has the right to be presumed innocent until proved guilty according to law in a public trial at which he[or she] has had all the guarantees necessary for his [or her] defence' (Art. 11).

In practice, most of the complaints about attacks on the rights of environmental defenders are filed in America. The American Convention on Human Rights,[9] in its Article 8, establishes that 'every person has the right to a hearing, with due guarantees and within a reasonable time, by a competent, independent, and impartial tribunal, previously established by law, in the substantiation of any accusation of a criminal nature made against him [or her] or for the determination of his [or her] rights and obligations of a civil, labour, fiscal, or any other nature'.

In 1982, the World Charter for Nature stated in its Principle 23:

> All persons, in accordance with their national legislation, shall have the opportunity to participate, individually or with others, in the formulation of decisions of direct concern to their environment, and shall have access to means of redress when their environment has suffered damage or degradation.[10]

The Declaration of Basic Principles of Justice for Victims of Crime and Abuse of Power[11] allows the circumstances to be determined of people criminalized and put on trial, who were first victimized when their collective rights had been violated, then later re-victimized when they were punished for complaining about the violation of their rights. This declaration establishes that '[v]ictims should be treated with compassion and respect for their dignity. They are entitled to access to the mechanisms of justice and to prompt redress, as provided for by national legislation, for the harm that they have suffered' (Art. 4). This declaration also establishes that 'informal mechanisms for the resolution of disputes, including mediation, arbitration and customary justice or indigenous practices, should be utilized where appropriate to facilitate conciliation and redress for victims' (Art. 7). Finally, in Article 19 the declaration refers to the idea that 'states should consider incorporating into the national law norms proscribing abuses of power and providing remedies to victims of such abuses. In particular, such remedies should include restitution and/or compensation, and necessary material, medical, psychological and social assistance and support'.

One of the key elements in the protection of defenders has been the official definition of the 'defence' of human rights as a right in itself and the recognition of the category of 'human rights defenders'. On 9 December 1998, Resolution 53/144 was adopted by consensus by the General Assembly of the United Nations, thus approving the Declaration on the Right and Responsibility of Individuals, Groups and Organs of Society to Promote and Protect Universally Recognized Human Rights and Fundamental Freedoms,[12] which establishes that 'everyone has the right, individually and in association with others, to promote and to strive for the protection and realization of human rights and fundamental freedoms at the national and international levels' (Art. 1) and each State 'shall adopt such legislative, administrative and other steps as may be necessary to ensure that the rights and freedoms referred to in the present Declaration are effectively guaranteed' (Art. 2). For these purposes, in accordance with Article 5:

> everyone has the right, individually and in association with others, at the national and international levels: (a) To meet or assemble peacefully; (b) To form, join and participate in non-governmental organizations, associations or groups; (c) To communicate with non-governmental or intergovernmental organizations.

And Article 9 recognizes, among other things, that:

> everyone has the right, individually and in association with others, *inter alia*: To complain about the policies and actions of individual officials and govern-mental bodies with regard to violations of human rights and fundamental freedoms, by petition or other appropriate means, to competent domestic judicial, administrative or legislative authorities or any other competent authority provided for by the legal system of the State, which should render their decision on the complaint without undue delay.

And Article 12 states that:

> 1. Everyone has the right, individually and in association with others, to participate in peaceful activities against violations of human rights and fundamental freedoms.
>
> 2. The State shall take all necessary measures to ensure the protection by the competent authorities of everyone, individually and in association with others, against any violence, threats, retaliation, de facto or *de jure* adverse discrimination, pressure or any other arbitrary action as a consequence of his or her legitimate exercise of the rights referred to in the present Declaration.
>
> 3. In this connection, everyone is entitled, individually and in association with others, to be protected effectively under national law in reacting against

or opposing, through peaceful means, activities and acts, including those by omission, attributable to States that result in violations of human rights and fundamental freedoms, as well as acts of violence perpetrated by groups or individuals that affect the enjoyment of human rights and fundamental freedoms.

Moreover, in 2000, the United Nations Human Rights Commission decided to create the figure of the United Nations Special Representative on Human Rights Defenders.[13] In August 2000, the Secretary General appointed Ms Hina Jilani. Her mandate was renewed by the Commission in 2003 and by the Human Rights Council in 2007,[14] after the adoption of a new resolution in support of human rights defenders by the General Assembly of the United Nations.[15]

In March 2008, the Human Rights Council decided to pursue their task even further and appointed Ms Margaret Sekaggya as new Special Rapporteur on the situation of human rights defenders.[16] Her mandate was renewed for three more years in April 2011.[17] In 2014, the Human Rights Council decided to continue the mandate on human rights defenders for a consecutive period of three years[18] and appointed, in June 2014, Mr Michel Forst as the new UN Special Rapporteur.

The mandate of the Special Rapporteur is:

(a) To promote the effective and comprehensive implementation of the Declaration on the Right and Responsibility of Individuals, Groups and Organs of Society to Promote and Protect Universally Recognized Human Rights and Fundamental Freedoms through cooperation and constructive dialogue and engagement with Governments, relevant stakeholders and other interested actors;

(b) To study, in a comprehensive manner, trends, developments and challenges in relation to the exercise of the right of anyone, acting individually or in association with others, to promote and protect human rights and fundamental freedoms;

(c) To recommend concrete and effective strategies to better protect human rights defenders through the adoption of a universal approach, and to follow up on these recommendations;

(d) To seek, receive, examine and respond to information on the situation and the rights of anyone, acting individually or in association with others, to promote and protect human rights and fundamental freedoms;

(e) To integrate a gender perspective throughout the work of his/her mandate, paying particular attention to the situation of women human rights defenders;

(f) To work in close coordination with other relevant United Nations bodies, offices, departments and specialized agencies, both at Headquarters and at the country level, and in particular with other special procedures of the Council;

(g) To report regularly to the Human Rights Council and the General Assembly.

The Organization of American States adopted its first Resolution on Human Rights Defenders in 1999.[19] Many others have subsequently approved, but with little impact.

In December 2008, the European Union Council adopted the EU Guidelines on Human Rights Defenders[20] to address specific concerns regarding human rights defenders, and also to contribute to reinforcing the EU's human rights policy in general.

Finally, as far as civil society is concerned, another document of interest related to the protection of environmental defenders is the Cartagena Declaration, adopted at the International Conference of Environmental Rights and Human Rights on 16 to 18 September, 2003 in Cartagena, Colombia, and organized by Friends of the Earth International, the Transnational Institute, and the Oilwatch network.[21] The declaration aims to safeguard the right of human rights defenders, environmentalists, and those who demonstrate against injustice and war to be free from criminalization and persecution.[22]

Protection of environmental defenders through the inter-American system of human rights

The African Commission on Human and Peoples' Rights adopted its first Resolution on the Protection of Human Rights Defenders in Africa on 4 June 2004.[23] But the regional system to protect human rights most sensitive to this issue has been the American, and the Inter-American Commission on Human Rights (IACHR) has expressed concern over clear instances of the persecution of environmental defenders.

For example, the Commission has received and processed claims of human rights violations against the leaders of African-descended communities in Colombia, and has requested that the Inter-American Court protect threatened African-descended leaders. In the case of the *Communities of Jiguamiandó and Curbaradó*, the Court ordered provisional measures to be taken in 2003[24] and renewed these measures in successive resolutions up to 2010.

Another more recent case is that of Teodoro Cabrera García and Rodolfo Montiel Flores against Mexico.[25] In 1998, Teodoro Cabrera García and Rodolfo Montiel Flores, along with other rural workers, established a civil association called the Campesino Environmentalist Organisation of the Sierra of Petatlán and Coyuca de Catalán (OCESP), to stop logging operations in the mountain forests in the state of Guerrero. In their opinion, these operations were threatening the environment and the livelihoods of those in the local rural communities. In 2010, the Court found Mexico guilty of infringing the Convention on a variety of counts: the right to personal freedom, the right to personal integrity, the requirement to investigate alleged acts of torture, the right to justice and judicial protection because information about the alleged tortures had been provided to the military criminal jurisdiction. However, the Court did not consider those aspects that were connected to their status of environmental defenders, insofar as the Inter-American Commission of Human Rights had not taken into account that the main reason for the case was the fact that they were environmental defenders and that they were being repressed.

On 25 October 2010, a general hearing was held before the Inter-American Commission on Human Rights, focused on the circumstances of environmental defenders in Mesoamerica.[26] Its aim was to denounce the pattern of violence against environmental defenders in Mexican and Central American mining areas. Environmental defenders from Mexico, Guatemala, Honduras, El Salvador and Panama denounced violence against environmental activists in the mining areas for two reasons. First, killings, kidnappings, torture, arbitrary detentions, and damage to the private property of environmental defenders have been verified in mining areas. Second, the countries in the region lack adequate legislative means to protect the effective enjoyment of human rights affected by the mining industry.

In his Second Report on the Situation of Human Rights Defenders in the Americas, the Inter-American Commission on Human Rights stated that:

> According to the information the Commission has received, the murders and other violations that environmental defenders have suffered are region-wide in the case of the Extractive industry, which is not just the mining industry as it includes other areas such as logging. The Commission has received information indicating that in Brazil, at least 125 activists and campesino leaders have received death threats; in the space of just five days (May 24 to 28, 2011) four persons were killed for their activities in defense of the environment. In El Salvador, in the short six-month period from June to December 2009, three environmental defenders opposed to the mining industries were murdered; another was killed in 2011. In Guatemala, four defenders working for a healthy environment died in just one month (January to February 2010). In Mexico, at least 12 persons were killed in the period from 2006 to 2012, among them public officials and civilian defenders of the right to a healthy environment.[27]

The case of *Kawas Fernández v. Honduras* is representative of the activity that the Inter-American system engages in with respect to environmental defenders.

On 4 February 2008, the Commission submitted a complaint before the Court against the Republic of Honduras, based on the pleading presented on 13 January 2003 by the Centre for Justice and International Law and the Team for Reflection, Research and Communication of the Company of Jesus in Honduras.

The case involved the murder of Blanca Jeannette Kawas Fernández, who was shot and killed in her home. Ms Kawas was the chair of the Foundation for the Protection of Lancetilla, Punta Sal, Punta Izopo, and Texiguat, an organization created for the purpose of 'improving the quality of life of the populations of the Bahía de Tela watershed (Department of Atlántida, Honduras)'. In her role as chair, Ms Kawas denounced, among other issues, attempts by individuals and entities to illegally assume power on the Punta Sal Peninsula, as well as the contamination of lakes and the degradation of forests in the region. After her death, serious omissions became evident, showing that State authorities failed to act with due diligence, as they did not adopt all of the measures necessary to support an investigation into the killing that could have resulted in a concrete outcome. As a consequence of the State's failure to perform its duties, the victim's family was denied their right to know the truth about what happened and to be compensated for the damages and losses they suffered.

According to the Inter-American Commission, 'the materials found in the file establish that there are indeed strong indications to suggest that the State is directly responsible in the alleged victim's loss of life'. Consequently, on 13 October 2005, the Commission approved Report No. 67/05,[28] by which the petition was declared admissible. Later, on 20 July, 2006, the Commission approved background report No. 63/06,[29] under the terms of Article 50 of the convention, which contained specific recommendations for Honduras.

The State was notified of this report on 4 August, 2006. Upon consideration of the information provided by the parties subsequent to the adoption of the background report, and upon 'the lack of substantive advances in effective compliance with' its recommendations, the Commission decided to submit this case to the jurisdiction of the Court.

The Commission alleged that:

> the effects caused by the impunity in the case and the failure to adopt the meas-
> ures needed to prevent the repetition of the acts have promoted a context
> of impunity in Honduras for acts of violence committed against defenders of
> human rights, the environment, and natural resources.' Furthermore, it stated
> that 'the case reflects the circumstances of defenders of the environment and
> natural resources in Honduras, attacks against such persons, and the obstacles
> to the investigation of acts of harassment and persecution.

The Commission concluded that the State was responsible for: violations of Article 4 of the American Convention (Right to Life) in compliance with the obligations established in its Article 1.1; violation of the rights recognized in Articles 8 (Legal Guarantees) and 25 (Legal Protections) of the American Convention in relation to its Articles 1.1 and 2.

The victim's representatives submitted their pleadings, motions and evidence under the terms of the Convention. The document alleged that Blanca Jeannette Kawas was a well-known Honduran environmental activist who promoted the protection of natural resources in her country, and that this activity was the motive for her murder on 6 February, 1995. The representatives reiterated that the death of Ms Kawas 'holds a special symbolism, because she was the first person killed in Honduras for defending natural resources and the environment. After her execution, and because of the impunity that characterised it, a series of assassinations of other environmental defenders occurred in Honduras.'

In its ruling of 3 April 2009,[30] the Inter-American Court states the following in relation to the issue at hand:

States have the duty to provide the necessary means for human rights defenders to conduct their activities freely: to protect them when they are subject to threats in order to ward off any attempt on their life or safety: to refrain from placing restrictions that would hinder the performance of their work, and to conduct serious and effective investigations of any violations against them, thus preventing impunity . . . Given the importance of the role that human rights defenders play in democratic societies, the free and full exercise of this right places a duty on States to create legal and real conditions in which they can freely carry out their activities.[31]

And it points out that:

The recognition of the work carried out by the defence of the environment and its relation to human rights is more valid in the countries of the region, where there is an increasing number of reports of threats, acts of violence and murders of environmentalists as a result of their task.[32]

The Court condemned Honduras for violating the rights to justice and judicial protection to the detriment of the family; the right to life, to the detriment of Blanca Jeannette Kawas Fernández; the right to personal safety, to the detriment of the family; and freedom of association, to the detriment of Blanca Jeannette Kawas Fernández.

In terms of reparation, the Court ruled that the State of Honduras must provide compensation to the representatives for material and immaterial damages and reimburse costs and expenses, as appropriate. It established the need for the State to conclude the criminal proceedings, or to initiate the appropriate ones, for the acts that generated the violations in the case and to resolve these proceedings under the terms established by law and within a reasonable period of time. It also declared the State's obligation to provide psychological assistance to members of Ms Kawas's family. Furthermore, it ordered the State to implement a national campaign to promote awareness and knowledge of the importance of the work performed by those who defend the environment in Honduras, and of their contributions to the defence of human rights. Finally, it determined that the State should hold a public event in recognition of international responsibility. And in particular:

it reiterates that the threats and attacks on the safety and life of the human rights defenders and the impunity of such events are particularly serious in a democratic society . . . the State has the duty to adopt legislative, administrative and judicial measures, or to improve existing measures, to guarantee the environmental defenders the freedom to carry out their activities; immediate protection for environmental defenders in the face of danger or threats that arise as the result of their work, and the immediate, serious and effective investigation of any acts that endanger the life or the safety of environmental defenders as a result of their work.[33]

Other mechanisms of protection

The ethics tribunal against the criminalization of defenders of nature, water and Pachamama

The Peoples' Tribunal against Criminalisation was held in Cuenca, Ecuador, on 22–23 June 2011. This Opinion Tribunal was organized on behalf of people, organizations, communities, and ethnic groups who have suffered some type of violation of their fundamental rights because of their defence of collective rights or the rights of nature, and who have been assigned – or who have been threatened with assignment of – criminal or formal administrative penalties after being accused of some type of crime, including in some cases terrorism.

Held within the context of the Continental Conference for Water and Pachamama, which took place on 21–23 June in Cuenca, Ecuador, the tribunal was organized by the associations Ecological Action (Acción Ecológica) , the Peoples' Ecological Network (Red de Ecologistas Populares), CEDHU and INREDH.[34]

The goal of the tribunal[35] was to hear testimony regarding the criminalization of protest, presented by Ecuadorian defenders of human rights and nature; then, based upon this testimony and the report drawn up by the public defender, to define the situation, expand awareness of it, and issue a verdict that could be applied in national and international cases.

The tribunal[36] ruled that the communities, peoples, and social and non-governmental organizations that have fought for collective rights and the rights of nature in Ecuador have been extensively and increasingly victimized by criminalization and punishment, encouraged by national and transnational companies – particularly in the extractive sector – and carried out by various judicial, police, military, and administrative authorities, as well as by private security forces. The tribunal therefore confirmed the existence of the 'systematic practice of criminalization as a means to punish and eliminate social protest', and that the justice system is used to criminalize the defenders of nature, while remaining passive against the human rights violations where these defenders and nature are the victims.

Networks of protection: the defenders who defend environmental defenders

Considering the current level of persecution against environmental defenders, a new generation of defenders has been born, that is, organizations which legally defend the environmental defenders and they are aimed to assist the victims of this persecution in order that they can freely develop its mission to protect the environment. The most relevant at the international level are the Environmental Law Alliance Worldwide (ELAW) and the Environmental Defender Law Center (EDLC).

In the Australian context, there is the National Environmental Defenders Office, which was established in 1996. It consists of nine independently constituted and managed community environmental legal centres located in each State and Territory of Australia, protecting the environment in the public interest. All together form the National EDO Network.

All of them have a core of common objectives, helping environmental defenders without charge. These objectives include not only assistance and support to environmental defenders, but also in giving strategic support, developing cases and strengthening environmental laws. Their main functions are: helping individual environmental defenders who are charged in politically-motivated criminal prosecutions, giving legal representation and advice, environmental law reform and policy formulation; finding top law firms to advocate, negotiate, or litigate on their behalf, advising, filing legal briefs, and providing resources, such as giving grants to fund cases; fighting against impunity for violations of the human rights of environmental defenders; ensuring that the community receives prompt advice and professional legal representation in public interest environmental matters; identifying deficiencies in the law and working for reform of these areas; and empowering the wider community, including indigenous peoples, to understand the law and to participate in environmental decision-making.

Final considerations: the defencelessness of the environmental defenders

Currently, environmental defenders are highly exposed to attacks on physical integrity, often by non-state actors, and many are killed by the work they do in relation to extractive industries and development projects, or in connection with the right to the land of indigenous peoples and minorities. Given all the considerations discussed above, and to cope with this situation of violence and vulnerability of environmental defenders, it would be appropriate to consider the following recommendations.

First, the recognition of their situation is necessary to protect them properly. In this regard, States should fully recognize the important work done by the environmental defenders and to try to strike a balance between economic development and respect for the environment, including the right to use land, wealth and natural resources and the rights of certain groups, such as minorities and indigenous peoples.

Second, after the recognition it would be desirable to strengthen institutional mechanisms to protect human rights defenders and environmental advocates. Often, attacks on environmental defenders go unpunished and exacerbate the lack of protection of rights, while generating suffering and distress in violation of the right to personal integrity. There is a pattern of criminalization against them as a means to punish and eliminate social protest. There is enormous inequality in the application of justice. While the justice system is used to criminalize the defenders of nature, it remains passive against the human rights violations of which these defenders and nature are the victims. Consequently, not only should the attacks be thoroughly investigated with due process, but the punitive role of the State should be strengthened to adequately respond to attacks against environmental defenders. Additional evidence is that administrators of justice in general have a lack of knowledge about the environmental defenders. Knowledge of the social and personal context help in research, knowledge and disposition of cases in which the defenders are immersed; not doing so would contribute its criminalization.

For this reason, national protection mechanisms should be promoted, such as ombudsmen, independent and investigative powers and effective recommendation; the development of national programs to protect defenders; and the adoption of protective measures such as rapid assistance services, urgent care and even relocation or protection of witnesses in cases of threats.

Third, and related to this, States must not allow the media or public officials to stigmatize the activity of these defenders, particularly in contexts of social polarization, as this can foster a climate of intimidation and harassment that could generate rejection and even violence against human rights defenders.

Fourth, the environmental regulatory framework may also contribute to the protection of environmental defenders, strengthening the supervisory and sanctioning capacity of the State. Also, improving the regulatory framework for industrial activities and control, would allow the State to ensure the effective enjoyment of the rights of communities affected by its impact, recognizing the link between environmental degradation and protection of human rights. The State, as guarantor of the rights protected by the international law of human rights, therefore must implement an appropriate regulatory framework, which effectively protects the rights of individuals, including the environmental defenders.

States must combat impunity for those who violate these defenders and their rights and facilitate a wider access to justice for environmental matters.

Finally, a last reflection: the defenders of nature and their families are defenceless and criminalized, putting them, their families and their surrounding communities in a situation of vulnerability, as well as any officials who dare to issue rulings in favour of such defenders. Not only them, but also the environment is rendered defenceless, when deprived of its environmental defenders.

Acknowledgements

This chapter was developed in the framework of the project Environmental Justice Organisations, Liabilities and Trade (EJOLT), FP7-Science in Society-2010–1 and of the project "Del desarrollo sostenible a la justicia ambiental: hacia una matriz conceptual para la gobernanza global", funded by the Spanish Ministry of Economy and Finance, for the period 2014–2016 (DER2013-44009-P).

Notes

1 Human Rights Defenders: Protecting the Right to Defend Human Rights, fact sheet no. 29, Office of the United Nations High Commissioner for Human Rights, United Nations Office at Geneva, Geneva, August 2004, p. 3. See also Agyeman (2002).

2 The organization Global Witness published in 2014 the result of their research in a report titled *Deadly Environment*, on the killing of activists working on environment and land issues. They report on 908 known cases of environmental defender killings, and provide key information on the issue, such as that 2012 recorded the highest number of deaths (147), three times more than in the previous 10 years; in last the last 4 years,

two environmental defenders have been killed every week; there are three main drivers (extractive industries and mining, land grabbing and land distribution, and illegal logging and deforestation); at international level most cases are reported from Central and South America; at regional level most cases are reported from the Philippines (67 cases), from Asia and Brazil (448 cases), and in the Americas, while Africa shows limitation in term of access to information; on perpetrators only 10 have been punished, most of them remain unknown, in 52 killings military or police have been identified and small groups of 1 to 6 people in other 171 cases. The report can be downloaded from www.globalwitness.org/deadlyenvironment.

3 Report of the Special Rapporteur on the situation of human rights defenders, Margaret Sekaggya, United Nations, Human Rights Council, Doc. A/HRC/19/55, 21 December 2011, para. 64. See also Friends of the Earth International, We Defend the Environment, We Defend Human Rights: Denouncing Violence against Environmental Defenders from the Experience of Friends of the Earth International, June 2014, available at www.foei.org/wp-content/uploads/2014/06/We-defend-the-environment-we-defend-human-rights.pdf.

4 See www.greenpeace.org/international/en/news/features/From-peaceful-action-to-dramatic-seizure-a-timeline-of-events-since-the-Arctic-Sunrise-took-action-September-18-CET.

5 The following information has largely been taken from the website of Goldman Environmental Prize; see www.goldmanprize.org/node/139.

6 Also, on 22 November 1998, the Sierra Club granted its Chico Mendes Award to Mr Nikitin for his activities directed at protecting the environment against the illegal dumping of nuclear waste in Russia.

7 Resolutions of 16 November 1995 (DO C 323 of 4 December1995, p. 112) on the harassment of the Bellona Foundation by Russian security forces; of 15 February 1996 (DO C 65 of 4 March 1996, p. 162) on the detention of Alexander Nikitin; and the Resolution on the case of Alexander Nikitin; Official Journal no. C 320 of 28 October 1996 p. 196, available at http://eur-lex.europa.eu/LexUriServ/LexUriServ.do?uri=CELEX:51996IP0995:ES:HTML.

8 Adopted and declared by General Assembly resolution 217 A (III) of 10 December 1948, available at www.un.org/es/documents/udhr (accessed 3 January 2012).

9 The Convention was adopted at the Inter-American Specialised Conference on Human Rights held in San Jose, Costa Rica, 7–22 November 1969, OAS Treaty Series No. 36, available at www.oas.org/juridico/english/treaties/b-32.html.

10 A/RES/37/7, 28 October 1982.

11 A/RES/40/34, 29 November 1985.

12 A/RES/53/144, 8 March 1999. See also Olagbaju and Mills (2004).

13 Resolution by the Human Rights Commission 2000/61, of 26 April 2000.

14 Resolution by the Human Rights Commission 2003/64, of 24 April 2003; Human Rights Council Resolution 5/1, Institution-Building of the United Nations Human Rights Council, 18 June 2007.

15 A/RES/62/152, 18 December 2007, Declaration on the Right and Responsibility of Individuals, Groups and Organs of Society to Promote and Protect Universally Recognized Human Rights and Fundamental Freedoms.

16 Human Rights Council, Resolution 7/8, Mandate of the Special Rapporteur on the Situation of Human Rights Defenders, 27 March 2008.

17 Human Rights Council. Resolution 16/5, Mandate of the Special Rapporteur on the Situation of Human Rights Defenders, 8 April 2011. The documents drafted by the Special Rapporteur can be consulted at www.ohchr.org/EN/Issues/SRHRDefenders/Pages/AnnualReports.aspx.

18 Human Rights Council, Resolution 25/18, Mandate of the Special Rapporteur on the Situation of Human Rights Defenders, 28 March 2014.

19 General Assembly of the Organization of American States, Human Rights Defenders in the Americas, Support for the Individuals, Groups, and Organizations of Civil

Society Working to Promote and Protect Human Rights in the Americas, AG/RES. 1671 (XXIX-O/99), 7 June 1999, available at www.oas.org/juridico/english/gares99/eres1671.htm.

20 The guidelines are available at http://europa.eu/legislation_summaries/human_rights/human_rights_in_third_countries/l33601_en.htm.

21 Also participating were 250 delegates from international organizations, NGOs and social movements from various parts of the world.

22 The text of the Declaration can be found online at http://wp.cedha.net/wp-content/uploads/2011/05/Declaraci%C3%B3n-de-Cartagena.pdf (accessed 14 January 2012).

23 African Commission on Human and Peoples' Rights, Resolution on the Protection of Human Rights Defenders in Africa, ACHPR/Res. 69 (XXXV), 4 June 2004, available at www.achpr.org/resolutions. Other resolutions on this subject were ACHPR/Res.119 (XXXII), 28 November 2007, and ACHPR/Res. 196 (L), 5 November 2011.

24 International Court of Human Rights, *Cases of the Jiguamiandó and Curbaradó Communities*, Provisional Measure, Resolution of 6 March 2003.

25 Inter-American Court of Human Rights, *Case of Teodoro Cabrera García and Rodolfo Montiel Flores v. Mexico*, Judgment of 24 June 2009.

26 General Hearing, 140th Regular Session, IACHR, The Situation of Environmental Defenders in Mesoamerica, 25 October 2010. See *Defensoras y defensores ambientales en peligro: Situación de defensores y defensoras del medio ambiente en Mesoamérica*, Report prepared by the Centre for International Environmental Law for the 25 October 2010 General Hearing during the 140th Regular Session of the Inter-American Commission on Human Rights, Available at www.miningwatch.ca/sites/miningwatch.ca/files/IACHR_Oct_10_Informe_CIEL.pdf (accessed January 2012).

27 Inter-American Commission on Human Rights, *Second Report on the Situation of Human Rights Defenders in the Americas, the Inter-American Commission on Human Rights*, OEA/Ser.L/V/II, Doc. 66, 31 December 2001, pp. 135–6 (notes omitted). Also see Amnistía Internacional (1999) and Amnistía Internacional and Club Sierra (2000).

28 In its admissibility report no. 67/05, the Commission decided to declare petition no. 61/03 admissible in relation to the alleged violation of articles 4, 8, and 25, in accordance with article 1.1 of the American Convention (documents in the appendices of the pleading, appendix 2, p. 683, para. 45). See Tanner (2011).

29 See Background Report No. 63/06, documents in the appendices of the pleading, appendix 1, page 672, para. 118.

30 Inter-American Court of Human Rights *Caso Kawas Fernández vs. Honduras* (Fondo, Reparaciones y Costas), Ruling of 3 April 2009, Series C No. 196.

31 Paras 145–6.

32 Para. 149.

33 Para. 213.

34 For more information, visit the conference's official website at www.aguaypachamama.org and also see http://movimientos.org/madretierra/pachagua/Convocatoria.pdf. See also Trujillo Orbe and Pumalpa Iza (2011).

35 The jury was made up of Elsie Monge from Ecuador, Raúl Zibechi from Uruguay, Lía Isabel Alvear from Colombia, and María Hamlin from Nicaragua, with Raúl Moscoso, Diana Murcia, and Carlos Poveda acting as Ecuadorian judges.

36 See Veredicto del Tribunal Ético ante la Criminalización a defensores y defensoras de los derechos humanos y de la naturaleza, Cuenca, Ecuador, 22 and 23 June 2011, available online at http://servindi.org/pdf/TribunalEtico23Jun2011.pdf (viewed 3 January 2012).

References

Agyeman, J. (2002) Constructing Environmental (In)justice: Transatlantic Tales. *Environmental Politics* 11(3): 31–53.

Amnistía Internacional (1999) *Defensores de los Derechos Humanos en Latinoamérica. Más protección, menos persecución.* Madrid: Editorial Amnistía Internacional.

Amnistía Internacional and Club Sierra (2000) *Ambientalistas Bajo Fuego.* Madrid: Editorial Amnistía Internacional.

Bellver Capella, V. (1996) El movimiento por la justicia ambiental: entre el ecologismo y los derechos humanos. *Anuario de Filosofía del Derecho* 3.

Bullard, R. (ed.) (2005) *The Quest for Environmental Justice: Human Rights and the Politics of Pollution.* San Francisco, CA: Sierra Club Books.

Coalición Internacional para el Acceso a la Tierra (2012) *Mecanismos internacionales para la protección de los defensores de los derechos humanos en riesgo por su trabajo en derechos de la tierra.* Rome: CINEP/PPP.

Coalición Internacional para el Acceso a la Tierra (2012) *Fondo de protección para defensores/ as del derecho a la tierra 'Defender la tierra'.* Rome: CINEP/PPP.

Comisión Interamericana de Derechos Humanos (2006) *Informe sobre la Situación de las Defensoras y Defensores de los Derechos Humanos en las Américas.* OEA/Ser.L/V/II. 124 Doc. 5 rev.1–7, March. Washington, DC: Comisión Interamericana de Derechos Humanos.

Comisión Interamericana de Derechos Humanos (2011) *Segundo Informe sobre la situación de las defensoras y los defensores de derechos humanos en las Américas,* OEA/Ser.L/V/II, Doc. 66, 31 December. Washington, DC: Comisión Interamericana de Derechos Humanos.

Gleason, J. M. (2009) Will the Confluence between Human Rights and the Environment Continue to Flow? Threats to the Rights of Environmental Defenders to Collaborate and Speak Out. *Oregon Review of International Law* 11(2): 267–99.

González Alcántara, J. (2001) Declaración de Defensores de Derechos Humanos de la ONU. *CIMAC Noticias* (4 November).

Manzini, E., and J. Bigues (2000) *Ecología y democracia: de la injusticia ecológica a la democracia ambiental.* Barcelona: Icaria Editorial.

Martín Quintana, M., and E. Eguren Fernández (2009) *Protección de defensores de derechos humanos: buenas prácticas y lecciones a partir de la experiencia.* Brussels: Protection International.

Martínez Alier, J. (2004) *El ecologismo de los pobres. Conflictos ambientales y lenguajes de valoración.* Barcelona: Icaria editorial.

Olagbaju, F. K., and S. Mills (2004) Defending Environmental Defenders. *Human Rights Dialogue* 2: 32–5.

Perazzi, R. J., and G. Orlandoni (1997) Sustentabilidad global, comercio internacional y política ambiental. *Economía* 22(13): 147–81.

Quintero, R. (2001) El acceso a la justicia ambiental, una mirada desde la ecología política. In *Justicia ambiental: Las acciones judiciales para la defensa del medio ambiente,* Jornadas Internacionales en Derecho del Medio Ambiente. Marzo de 2001, Bogotá (Colombia). Bogotá: Universidad Externado de Colombia.

Tanner, L. R. (2011) *Kawas v. Honduras* – Protecting Environmental Defenders. *Journal of Human Rights Practice* 3(3): 309–26.

Trujillo Orbe, R. (2010) *Manual para Defensores y Defensoras de Derechos Humanos y la Naturaleza.* Quito: Fundación Regional de Asesoría en Derechos Humanos (INREDH).

Trujillo Orbe, R., and M. Pumalpa Iza (2011) *Criminalización de los Defensores y Defensoras de Derechos Humanos en Ecuador.* Quito: Fundación Regional de Asesoría en Derechos Humanos, INREDH.Wakild, Emily, "Environmental Justice, Environmentalism, and Environmental History in Twentieth-Century Latin America", en *History Compass,* 11: 163–76. doi: 10.1111/hic3.12027, 2013.

22 Mind the gap

State governance and ecological integrity

Klaus Bosselmann

Introduction

This final chapter – the 'last word' – covers terrain that will be familiar to members of the Global Ecological Integrity Group. After all, 'our mandate is to push the boundaries of scholarly endeavor'.[1]

In pushing the boundaries a little further, my assertion is that ecological integrity is a fundamental concept of international law and governance, however, with the proviso that this has not yet been recognized by states.

If recognized by the international community, the dynamics of global governance would completely change as states would begin to function as guardians or trustees of the planet's commons. At present, states behave more like profit-oriented corporate organisations with an unstoppable appetite for everything that can be exploited. So expecting states to act as guardians of the planet seems a bit like asking the fox to look after the chicken.

But, unlike foxes, states are cultural arrangements. They are creations of human intelligence to allow governance of people and, in a democracy, by and for the people. Legitimacy of state power rests, to a considerable extent, on performing fiduciary obligations towards its citizens.[2] Such fiduciary obligations are recognized typically in public law,[3] exist in common law and civil law (although in varying forms and degrees[4]) and are also known in international law.[5]

Ultimately, state power can only be justified in so far as it does not violate the fundamental rights of individuals. And *vice versa*: fundamental rights such as the right to life or the right to dignity or the right to freedom are unalienable pre-state guarantees that no state can ignore.[6] This fiduciary function of the state can also be described as a trusteeship function.[7]

Given that states act, or should act, as trustees with respect to the fundamental rights of their citizens, it is not too difficult to include ecological integrity here. What could be more fundamental than the integrity of the planetary ecological systems that all human life depends on?

In the following I want to show the fundamentality of ecological integrity, both in terms of law and in terms of governance in order to close the gap between state governance and ecological integrity. In legal terms, the preservation of ecological integrity can be seen as a fundamental principle, or *grundnorm*.

In governance terms this calls for guardianship or trusteeship concepts forming part of institutional arrangements including states.

Closing the gap is not an exercise of mere jurisprudential argument, of course. It takes a lot more to overcome the ecological ignorance of the state which in turn has its roots in the ecological ignorance of Western civilization.[8] However, given the magnitude and dynamics of the current global crisis, change may happen unexpectedly and more quickly than we think. In some strange way time is on our side; we need to have the models ready when they are needed.

One such promising model is systems integrity (i.e. the integrity that each system including ecological and social systems needs to maintain in order to function properly). The title of this book and much of our work explores this model: as social systems, governance and law need integrity at their core (i.e. social, moral and ecological integrity). This seems obvious considering their need to function properly, yet we are seeing the signs of their collapse and demise. Some 'constitutional moment' may now be necessary to shake us up and begin the restoration of properly functioning governance and law.

Ecological integrity in domestic constitutional and international law

A 'constitutional moment' can occur when there are unusually high levels of sustained popular attention to questions of constitutional significance.[9] History has seen many constitutional moments, usually triggered by revolutions such as in England in 1688, United States 1787, France in 1789 or in recent times by peaceful revolutions such as in Russia, Poland, Germany or South Africa. A constitutional moment arises in situations of crisis. When the gap between promise and reality becomes too wide, then the 'will of the people' may articulate a new narrative or ideology demanding its expression in constitutional terms.

Such a situation arose during the 1980s in West Germany when the Green movement began to articulate the legal implications of ecological ethics. Following academic critique of anthropocentric reductionism in environmental law,[10] a group of eco-lawyers established the Institute für Umweltrecht (Institute for Environmental Law) and Verein für Umweltrecht (Association for Environmental Law) to promote the ecocentric approach to law. Through lobbyism of oppositional political parties and with the support of governments of some *Bundesländer* (federal states) a constitutional moment occurred when the Federal Government decided to conduct a major review of the *Grundgesetz* (Basic Law).

The constitutional review (1985–9) considered a shift from traditional anthropocentrism to an ecocentric value system underlying the constitution. Of central importance was the question of whether ecological realities require a redefinition of human rights to accept 'ecological limitations' and a special obligation of the state to protect the environment 'for its own sake' as well as for future generations.[11] In the end, the Joint Constitutional Commission did not resolve these issues, but called for a wider public dialogue precisely because they are so important: 'The question of either an anthropocentric or ecocentric approach to the constitution is of such fundamental importance, that the Commission did not see itself as mandated to answer it. Instead the Commission calls for a wide expert and public dialogue before considering such a change.'[12]

One step in this direction was the draft constitution of 1991,[13] which defined a 'socio-ecological market system' and included ecological limitations of human rights and property rights as well as a state obligation to protect the environment for its own sake. While this draft was rejected by the Federal Government, the 1994 and 2002 amendments to the *Grundgesetz* reflected a certain move away from anthropocentrism. Article 20a, for example, established a new state obligation: 'Mindful also of its responsibility toward future generations, the State shall protect the natural bases of life' – not just 'human' life following parliamentary debate on anthropocentrism versus ecocentrism.[14] In 2004, a further amendment added 'the animals' to the state obligation in response to uncertainties surrounding the constitutional status of animals.[15]

These developments suggest that fundamental change can occur when the time is right. At the end of 1980s the time was right for the peaceful revolution in East Germany. The time was not right, however, for an ecological revolution. The initial constitutional debate of the 1980s lost its momentum following German unification; instead some minor amendments were made. Undoubtedly, protecting the integrity of ecological systems and thereby human survival remains of utmost importance and will be constitutionalized when the time is 'right'.

Like the German *Grundgesetz*, virtually all European constitutions have been amended in the light of sustainability challenges. Worldwide, 147 countries have enacted constitutional provisions for environmental rights and responsibilities. Approximately 100 countries have incorporated a state obligation to protect the environment; of those, 60 national constitutions also recognize a human right to a decent environment, while 60 constitutions include collective and individual responsibilities for the environment.[16] The most ambitious of all is the Pachamama approach of some Latin American constitutions, especially Bolivia's (2009), which made ecological integrity a *grundnorm* of society.

The concept of ecological integrity is, of course, well-known in international environmental law. No less than 23 international soft and hard law agreements contain specific reference to it. The first such agreement was the Convention on the Conservation of Antarctic Marine Living Resources adopted in 1980, which recognized in its preamble the importance of 'protecting the integrity of the ecosystem of the seas surrounding Antarctica'. Another example is the preamble of the 1992 Rio Declaration on Environment and Development which calls for 'working towards international agreements which respect the interests of all and protect the *integrity* of the global environmental and developmental system'. Then Principle 7 of the Rio Declaration: 'States shall cooperate in a spirit of global partnership to conserve, protect and restore the health and *integrity* of the Earth's ecosystem.' This is repeated in key documents such as Agenda 21 or the 2002 Johannesburg Declaration. Even the 2012 Rio+20 outcome document, *The Future We Want*, widely perceived as unambitious and weak, calls for holistic and integrated approaches to sustainable development to guide humanity for restoring the health and *integrity* of the Earth's ecosystem (II. 40).

Finally, the Brundtland Report itself described the 'integrity of the natural system' as the basic condition for 'the survival of life on Earth' and in this way described the core idea behind sustainability as a prerequisite for development.[17]

Applying the usual standards for the recognition of concepts as international law, we can say that the repeated and consistent references to ecological integrity amount to an emerging fundamental goal or *grundnorm* of international environmental law.[18] But even independently of any legal status under international law, the *grundnorm* quality of ecological integrity should be without dispute. Surely, no one would doubt the fundamental importance of keeping the Earth's life-supporting systems intact.

Failure of states and response of legal scholars

Yet states have consistently ignored their own commitment to ecological integrity. This is particularly visible in the way states interpreted the formula that was meant to capture governance for sustainability (i.e. the concept of sustainable development).

I recently met Jim MacNeill, former Secretary-General of the UN Commission for Environment and Development and lead author of the 1987 Brundtland Report.[19] He confirmed my suspicion that, right from the start, governments and multinational companies deliberately rejected the key recommendation of the Brundtland Report, namely to organize social and economic development within the limits of nature. Instead, they cherry-picked one sentence of the report ('development that meets the needs of the present without compromising the ability of future generations to meet their own needs') to reconcile it with the traditional paradigm of economic growth with environmental protection as an add-on. Understood in this way, sustainable development made its world career precisely because it is so meaningless.

It was a great mistake of legal scholars to not criticize this ideologically-motivated behaviour of states in clear terms. In the main, legal scholarship underestimates the importance of ecological integrity. One of the pioneers of environmental law, the late Staffan Westerlund, University of Uppsala, went so far as to say that the entire academic discipline of environmental law over the last thirty years has failed. Rather than looking at sustainability from the perspective of law, we should have looked at the law from the perspective of sustainability. According to Westerlund, the central reference point of environmental law is not some undefined 'environment' or 'sustainable development', but sustainability with ecological integrity at its core.[20]

Douglas Fisher, one of Australasia's pioneers of environmental law, makes the same point: in his new book, *Legal Reasoning in Environmental Law* (2013), Fisher has analyzed the methodology of politicians, administrators and judges when dealing with environmental matters. They reach their decisions without a specific environmental 'point of commencement' (as Fisher calls it). Rather, they readily employ the well-trodden assumptions about human well-being, economic prosperity, cost–benefit analysis, and so on. The environment appears as an unknown entity, too abstract and not nearly as well defined as human rights or property rights. As a consequence, vague environmental interests are bound to lose against hard economic interests. The book concludes with a plea for 'processes of legal reasoning which reflect the fundamental *grundnorms* of the system – the rule of law in general and sustainability in the context of environmental governance'.[21] Spot on!

To me, the 'forgetfulness' (if one wants to call it that) of states, politicians, bureaucrats, lawyers and so many other professional groups is systemic. The current system of state governance is oblivious to the most fundamental requirement of human survival, namely that any action impacts on space and time. The cumulative effects of individual actions ignoring the wider spatial and temporal dimensions are visible in runaway climate change, irreversible biodiversity loss, collapsing ecological systems and a general decline of human well-being, life quality, living standards and wealth, eventually causing the collapse of civilization.

Rebuilding governance

This leads me to conclude that a fundamental rebuild of the governance system is needed. With respect to global governance, the shortcomings of current state-negotiated compromises are well known, but not easily overcome. The idea of creating new institutions for safeguarding the global commons is ambitious, but aims for a middle-ground. States are still assumed to play a pivotal role even though shared with members of global civil society who represent the interests of today's and future generations more credibly than state delegates.

Whatever the prospects, two preconditions must be met before even contemplating the feasibility of a partnership between states, the United Nations and civil society (as proclaimed by the Earth Charter).

The first is to recognize the severity of global threats and not to be side-tracked. While we can assume that most political leaders are concerned about at least some of these threats, they continue to isolate them from each other and from their underlying causes. Instead, they focus on issues that can be managed prag-matically and within the dominant paradigm of economic growth with environ-mental protection at its periphery. In other words, states and their representatives continue to ignore the severity of the problem. The first step, therefore, is to halt pragmatism, accept the systemic nature of global threats and start to think about solutions from there. This will in itself instil a sense of urgency that actions under the current system of international law and governance appear to be missing.

An equally important second precondition is to realize, and act upon, the inherent limitations of the current governance system. Historically, international law has evolved around the recognition of absolute state sovereignty. This continues to be relevant with respect to 'traditional' challenges of safeguarding peace and the autonomy and welfare of peoples and citizens. However, today's ecological, social, cultural and economic interdependencies require a more sophisticated concept of state sovereignty. States are not factually sovereign, but utterly dependent on other states to meet their common responsibilities and responsibilities for the commons. Notions such as *smart sovereignty* and *responsible sovereignty*[22] or *erga omnes* and *ius cogens* duties aim for defining the relationship between the national and the global in a new way. They have emerged from the experience of the last 20–30 years of dismal states' behaviour (ignoring, for example, the ethical foundations of the United Nations or the IUCN).

Putting national interests over global interests is profoundly wrong as it overlooks the mutual dependency of states and their citizens from the well-being of the planet as a whole. While rich and powerful states tend to defend their national interests stronger than poor and less powerful states, there are no real winners. Ultimately, we are all facing the same risks and challenges. The only credible solution for a dismally failing system of international law and governance is to renew it in a way that gives full weight to global interests alongside any national interests.[23]

In institutional terms, the renewal can be described as a shift away from the current state-centred model of international governance to an Earth-centred model of global governance, a model based on the partnership of nation-states, the United Nations and global civil society. The term *Earth governance* tries to capture institutional arrangements around this new partnership.[24]

Earth governance starts from the viewpoint of 'one Earth for all people'. It recognizes the (borderless) interconnectedness of planetary systems ('one Earth') and the right of each person and each community to benefit from their bounty and life-enabling services ('for all people').

Recognition of planetary systems involves two aspects. One is that planetary systems are finite and irreplaceable. They provide objective limits to the 'safe operating space for humanity'.[25] We can call this effect the importance and conditionality of planetary boundaries for human life. The other aspect concerns the protection of planetary systems themselves. How do we ensure that they are protected from human overconsumption? This aspect is different from mere recognition of ecological realities as it requires a positive concept of action. Living within planetary boundaries is only possible if we know how to recognize *and* positively protect them. As we identify, describe and promote the various – currently nine – planetary boundaries, we are implying the possibility of respecting them. Just how this respect can be defined in scientific, ethical and legal terms is a key consideration for tangible Earth governance.

The other aspect of Earth governance is, therefore, a positive definition of what keeps planetary systems alive and intact. Fundamentally, Earth and its ecological systems, large or small, possess a certain integrity understood as a dynamic equilibrium between stability and change. Any living system owes its existence to its ability to maintain such a dynamic equilibrium or steady state. In fact, we could say that any system, including, for example, an economic system, requires steady state qualities in order to not threaten itself or other systems, particularly ecological systems. With respect to ecological systems science, ethics and law have described such ability as ecological integrity.

As mentioned, the integrity of the Earth's systems has often been referred to in international agreements and described as a general duty of states. It is also the central concept of the Earth Charter, the only document to-date reflecting a global consensus across cultures, religions and civil society groups. The Earth Charter defines ecological integrity as fundamental to humanity, however, in doing so it describes it as inextricably linked to social and economic justice, democracy and peace. There is no prospect for humanity without realizing ecological integrity as a prerequisite *and* integral part of social integrity.

Ethically, caring for the integrity of ecological systems requires a sense of trusteeship, guardianship or stewardship. Politically, preserving and restoring ecological integrity is an overarching goal of public policy. Legally, this goal can be formulated as a central objective or underlying *grundnorm* and defined in a quantifiable and enforceable manner (e.g. in legislation or treaties). And institutionally, respecting planetary boundaries and ecological integrity requires governance based on trusteeship or guardianship.

The Earth Governance project

In 2014, a collaboration began between a number of global networks including the Planetary Boundaries Initiative (PBI), the Global Ecological Integrity Group, Earth Systems Governance Project, the Stockholm Resilience Centre, the IUCN World Commission on Environmental Law, the Earth Condominium Project, and the UN Scientific Advisory Board to the Secretary-General. The collaboration aims for investigating some of the normative foundations of more effective global governance (as indicated above) and for developing a proposal to institutionalize trusteeship governance at international level.

Institutional reform within and around the UN system has been of interest for some time and a range of proposals have been presented. For example, there is a long-standing discussion over the possibility of a World Environment Organization (WEO) or Global Environmental Organization (GEO).[26] An early proponent of this idea, Renato Ruggiero, then executive director of the World Trade Organization (WTO) drew much attention when he advanced the idea of a world environment organization as a counterweight to WTO.[27] Before that, in 1997, the German government had already proposed – together with Brazil and others – the establishment of a 'global umbrella organization for environmental issues with the United Nations Environment Programme (UNEP) as its major pillar.'[28] Recently, UNEP has been 'upgraded' to an environmental organization with universal membership. There is also renewed discussion about reviving the defunct UN Trusteeship Council to function as a central trusteeship organization of the global commons.[29]

Over the last few years, the interest in trusteeship governance has considerably increased. On the one hand, there is a rapidly growing body of literature advocating new forms of governance for the commons including the global commons.[30] On the other hand, the United Nations itself is undergoing a process of review and reform triggered by the Rio+20 outcome document *The Future We Want* (2012). Current developments include several official UN publications such as, for example, a 'Thematic Think Piece' entitled *Global Governance and Governance of the Global Commons in the Global Partnership beyond 2015*, compiled by the UN System Task Team on the Post-2015 UN Development Agenda.[31]

With the recent establishment of a 'UN Secretary General's Scientific Advisory Board to strengthen the connection between science and policy',[32] an opportunity exists to directly feed into the deliberations of the UN Secretary General and influence further development of UN policies.

It is important to note that there is currently not sufficient support among UN member states for such far-reaching reforms or even for charging a UN-authorized working group with developing a proposal as envisaged here. But this may change as the dynamics of the global governance crisis unfold. There is a general sense of urgency within global civil society and increased frustration with the failures of state-dominated international law and governance. This may well have spill-over effects on UN organizations and, at least, some UN member states known for their past support of trusteeship-type institutions and likely to respond to a new initiative more favourably.[33]

Another objective is the organization of a major international conference around the theme of this project. A consortium of environmental law networks, under the leadership of the IUCN World Commission on Environmental Law (WCEL) and Earth Charter International (ECI), has developed a proposal for a conference on 'Democracy and Sustainability' to be held at the Institute of Advanced Sustainability Studies in Potsdam/Germany in 2016 as part of the 'Earth Democracy' project that the WCEL Ethics Specialist Group is coordinating. This conference aims for bringing together some of the world's leading democracy and sustainability theorists as well as representatives of the UN, governments and NGOs. The theme of the conference reflects the topic of this project.

The timeframe is for an initial period of three years. For us as the Global Ecological Integrity Group, I see this project as a great opportunity to not only link up with some of the world's largest professional networks, but also to get a foot into the door of the UN system.

Final remarks

I am closing with what I believe is the most pressing issue right now. This concerns the way or methodology under which decisions are currently being made. We should be mindful that any new institutions and laws can only ever be signposts for better decision-making. What really matters are people and how they think and act.

We need to get into the habit of observing what the people who govern us are actually doing. Politicians and decision-makers need to be directly confronted with the ever-increasing gap between morality and public policy. We should look politicians over the shoulders, remind them of the ethical implications of their decisions and hold them accountable, morally and, where possible, legally.

According to Don Brown and his rich experience with the US government system, public policy has become so dominated by neo-classical economics that questions of ethics and morality are consistently kept off the table. For the most part, this is done through two kinds of arguments. First, the goal of all policy is welfare maximization within given markets, in which all values are determined by the 'willingness to pay' measured in money. Second, no government involvement in free markets is ever justifiable unless there are high levels of scientific proof of some serious market failure with undeniable serious environmental impacts. This puts the stakes so high that only overwhelming fly-in-your-face evidence may result in some action. And even if there is overwhelming evidence – think of climate change – any action has to pass a utility test, usually in the form of cost–benefit analysis. Instead of respecting non-negotiable ethical bottom-lines, governments aim for a middle-ground between market needs and environmental needs, no matter what this may mean for our prospect of sustainability.

Ecological integrity as a *grundnorm* would completely turn this around. It found its best expression in the Earth Charter with its covenantal plea to care for the community of life and to protect and restore the integrity of Earth's ecological systems. The Earth Charter has been endorsed by a number of states and many thousands of organizations, local communities, universities and so on. And yet, nothing will ever change as long as the heads and hearts of real people don't change.

The problem, as Don Brown, Ron Engel and others have shown, is the dominance of the liberal pretence of separating ethics from the concrete specifics of social, political and economic reality. The separation between morality and legality seems to be defining feature of our time. This separation, for example, allowed a mass murderer like Adolf Eichmann to appear perfectly 'normal' in his conduct as human being (citing his duties as public servant). Hanna Arendt famously described this phenomenon as the 'banality of evil',[34] but she did not just refer to Nazis and their followers here. Her overarching concern was that modern society increasingly creates new realities that move further and further away from the reality experienced through personal encounters with human beings – the gap between the 'systems world' and the 'life world', as Jürgen Habermas analyzed it.[35]

It is astonishing to see how realities created by mainstream media, corporate managers or Wall Street traders exist without any morality; in fact, they function best if they deliberately reject morality as in any way relevant to their business.

The challenge therefore is to see each and every aspect of our lives as being guided, or at least influenced, by ethics. To me, the ethical aspirations of the Earth Charter and our Global Ecological Integrity Group are a source of great personal strength, but are also limiting in their implicit assumption that their existence makes a crucial difference. Unless we live by the values expressed there, and unless we test ourselves every day and on all occasions, the Earth Charter and our group will not succeed.

I have no solution to offer, but would like to end with quoting Ron Engel, who, in his critical reflections on the Earth Charter, offers encouragement:

What could be more motivating, interesting, ennobling, more inherently worthwhile, than engaging with colleagues in discussions regarding the ethical responsibilities incumbent upon our species if the world is to enter a new axial age? What greater satisfaction than contributing to the creation of a new order of existence that brings us into alignment with what is truly and everlastingly good?[36]

Engagement with each other and truthfulness to our values may not change the world, but it will make the change just a little more likely.

Notes

1 Global Ecological Integrity Group, Our Mandate, www.globalecointegrity.net
2 For example, John Locke: 'government is not legitimate unless it is carried on with the consent of the governed'; Richard Ashcraft (ed.), *John Locke: Critical Assessments* (London: Routledge, 1991), 524. See generally E. Fox-Decent, *Sovereignty's Promise: The State as a Fiduciary* (Oxford: Oxford University Press, 2012); T. Frankel, 'Fiduciary Law' *California Law Review* 71 (1983), 795.
3 Including constitutional law, administrative law, tax law, criminal law and environmental law.
4 For example, the United States, Canada, Australia and New Zealand recognize them with respect to indigenous peoples, ratepayers and (with the exception of New Zealand) in the form of public trusts, whereas continental European countries more fundamentally rely on public law to assume fiduciary relationships between individuals and governments.
5 M. Blumm and R. Guthrie, 'Internationalizing the Public Trust Doctrine', *University of California Davis Law Review* 45 (2012), 741; H. Perritt, 'Structures and Standards for Political Trusteeships', *University of California Los Angeles Journal of International Law and Foreign Affairs* 8 (2004), 391; E. Brown Weiss, 'The Planetary Trust: Conservation and Intergenerational Equity', *Ecology Law Quarterly* 11 (1984), 495.
6 E. Criddle and E. Fox-Decent, 'A Fiduciary Theory of Jus Cogens', *Yale Journal of International Law* 34 (2009), 369.
7 P. Finn, 'The Forgotten "Trust": The People and the State', in M. Cope (ed.), *Equity: Issues and Trends* (Sydney: The Federation Press, 1995), 131–51.
8 Or, more precisely, what can be termed the European cosmology: **d**ualism (of humans and nature), **a**nthropocentrism, **m**aterialism, **a**tomism, **g**reed (individualism gone mad) and **e**conomism (the myth of no boundaries and limitless opportunities), or 'damage' for short; K. Bosselmann, 'From Reductionist Environmental Law to Sustainability Law', in P. Burden (ed.), *Exploring Wild Law: The Philosophy of Earth Jurisprudence* (Adelaide: Wakefield Press, 2011), 205.
9 Bruce Ackerman, *We the People, Vol.1: Foundation* (Cambridge, MA: Harvard University Press, 1991); Anne-Marie Slaughter and William Burke-White, 'An International Constitutional Moment', *Harvard International Law Review* 43 (2002), 1.
10 K. Bosselmann, 'Wendezeit im Umweltrecht. Von der Verrechtlichung der Ökologie zur Ökologisierung des Rechts', *Kritische Justiz* 18 (1985), 345–61; K. Bosselmann, 'Eigene Rechte der Natur? Ansätze einer ökologischen Rechtsauffassung', *Kritische Justiz* 19 (1986), 1–22.
11 K. Bosselmann, *Im Namen der Natur: Der Weg zum ökologischen Rechtsstaat* (Munich: Scherz, 1992), 196–202.
12 K. Bosselmann, *Ökologische Grundrechte: Zum Verhältnis zwischen individueller Freiheit und Natur* (Baden-Baden: Nomos, 1998), 81–2.

13 Drafted by approximately one hundred professors of law and social sciences. See Kuratorium für einen demokratisch verfaßten Bund deutscher Länder, *Vom Grundgesetz zur Deutschen Verfassung: Denkschrift und Verfassungsentwurf* (Baden-Baden: Nomos, 1991); Bosselmann, *Im Namen der Natur*, 201.

14 K. Bosselmann, *The Principle of Sustainability: Transforming Law and Governance* (Aldershot: Ashgate, 2008), 139–40.

15 E. Evins, 'The Inclusion of Animal Rights in Germany and Switzerland: How Did Animal Protection Become an Issue of National Importance?', *Society and Animals* 18 (2010), 236.

16 Bosselmann, *The Principle of Sustainability*, 125–7.

17 World Commission on Environment and Development (WCED), *Our Common Future* (Oxford: Oxford University Press, 1987; available at www.un-documents.net/our-common-future.pdf), part I.1.A.I: 'Nature is bountiful, but it is also fragile and finely balanced. There are thresholds that cannot be crossed without endangering the basic integrity of the system. Today we are close to many of these thresholds; we must be ever mindful of the risk of endangering the survival of life on Earth.'

18 R. Kim and K. Bosselmann, 'Towards a Purposive System of Multilateral Environmental Agreements', *Transnational Environmental Law* 2 (2013), 285–309; R. Kim, and K. Bosselmann, 'Operationalizing Sustainable Development: Ecological Integrity as a *Grundnorm* in International Law', *Review of European Community and International Environmental Law* (forthcoming 2015).

19 WCED, *Our Common Future.*

20 S. Westerlund, 'Theory for Sustainable Development Towards or Against?', in H. C. Bugge and C. Voigt (eds), *Sustainable Development in International and National Law* (Groningen: Europa Law Publishing, 2008), 48–65.

21 D. Fisher, *Legal Reasoning in Environmental Law: A Study of Structure, Form and Language* (Cheltenham: Edward Elgar, 2013), 433.

22 I. Kaul, 'Meeting Global Challenges: Assessing Governance Readiness', in Hertie School of Governance, *The Governance Report 2013* (Oxford: Oxford University Press, 2013), 33–58.

23 C. Kegley and G. Raymond, *The Global Future: A Brief Introduction to World Politics*, 5th edn (Boston, MA: Wadsworth, 2010), 300–330.

24 K. Bosselmann, *Earth Governance: Trusteeship for the Global Commons* (Cheltenham: Edward Elgar, 2015).

25 J. Rockström, W. Steffen, K. Noone, Å. Persson, F. S. Chapin, E. Lambin, T. M. Lenton, M. Scheffer, C. Folke, H. Schellnhuber, B. Nykvist, C. A. De Wit, T. Hughes, S. van der Leeuw, H. Rodhe, S. Sörlin, P. K. Snyder, R. Costanza, U. Svedin, M. Falkenmark, L. Karlberg, R. W. Corell, V. J. Fabry, J. Hansen, B. Walker, D. Liverman, K. Richardson, P. Crutzen, and J. Foley, 'Planetary Boundaries: Exploring the Safe Operating Space for Humanity', *Ecology and Society* 14(2) (2009): article 32, available at www.ecologyandsociety.org/vol14/iss2/art32.

26 For example, F. Biermann, 'The Emerging Debate on the Need for a World Environment Organization: A Commentary', *Global Environmental Politics* 1 (2001), 45; D. Esty, 'Toward a Global Environmental Organization', in C. Bergsten et al (eds), *Toward Shared Responsibility and Global Leadership. A Report to the Leaders of the G-8 Member Countries* (Washington, DC: Institute of International Economics, 2001), 30; J. Whalley and B. Zissimos, 'What Could a World Environmental Organization Do?', *Global Environmental Politics* 1 (2001), 29–34; S. Charnowitz, 'A World Environmental Organization', *Columbia Journal of Environmental Law* 27 (2002), 329. Compare with: K. van Moltke, 'Clustering International Environmental Agreements as an Alternative to a World Environment Organization', in F. Biermann and S. Steffen Bauer (eds), *A World Environment Organization* (Aldershot: Ashgate, 2005), 176.

27 R. Ruggiero, 'A Global System for the Next Fifty Years', Address to the Royal Institute of International Affairs (30 October 1998).

28 F. Biermann, 'The Case for a World Environment Organization', *Environment* 42 (November 2000), 24.

29 K. Gautam, 'Ensuring United Nations Trusteeship of the Global Commons', available at www.kulgautam.org/2013/11/ensuring-united-nations-trusteeship-of-the-global-commons.

30 See the bibliography of the Commons for the United Nations network at www.commonsactionfortheunitednations.org/commons-resources-2/articles-and-documents. See also the booklet 'Growing the Commons Top Down and Bottom Up', available at www.commonsactionfortheunitednations.org/wp-content/uploads/2013/03/Commons-Booklet-March-21-2013.pdf.

31 See UN System Task Team on the Post-2015 UN Development Agenda, *Global Governance and Governance of the Global Commons in the Global Partnership beyond 2015*, available at www.un.org/en/development/desa/policy/untaskteam_undf/thinkpieces/24_thinkpiece_global_governance.pdf.

32 See UNESCO, 'UN Secretary-General's Scientific Advisory Board to Strengthen Connection between Science and Policy', available at http://en.unesco.org/post2015/news/un-secretary-general's-scientific-advisory-board-strengthen-connection-between-science-and-policy.

33 For an analysis of international trusteeship-type organizations within, or related to, the UN system, see Bosselmann, *Earth Governance*.

34 H. Arendt, *Eichmann in Jerusalem: A Report on the Banality of Evil* (New York: Viking Press, 1963).

35 J. Habermas, *The Theory of Communicative Action, Volume 2: Lifeworld and System: A Critique of Functionalist Reason* (Boston, MA: Beacon Press, 1985); K. Bosselmann, R. Engel and P. Taylor, *Governance for Sustainability: Issues, Challenges and Successes*, Environmental Law and Policy Series vol. 70 (Bonn: IUCN, 2008) available at http://cmsdata.iucn.org/downloads/governance_final_fur_web.pdf.

36 R. Engel, 'Prologue Summons To a New Axial Age: The Promise, and Future of the Earth Charter', in L. Westra and V. Mirian (eds.), *The Earth Charter, Ecological Integrity and Social Movements* (Abingdon: Earthscan/Routledge, 2014), xv, at xxviii–xxix.

Index

Note: References in **bold** are for tables, and figures are shown in *italics*.

link with environmental protection 256–7; official definition of the 'defence' of human rights 260; overview of 163; remedies for citizens to fight climate change 171; term 'land' 50–1; Universal Declaration of Human Rights (UDHR) 163; violation by climate change 166

human rights defenders: international legal framework for the protection of 258–62; official definition of the 'defence' of human rights 260; term 256; UN Special Rapporteur, mandate 261

hydrofracking: and civil unrest 245; investments in and climate change vs. environmental issues 100; Jessica Ernst case, Canada 13; opposition to, Canada 134; process of 12–13, 134

income inequality: and access to political processes 237; in Canada 204–5, 210; global 220; income level and political participation rates, Canada 210–11

India: biopiracy protection and national legislation 42–3; genetic resource disclosure proposal 41; ground water as national wealth 72; loss of indigenous lands for resource extraction 245; water scarcity 65

Indian Residential Schools Settlement, Canada: Canadian nation's apology for 176; conflict with adversarial torts-based legal system 178, 179, 182–3; consultation with First Nation communities 178; forgiveness within 184; former students initial legal actions 177; future relationship between Aboriginal and non-Aboriginal Canadians 180–2; historical policy of assimilation 177; importance of victims' perspectives 181–2; negotiation process 177; reconciliation mandate 178, 183–4; restorative justice principles 183; Truth and Reconciliation Commission 179–80

indigenous peoples; *see also* Indian Residential Schools Settlement, Canada: cultural and spiritual land values 50–1, 54; land rights 50–1; loss of lands for resource extraction 245; protection under the Nagoya Protocol 38

intellectual property rights 39; *see also* patents; TRIPS Agreement (Agreement on Trade-Related Aspects of Intellectual Property Rights)

Inter-American Commission on Human Rights (IACHR) 262–5

Intergovernmental Panel on Climate Change (IPCC) 96

internal environmental conflicts: environmental impact assessment (EIA) 103–4; holistic impact assessment (HIA) 103–4; overview of 99–102; strategic impact assessment (SIA) 103–4

International Center for Human Rights (ICHR) 67

International Court of Justice (ICJ) 66–7

International Covenant on Economic, Social and Cultural Rights (ICESCR) 51

International Emissions Trading (IET) 98–9

International Labour Organization (ILO) Convention no. 169 50

international laws: customary international water law 80–1; influence of soft laws on 54, 59; on the right to water 65–8, 79–81

internet; *see also* media violence: cyberbullying 224; dangers to young people 224–5; incidence of hate crimes 226–7; networking power of 221; online communications interception 227; pornography 224–5; and radicalisation of terrorists 227; relationship with the global economy 221–2; sexting 224

investment: climate change measures in conflict with environmental issues 99–102; ecological sustainability paradigm 102–3; Holistic Impact Assessment 103

investor-state dispute settlements (ISDS) 141–4, 145

islands, classification of 110–12

Italy: palm trees and biological invasion threat 26; qualification of water as a common good 64

Jacobs, J. 76
Jeffrey, P. 65
Jenkins, B. 150
Jongman, 150
Judt, T. 203, 210